T0399145

World and Hour in Roman Minds

Center panel of three floor mosaics, unearthed in 2013 at Syrian Antioch (modern Antakya, Turkey), probably in a dining-room or its entrance vestibule, dating to around 300 CE. A cleanshaven man whose right sandal has dropped off looks up – right hand raised, index finger extended – at the sundial on top of a column; above him, in Greek, the description ΤΡΕΧΕΔ[Ε]ΙΠΝΟΣ, "dasher to dinner." His evident lateness is signified by the half-figure of a bearded man in black – described above as ΑΚΑΙΡΟΣ, "bad timer" – who tugs at his cloak from behind. Over the crescent of the sundial appears the letter Θ with a macron (bar) superscript, denoting that here it represents a numeral, nine, signifying the ninth hour; elsewhere, without macron it commonly abbreviates ΘΑΝΑΤΟΣ, "death." Cf. figs. 19.1 and 6 below. Photo: Pamir and Sezgin (2016), 275, 277.

World and Hour in Roman Minds

Exploratory Essays

RICHARD J. A. TALBERT

OXFORD
UNIVERSITY PRESS

Oxford University Press is a department of the University of Oxford. It furthers the University's objective of excellence in research, scholarship, and education by publishing worldwide. Oxford is a registered trade mark of Oxford University Press in the UK and certain other countries.

Published in the United States of America by Oxford University Press
198 Madison Avenue, New York, NY 10016, United States of America.

Library of Congress Cataloging-in-Publication Data
Names: Talbert, Richard J. A., 1947– author.
Title: World and hour in Roman minds : exploratory essays / Richard J.A. Talbert.
Description: New York : Oxford University Press, [2023] |
Includes bibliographical references and index.
Identifiers: LCCN 2022027369 (print) | LCCN 2022027370 (ebook) |
ISBN 9780197606346 (hardback) | ISBN 9780197606360 (epub)
Subjects: LCSH: Rome—Maps. | Rome—Historical geography. |
Cartography—Rome—History. | Roads, Roman. | Time perception—Rome. |
Rome—Geography. | Geographical perception—Rome. | Rome—Civilization.
Classification: LCC DG31 .T35 2023 (print) | LCC DG31 (ebook) |
DDC 911/.3—dc23/eng/20220629
LC record available at https://lccn.loc.gov/2022027369
LC ebook record available at https://lccn.loc.gov/2022027370

DOI: 10.1093/oso/9780197606346.001.0001

1 3 5 7 9 8 6 4 2

Printed by Integrated Books International, United States of America

Dedicated with love and optimism to my grandchildren

Alastair

Isla

Lazslo

Luisa

Contents

Preface and Acknowledgments

This collection fulfills a long-standing wish to offer a match for my earlier one, *Challenges of Mapping the Classical World*, handsomely published by Routledge in 2019. Covid accelerated the preparation this time, after diversions demanding travel and access to libraries were blocked from spring 2020 onward. Even so, the need for extensive help remained, and I am more than grateful to those who provided it in difficult circumstances. Once again, Peter Raleigh ably converted printed texts to formats that could be edited, and Lindsay Holman at the Ancient World Mapping Center proved the most skilled manipulator of files and images. In Chapel Hill's Davis Library, the tireless efficiency of the Inter-Library Borrowing office headed by Sellers Lawrence has been especially appreciated. The University of North Carolina was generous in meeting the considerable expense of fees for the reproduction of certain images.

The thanks to colleagues whose help I acknowledged at the time of original publication are of course retained and reaffirmed in the fresh presentation of each text. There are now further debts to record, due to others who either permitted the reproduction of one of the additional images, or specially helped to secure the necessary authorizations for their inclusion. While naming Hatice Pamir first for her provision of the splendid jacket image (in color) and frontispiece, I also thank heartily: Işık Adak Adıbelli, Aït Allek Chafiaa, Nacéra Benseddik, Claude Briant-Ponsard, Katherine Dunbabin, Lien Foubert, Francisco González Ponce, Ekaterina Ilyushechkina, Anne Kolb, Sergei Maslikov, Ma Menglong, Fatih Onur, Tom Parker, Jesús Rodríguez Morales, Benet Salway, Zhongxiao Wang, and Juping Yang.

This is the third book of mine brought to publication with Stefan Vranka's shrewd and supportive guidance as editor. I thank Stefan and his colleagues at Oxford University Press, as well as the experts recruited by him to referee the manuscript. The final tribute goes to the kindness of my wife Zandra: this book is my first that had to be prepared entirely at home, never on campus as usual—a demanding intruder therefore, accommodated with equanimity. We shall both be glad to see its successful launch.

Chapel Hill, North Carolina
August 2021

Essays originally published in:

1 *JRS* 77 (1987), 210–12
2 *Studies in Chinese and Western Classical Civilizations: Essays in Honour of Professor Lin Zhi-Chun on His Ninetieth Birthday* (1999), 413–24. Changchun: Jilin People's Publishing House
3 K. A. Raaflaub, ed., trans., *The Landmark Julius Caesar* (2017), Web Essay W. New York: Pantheon
4 *BMCR* 2004.12.23
5 *Shagi/Steps* 6.1 (2020), 214–28
6 *Caesarodunum* 39 (2005), 93–101
7 Luuk De Ligt, E. A. Hemelrijk, and H. W. Singor, eds., *Roman Rule and Civic Life: Local and Regional Perspectives* (2004), 21–37. Amsterdam: Gieben
8 Klaus Geus and Michael Rathmann, eds., *Vermessung der Oikumene* (2013), 163–70. Berlin: de Gruyter
9 Rudolf Haensch and Johannes Heinrichs, eds., *Herrschen und Verwalten: Der Alltag der römischen Administration in der Hohen Kaiserzeit* (2007), 256–70 [Festschrift Werner Eck] Cologne: Böhlau
10 *BMCR* 2007.02.31
11 Claudio Gallazzi, Bärbel Kramer, and Salvatore Settis, eds., *Intorno al Papiro di Artemidoro, II. Geografia e Cartografia* (2012), 185–96. Milan: LED
12 F. J. González Ponce, F. J. Gómez Espelosín, and A. L. Chávez Reino, eds., *La Letra y La Carta: Descripción Verbal y Representación Gráfica en los Diseños Terrestres Grecolatinas* (2016), 313–20. [Festschrift Pietro Janni] Sevilla: Universidad de Sevilla, Universidad de Alcala
13 Richard Talbert and Kai Brodersen, eds., *Space in the Roman World: Its Perception and Representation* (2004), 113–31. Münster: LIT
14 Michael Rathmann, ed., *Wahrnehmung und Erfassung geographischer Räume in der Antike* (2007), 221–30. Mainz: von Zabern
15 Unpublished (conference presentation 2019)
16 S. E. Alcock, John Bodel, and R. J. A. Talbert, eds., *Highways, Byways, and Road Systems in the Pre-modern World* (2012), 235–54. Malden, MA: Wiley-Blackwell
17 Anne Kolb, ed., *Roman Roads: New Evidence—New Perspectives* (2019), 22–34. Berlin: de Gruyter

18 F. S. Naiden and R. J. A. Talbert, eds., *Mercury's Wings: Exploring Modes of Communication in the Ancient World* (2017), 340–62. Oxford: Oxford University Press

19 Alexey Belousov and Ekaterina Ilyushechkina, eds., *Homo Omnium Horarum: Symbolae ad Anniuersarium Septuagesimum Professoris Alexandri Podosinov Dedicatae* (2020), 534–55. Moscow: Academia Pozharskiana

Figures and Tables

Figures

Tables

Abbreviations

AE	*L'Année Epigraphique*
AntAf	*Antiquités Africaines*
Arh Vest	*Arheološki Vestnik*
BAtlas	R. J. A. Talbert, ed., *Barrington Atlas of the Greek and Roman World* and *Map-by-Map Directory* (Princeton, NJ: Princeton University Press, 2000)
BMCR	*Bryn Mawr Classical Review*
BNP	*Brill's New Pauly*
CAH^{2}	*The Cambridge Ancient History* (2nd ed.)
CIL	*Corpus Inscriptionum Latinarum*
CQ	*Classical Quarterly*
CR	*Classical Review*
DarSag	Charles Daremberg and Edmond Saglio, eds., *Dictionnaire des Antiquités Grecques et Romaines* (Paris: Hachette, 1875–1919)
FIRA^{2}	*Fontes Iuris Romani Antejustiniani* (2nd ed.)
FGrHist	Felix Jacoby et al., eds., *Die Fragmente der griechischen Historiker*
FRHist	T. J. Cornell, ed., *The Fragments of the Roman Historians* (Oxford: Oxford University Press, 2013)
G&R	*Greece and Rome*
GLM	Alexander Riese, ed., *Geographi Latini Minores* (Heilbronn: Henninger, 1878)
IK	*Inschriften griechischer Städte aus Kleinasien*
ILBulg	*Inscriptiones Latinae in Bulgaria Repertae*
ILS	*Inscriptiones Latinae Selectae*
IScM	*Inscriptiones Scythiae Minoris Graecae et Latinae*
JHS	*Journal of Hellenic Studies*
JRA	*Journal of Roman Archaeology*
JRS	*Journal of Roman Studies*
LTUR	E. M. Steinby, ed., *Lexicon Topographicum Urbis Romae* (Rome: Quasar, 1993–2001)
MDAI(R)	*Mitteilungen des deutschen archäologischen Instituts: Römische Abteilung*
MEFRA	*Mélanges d'archéologie et d'histoire de l'Ecole Française de Rome*
OBO	*Oxford Bibliographies Online*
OCD	*Oxford Classical Dictionary*
ODCC	*Oxford Dictionary of the Christian Church*
ODLA	*Oxford Dictionary of Late Antiquity*

OLD	*Oxford Latin Dictionary*
Peut.	Peutinger map as presented online at www.cambridge.org/9780521764803: Map A (photographed in color); Map B (photographed in monochrome, 1888). See further Talbert (2010), 8–9, 196–98
PIR[2]	*Prosopographia Imperii Romani* (2nd ed.)
PL	*Patrologia Latina*
P.Lond.	*Greek Papyri in the British Museum*
P.Oxy.	*Oxyrhynchus Papyri*
P.Ryl.	*Greek Papyri in the John Rylands Library*
RE	*Real-Enyclopädie der classischen Altertumswissenschaft*
SEG	*Supplementum Epigraphicum Graecum*
TAVO	*Tübinger Atlas des Vorderen Orients*
TLL	*Thesaurus Linguae Latinae*
TLS	*Times Literary Supplement*
ZPE	*Zeitschrift für Papyrologie und Epigraphik*

Introduction

This collection of nineteen varied writings, or "essays," records stages of an ongoing intellectual journey into uncharted territory. It presents in three parts a "world" quest launched during the 1980s by no more than a succession of accidents. Yet the explorations that they prompted in various directions turned out surprisingly rewarding and cohesive. Eventually, from these expanding ventures—written up as progress was made—a further "hour" path emerged. Its new direction I have only recently begun to follow. So there is good reason to pause at this point and to draw together the main findings to date.

The two essays placed first were indeed the earliest in the collection to be written, and four of the final five are among the latest. In between, however, the order is loosely by topic rather than by date of writing. Each essay was intended to be self-standing without thought of a collection being made, so they can be browsed at random; but to read them in order should add to their value and impact. I have in any case frequently added pointers to themes that link them. Because several of the nineteen have only appeared in publications with limited circulation, I expect few potential readers to be already familiar with them all, and obviously not with the one essay published here for the first time (essay 15). Here and there, passages have been cut to eliminate repetition or dense technical detail. On some aspects, brief reference is now made to more recent scholarship. Various slips have been quietly corrected. Most often, the welcome opportunity has been taken to insert illustrations or maps that for one reason or other could not be used originally. Some of these additions are reproduced from publications of mine not selected for inclusion.

In retrospect, a random encounter at a conference reception in 1981 can be seen as the initial accident to trigger my "world" quest, the focus of the first part *World and Empire in Mind's Eye*. Here I met Richard Stoneman, a talented editor seeking to expand his classics list, who soon afterward (and quite unexpectedly) invited me to produce an atlas of classical history. This modest textbook—for which I recruited twenty-four collaborators, all in the British Isles—appeared from Croom Helm in 1985. With it in mind presumably, the following year John North, *Journal of Roman Studies* review editor, approached me to tackle Oswald Dilke's new book *Greek and Roman Maps*. My work for the atlas hardly matched Dilke's focus, but North was no doubt at a loss given the general lack of interest in ancient cartography at that date, and his request was for merely a 350-word notice. In the event, after I accepted and awoke to the value of a far lengthier appraisal, he generously granted my plea for ten times as many words. For all the book's merits, Dilke in the traditional way took it for granted that Greeks and Romans perceived the world and used maps as modern Westerners do. My review (essay 1) begged to differ, daring to question the nature and degree of map consciousness in classical antiquity. I urged rather that alternative ways of perceiving and organizing geographical space mattered more, and that wide-ranging investigation of such elusive "mental mapping" or "informal geography" should be attempted.

A remarkable feature of this review that has become apparent with time is the extent to which it envisages lines of inquiry subsequently pursued by myself and others. In fact one colleague in Russia (Alexander Podossinov) and another in Italy (Pietro Janni) had already made a start, although at that date I was barely aware of the latter's efforts, and not at all of the former's. So years later, and now better informed, I especially appreciated invitations to contribute an essay to volumes in honor of each, included as essays 19 and 12 here. Earlier, however, my own efforts were slowed by a call from the American Philological Association in 1988 to rethink and restart its ambitious, but stillborn, plan for a definitive classical atlas. The major international collaborative project that I developed in response, and its fulfillment with publication of the *Barrington Atlas* in 2000, are explained elsewhere in another collection of writings, *Challenges of Mapping the Classical World* (2019). The one line of inquiry I could at least embark upon was some cross-cultural comparison of Roman and ancient Chinese mapping and awareness of space. A paper on this revealing topic made a timely contribution to China's first-ever conference of Chinese and Western ancient historians, held at Nankai University, Tianjin, in 1993, included here as essay 2.

Essays 3 and 4 seek to identify and appraise the approaches taken by two very different prominent Romans who both confront the world without (it seems) recourse to maps: Julius Caesar at the head of an army intent upon conquering Gaul and possibly even Britain; Pliny the encyclopedist daring in books 3–6 of his *Natural History* to encompass physical, ethnic, and political geography on a global scale from the novel perspective of Rome as center-point. This geographical record lays the foundation on which Pliny's entire monumental work rests. It was the inspiring study by Trevor Murphy reviewed in essay 4 which prompted me to observe how long overdue the sole (in effect) English translation of Pliny's geographical books is for replacement. At the time this was not at all a challenge I envisaged undertaking myself, but a decade or so later the invitation from a former pupil, Brian Turner, to join him in meeting it—even extending the scope to book 2 on the universe, as well as further geographical passages—proved irresistible. Essay 5 analyzes the missteps and shortcomings of the previous translation, and then explains the principles that Turner and I formulated for our rendering of this exceptionally important and influential Roman account of the world. We aim for a translation that is accurate, scholarly, and above all accessible to nonspecialist readers who will never refer to the original Latin: in short, a taxing combination of goals, but attainable ones.

In the next four essays (6–9) the focus turns to noncartographic means by which Romans visibly demarcated space and mentally perceived it. Essay 6 calls for scholars' preoccupation with the nature of the Roman empire's external boundaries to be balanced by matching consideration of the reality that in its interior there were boundaries everywhere demarcating provinces, communities, individuals' landholdings, and more. The frequent occurrence of boundary disputes and their tendency to become protracted can only have heightened awareness of these lines crisscrossing the landscape. An investigation of the surprisingly neglected category of boundaries between Roman provinces follows in essay 7. The question of whether a traveler would be made aware of leaving one province and entering another, and by what means, had evidently never been raised. Still less had thought been given to how far, if at all, provinces might become units in the Roman imagination that were visualized in order to comprehend the broad spatial relationships of the empire's territories to one another as well as to Italy and the Mediterranean—on the pattern of today's jigsaws where the counties of England, say, or Canada's provinces or US states are to be pieced together.

Essay 8 tests the visualization just outlined—a plausible one in my view—by making a novel attempt to penetrate the worldview of long-serving noncitizen soldiers and naval crew in the first to third centuries CE, a large, distinct group with minimal education. I contend that glimpses of these men's outlook may be captured from a type of standard document inscribed on bronze that has been found in increasing numbers ever since metal detectors came into widespread use around 1990. Because in each instance the man's name and his "origin" are recorded, every such document is by definition unique and personal. Each also includes a record of the province where the document was issued, often somewhere distant from the recipient's place of origin (be it province, city, village, people, or more than one of these). The reaction offered by Michael A. Speidel (2017) that my inferences may be rash deserves attention, but still the attempt itself provides an example of exploratory method that bears repeating. As will emerge, essay 18 outlines a further effort on my part to gain insight into geographical awareness by again exploiting overlooked evidence (from portable sundials) not tapped hitherto for the purpose.

The springboard for writing essay 7 was to caution against the claim—originating with Pietro Janni, but later articulated more sweepingly by Charles (Dick) Whittaker—that Romans' conception of space was merely linear, one molded in particular by a distinctive habit, the compilation and use of itineraries for journeys along the empire's highways. However, closer reading of the many surviving itineraries finds references to provinces, seas, rivers, and mountain ranges. Tangential these references may be, but they still point to an awareness by no means one-dimensional. My engagement with this material gave rise to basic questions posed in essay 9 about the largest ancient collection, the so-called *Antonine Itinerary*: does it deserve the immense respect paid to it ever since the Middle Ages? Is it well organized, as well as genuinely useful to ancient travelers on the ground? Having reached negative conclusions about each of these aspects, I proceed to inquire further what kind of individual would make such a defective compilation, and with what audience in mind, if any, especially if the work fits no established genre. In my revisionist opinion, the compiler is no educated intellectual but instead middle-ranking at most, a purely self-centered hobbyist who never intended to put his patchy, unedited assemblage into circulation. To be sure, since antiquity it has become a precious, even evocative, chance survival, but its genesis and place in contemporary context still call for the more rigorous assessment that essay 9 puts forward. It was written for a conference on the Roman empire's day-to-day administration.

There follows (essay 10) a review of John Matthews' book about one of classical antiquity's most meticulously recorded journeys: a return trip by Theophanes—a high-ranking lawyer—and his entourage, made in the 320s CE by Nile riverboat and then overland between Egypt and Syria. Not only do itineraries for it written on papyri by unnamed staff survive, but also ledgers sufficiently detailed to reconstruct the meals prepared day by day, thus furnishing insight into diet and expenditures. It is clear too that, after a two-and-a-half-month stay in Syrian Antioch, an "agreement" was made and written up on specially purchased papyrus, but otherwise—ironically—there is no clue to the trip's purpose, let alone to the worth of its outcome. In terms of method, the value of this study lies in its creative attention to hitherto ignored minutiae of cultural and social history, and in its ability to tease out information and draw instructive conclusions from the most rebarbative of documents. These—again in exemplary fashion—Matthews translates into English, thereby opening them to nonspecialists for the first time in the way mentioned above for books by Pliny.

In the second part *Maps for Whom and Why*, the next five essays (11–15) address issues relating to Romans' production and use of maps; attention here centers on the single substantial surviving Roman map with more than localized scope. Once I began to follow Dilke's lead in order to master the various dimensions of this large topic, the mass of difficulties became only too plain. Above all, evidence for use of a map is seldom to be found in any kind of source, and the single substantial map just mentioned—the so-called Peutinger map—survives only as an incomplete medieval copy. How well this copy reproduces the original drawn centuries earlier, without being unduly impaired or altered by successive copyists' slips and modifications, is a fundamental but uncomfortable dilemma laid out in essay 15 (written for a 2019 conference mentioned further below). I use the adjective "uncomfortable" quite deliberately, because anyone with an interest in Roman maps is predisposed to resist endorsing the suspicion that our sole copy may have become too much changed to be regarded as Roman work any longer. Be that as it may, possible reasons for reaching this extreme conclusion (which I do not), so damaging for the study of Roman mapping, ought at least to be considered, and the exceptional difficulties of making an accurate copy of the Peutinger map better appreciated. Even copies made as late as the nineteenth century, and claimed to be flawless, are demonstrably not perfect.

However, despite the many causes for frustration, I am confident that rewarding avenues to insight can be found. Essay 12 identifies an overlooked instance of the use of a map in Roman public life, an initiative by the emperor Claudius to illustrate a speech to the senate with a visual aid. The identification rests on adopting a neglected perspective to interpret allusions in the much-studied verbatim text of the speech. An untypical instance it may be in some respects, but still credible and suggestive. More excitingly, during the 1990s tangible fresh testimony surfaced in the form of an actual map: it is the subject of essay 11, one of several components on a papyrus roll reckoned to date to the first century CE. Because the papyrus also includes—adjacent to the map—text about the Iberian peninsula attributable to the geographer Artemidorus around 100 BCE, it has been named after him. Even so, for all that the map is a unique and thrilling find, initial expectations about its value must be tempered by the multiple obstacles to interpreting it with confidence. In general, expert opinion still remains divided on whether the contents of the papyrus are ancient at all (as I take them to be) or instead the work of a forger in, say, the nineteenth century. More specifically, the map gives the impression (false perhaps) of having been abandoned unfinished for no obvious reason. For certain, it lacks color and lettering, and its puzzling linework and symbology are hard to relate convincingly to an outline of the Iberian peninsula in whole or in part; but clues to where else it may represent, and at what scale, seem absent. At least its present state does illuminate the stages by which the mapmaker or copyist proceeded; there may be useful pointers here to how the Peutinger map was produced.

In struggling to understand this fundamental aspect of both the Artemidorus and Peutinger maps, I came to appreciate the value of experience gained in designing the *Barrington Atlas*, work undertaken around 1990 when no stage of mapmaking was yet a digital process. In addition, *Barrington*'s comprehensive, uniform, and relatively detailed coverage of the entire Roman empire and far beyond to the East was to prove an indispensable resource for understanding the Peutinger map, given that both atlas and map have comparable scope (and, as it happens, North orientation). When a grasp of the map became my goal around 2000, no classical atlas of *Barrington*'s caliber had appeared since the 1870s, nor had any comprehensive study of the map since World War I. To the end of the twentieth century, scholarship on it remained almost exclusively preoccupied with the routes shown, a conspicuous and important feature without doubt, but in fact only one of many. Even so, the map was commonly categorized as a route diagram, with its other landscape features quickly dismissed as mere accompanying decoration. In addition, its remarkable shape was liable to be ridiculed: a "ribbon" only just over 30 cm tall, but approaching seven meters in width even with the western end lost. Minimal thought was given to its design and production, and the information it offers was described as "hopelessly crowded" (Whittaker [2002], 102).

Having been recently so preoccupied with both these aspects of map-making, I was struck by the persistent failure of scholars (Dilke among them) to consider this map seriously from a contemporary Roman perspective rather than judging it by largely irrelevant modern standards and ignoring how it was produced. Just why should its scope so ambitiously span the Atlantic to Sri Lanka, albeit with marked unevenness of scale? And why should so extreme a shape be chosen, for what context? Altogether what is featured and named, and how? Equally, what is omitted and why? Who are the map's intended viewers? Since answers to these and many other related questions can only be deduced from the map itself, speculation is unavoidable and much will have to remain obscure. Nonetheless a thoughtful approach of this type ought surely to be ventured rather than left untried for fear of provoking ridicule.

An invaluable aid to study that I commissioned was one out of reach before the twenty-first century: an online full-size image of the map reassembled as the single piece it was designed to be. The eleven successive parchments on which the surviving copy (in Vienna) was produced had been separated in 1863. The color photographs of all eleven published in 1976 could at least be joined, it is true, but there would then loom the problem of securing use of a table long enough to inspect the single piece. Not even the great map room of the New York Public Library had one; nor was one found for the extraordinary occasion in 2019 when the entire map was displayed to participants in a conference about it held at the Austrian National Library—the only time I have ever enjoyed the privilege of seeing its eleven segments together. Needless to stress, the ability to move through the map at will on a computer screen, to pan, zoom, create removable layers, link to a database, and more, is transformative for study purposes.

My full reappraisal of the map was published as the book *Rome's World: The Peutinger Map Reconsidered*, together with extensive materials online (2010). However, an advance statement of my main avenues of approach appeared earlier as they were being developed. This is essay 13 here. An equally fundamental advance publication follows (essay 14): it identifies the choices that I envisage were made to create the map's entire basis, namely its physical landscape, and considers the reasoning. Elaboration of my provocative conclusion that the map's main purpose was hardly cartographic at all, but rather a triumphalist celebration of Roman imperial reach, power and values, is postponed to essay 18, which includes a wider-ranging discussion of ancient propagandistic maps.

In the third part *From Space to Time*, essay 16 builds on my observation that the Peutinger map is exceptional for the prominence it accords roads almost everywhere in the empire and beyond to the East. Otherwise, in surviving source material of all types, Romans in general and the imperial authorities in particular are seldom found opting to draw attention to them. The essay explores the resulting puzzle: does such a low profile indicate that emperors deliberately restrained themselves from publicizing this huge asset both for their own control of the empire and for enemies' penetration overland? In addition, or instead, are we to infer that Romans' spatial perception lacked the capacity or the interest to conceive their roads as an interconnected system at all, let alone one with formidable strategic potential? Here again it proves advantageous to inquire whether there is useful insight to be gained from comparing Romans' attitudes with what may be learned of typical thinking in other premodern states likewise well known for their roads, such as Achaemenid Persia, China, Japan, and the Abbasid caliphate. A conference I proposed as a stimulus to such comparative thinking more than fulfilled its potential, moreover striking almost all participants as a novel approach.

Essay 17, a keynote address to a later conference inviting new evidence and perspectives on Roman roads, exploits this approach further with reference to the Inca empire, China, and Japan. These comparisons throw into sharp relief the superior qualities of Roman roads and the minimal restrictions on their use. Rome built for wheeled traffic, funded major bridges and viaducts, and normally permitted almost all free persons to travel at will without charge on every road. Such generosity or indifference emerges as far from typical elsewhere. The address closes with a plea for more efficient integrated dissemination of Roman milestone data, online rather than in print, and with commentary in a widely understood modern language rather than in Latin—still used for the definitive series in which milestones are currently published. In this subfield too, it should become a priority to offer nonspecialists fuller access.

Cross-cultural comparison remains relevant in essay 18, written as a chapter on Roman maps in a pathbreaking volume I coedited with F. S. Naiden which explores modes of communication in the ancient world. This essay rates Roman society as unmatched there for its imaginative recasting of factual, scientific maps (made and used only by intellectuals and some officials) into arresting artwork for the communication of Rome's might to a wide, less educated public. As already noted, I elaborate here upon my view that the Peutinger map was designed primarily for such promotion, and I also contend that the same may be claimed for the gigantic "Marble Plan" of the city of Rome erected there around 200 CE. Its generous scale permitted the incorporation of astonishing details—even individual pillars, steps, doorways—yet the map was positioned too far above floor-level for viewers to distinguish them.

Again as already noted, this essay introduces a further instance where I seek to gain insight into awareness of geography, and perhaps maps, by exploiting overlooked evidence. This data is inscribed on small portable sundials, and comprehension of it has been immensely improved in recent years by use of Reflectance Transformation Imaging to recover linework and lettering invisible to the naked eye. Roman users understood that in order to establish the hour in any location satisfactorily, the instrument had to be adjusted to take account of the latitude (a Greek concept) and the time of year. Hence each portable sundial incorporates a list of cities and regions for reference, each with its latitude figure in the form devised by Ptolemy in the mid-second century CE. Because the choice of locations seems typically to be that of a particular designer or owner rather than some workshop's standard selection, there opens up a means to penetrate individuals' worldviews and compare them. It is intriguing that in some lists the names are arranged by latitude, while in others they evidently outline a long-distance journey, albeit perhaps one only in the imagination. Also apparent is spatial perception of the Roman empire on the jigsaw model proposed in essay 7 above. Attachment to these remarkable gadgets as status symbols can be suspected too. Regardless of the patent flaws in design or data on some, all were enviable possessions which boosted the owner's sense of superiority as well as (again) Roman might and Romans' freedom to travel. The enthusiasm which such "scientific" sundials aroused had some potential—we might think—to overcome Romans' disinterest in maps, but no such intellectual shift would occur.

When my full findings on the cultural and social significance of this data were published in the book *Roman Portable Sundials: The Empire in Your Hand* (2017), disappointment was expressed that I had declined to extend its scope—deliberately confined to geography and worldview—into consideration of Romans' related interest in the time of day (always a twelve-hour period from sunrise to sunset, hours consequently varying in length according to season). Essay 19, which opens up my new "hour" path of exploration emerging from essay 18, is my initial response. This path proves promising, in part because recent scholarship has focused mainly on periods longer than the day. To be sure, this imbalance is now changing, as is well demonstrated, for example, by the volume *Down to the Hour: Short Time in the Ancient Mediterranean and Near East,* edited by Kassandra Miller and Sarah Symons (2020), which appeared too late for me to take into account. At least its contributors are not inclined to minimize the Roman concern to know the hour—a misjudgment in my view, although still expressed—but fuller recognition of the epigraphic and material evidence may be called for. I find cause for enlisting social psychology to explain not only why Roman society at all levels became more concerned than any other in the ancient Mediterranean and Near East to know the hour, but also why it maintained this fixation for centuries. It is one that at least can be seen to match Roman concern for well-defined boundaries, as well as for marking mile-intervals on roads, and for recording the length of journeys even to the nearest half-mile. Cross-cultural comparison with the attention paid to similar seasonal hours in Japan—persisting into the 1870s—could prove another rewarding new approach.

By its very nature, a collection of the present type demonstrates the contribution that a single individual has made to penetrating an area of inquiry and shaping it further. Even so, in the present case I trust it is repeatedly apparent that my pioneering ventures have by no means been pursued in isolation. Rather, at every stage I have gained invaluable stimulus from interchange with generous colleagues worldwide, as well as with the participants in four memorable series of seminars I cotaught with Hugo Meyer, Michael Maas, Grant Parker, and Patrick Gautier Dalché, respectively, in Princeton, Rome, Chapel Hill, and Paris between 1997 and 2011. I hope to have enlarged the thinking of all, as they have enlarged mine. Many colleagues—Kai Brodersen, James Akerman, Kurt Raaflaub, Elizabeth Wolfram Thill, and Richard Unger especially—have been my collaborators in initiatives of various kinds, several of these extending far beyond classical antiquity in space and time. I am equally indebted to the experts—in particular Tom Elliott, Ryan Horne, and Daryn Lehoux—who have so ably applied digital techniques and tools to questions and materials I have sought to investigate. These novel resources have opened up approaches that were out of reach, not to say out of mind, back in the 1980s. Their creative potential will again prove vital post-Covid, when my immediate aim is to rethink Romans' use of water-clocks. From this next phase, too, of the exploratory journey I anticipate revealing insights. Here, however, its unfolding course to date is encapsulated, beginning over thirty years ago with my review of Oswald Dilke's *Greek and Roman Maps*.

PART I

WORLD AND EMPIRE IN MIND'S EYE

1
Oswald Dilke's *Greek and Roman Maps* (1985)

As Oswald Dilke's preface explains, he was prompted to write this fascinating book in the course of research which originally resulted in *The Roman Land Surveyors: An Introduction to the* Agrimensores (1971). The overlap between the two works is acknowledged. The earlier one included discussion of maps, while the present one incorporates updated treatment of land surveying; there is some repetition of plates and figures. However, what is different, and very striking, about the present work is its immense scope in both chronological and material terms, a feature not fully conveyed by the title *Greek and Roman Maps*. In fact Dilke begins with one chapter (perhaps unduly brief) on maps in other early Mediterranean and Near Eastern civilizations, and concludes with two on the development and use of ancient maps among Arabic, Byzantine, and Western scholars up to the rebirth of European cartography and the production of the first printed maps in the fourteenth and fifteenth centuries respectively. In between come nine chapters on evidence from Greece, mapmakers (especially Agrippa), geographical writers (especially the Elder Pliny and Ptolemy), land surveying, Roman stone plans, road maps and itineraries, accounts of sea voyages (*periploi*), and maps (mainly postclassical) in art form.

The result is a splendid and valuable survey of every type of "map" from thumbnail sketch to world panorama. Throughout, Dilke writes with clarity, enthusiasm, and learning about important authors and topics too often unappreciated because of their obscurity and technical difficulty. Though he has almost always met the need to be both judicious and concise, inevitably perhaps there remain occasional repetitions as well as certain dubious claims inadequately documented—about Tarshish/Tartessus, for example, and Carthaginian policy in the western Mediterranean. The "general reader interested in classical antiquity," who is supposedly included in Dilke's target audience, might surely wonder whether chapter 3's detailed appraisal of references to Agrippa in the Elder Pliny could not more appropriately have been relegated to an appendix. It might also be felt that appendix IV, listing the manuscript variants on British place-names in Ptolemy, is a little too learned to merit inclusion in a volume of this size and character. The recommendations of further reading are mostly excellent, though I missed any mention of David Hunt's *Holy Land Pilgrimage in the Later Roman Empire* (1982) or Jonathan Sumption's *Pilgrimage: An Image of Mediaeval Religion* (1975).

Some broader reservations may be entered. First, the meager stock of evidence and Dilke's understandable desire not to see it reduced hardly encourage objective appraisal of all doubtful cases. For instance, no hint is given that very considerable controversy attaches to the provenance, date, and stippled design of the coins that he simply states (31, 146) to be issues by Memnon of Rhodes, portraying "some sort of relief map of the hinterland of Ephesus" on the reverse. In fact, when these pieces are viewed in the context of the Achaemenid issues in the vicinity, this identification of the design appears quite fantastic. Dilke's acceptance of Mordechai Gichon's suggestion that a fourth-century Palestinian oil lamp depicts the outline of a Roman camp (148) does happen to be more convincing. However, he still passes over the related points in Gichon's admirably cautious publication (1972), that such designs do not appear to have gained any popularity in Palestine, and that they are unparalleled in the Greek and Roman worlds. Dilke himself is cautious enough concerning what he believes might be a Roman map of Gaul possibly associated with Caesar's campaigns, discovered in 1976 (102–3); but if such a remarkable claim is to carry conviction, publication with a photograph is still essential. The inference (167) from the mention of a *comes formarum* in the *Notitia Dignitatum* that there existed in the Late Empire "a branch of the civil service in Rome . . . which dealt with maps" is seriously misleading, since other evidence proves that this *comes* was no more than a local official with responsibility for the city's aqueducts; a map department in the backwater of late imperial Rome would have been useless to the administration anyway.

Secondly, the comprehensiveness of Dilke's survey strengthens the impression that the majority of Greeks and Romans had only the most limited use for maps. Dilke himself draws no such conclusion, and to my knowledge it has never been more than hinted at except by Pietro Janni, *La Mappa e il Periplo: Cartografia Antica e Spazio Odologico* (1984), a provocative pioneer study, though its treatment of the question is still circumscribed. Other scholars remain sufficiently bound by twentieth-century preconceptions to insist that there must have been "really practical, detailed maps . . . particularly for fiscal, military and navigational purposes,"[1] and can assume that conquerors enjoyed the benefit of advanced cartographic expertise.[2]

As Dilke's work demonstrates, there is ample evidence for interest in cartography on the part of ancient thinkers, and it is clear that they made remarkable progress in their techniques. More broadly, a considerable amount of research (albeit rather uncoordinated, and only touched upon by Dilke) has been done by modern scholars to elucidate ancient geographical writers' perspective of the world. But in this sphere throughout antiquity the usual divorce of scientific investigation from its practical applications is blatant. It looks as if the only type of map ever widely valued was the survey of landholdings developed by Romans and wanted for legal reasons more than geographical ones. In contrast there is a marked absence of plans in Greek papyri relating to property. Once an empire-wide network of main roads was established by the Romans, their invention of "strip" maps met all the practical needs of most long-distance travelers, who rode or drove just from one city to another. It was duly recognized that to go further and prepare accurate maps of wide expanses (let alone reproduce them) presented immense obstacles—witness the jurist Marcian, quoted by Dilke (142–43).

Even so, such difficulties were hardly the decisive influence; it was the necessary attitude of mind that was missing. Seafarers, soldiers, and historians concentrated upon written accounts, not maps, while even among so-called experts there was regularly complacent repetition of outdated, muddled, or inadequate geographical information. In this connection there is instructive comparison to be made with studies like M. J. Bowden, "The Great American Desert in the American Mind: The Historiography of a Geographical Notion" (1976). J. J. Coulton (1977, cited by Dilke 102 n. 3) has argued with conviction that even Greek architects working on major projects as late as the classical period did not draw up general plans and elevations to scale. One instance of what might seem exceptional behavior (if it can be credited at all) is Plutarch's account of Athenians in 415 sketching the outline of Sicily and placing it in relation to North Africa.[3] But sufficient Athenians to man upward of sixty triremes had visited Sicily within the previous decade, so that

[1] Johnston (1967), 92; for geographers' concurrence, Moreland and Bannister (1983), 3–4.

[2] Note Sherwin-White (1984), 176–85, on Lucullus' invasion of Armenia.

[3] *Nicias* 12; *Alcibiades* 17, both passages unjustifiably embroidered in the Loeb and Penguin translations.

it was really just an area familiar to them from firsthand acquaintance which was being sketched.

The general absence of a "map consciousness" among Greeks and Romans is highlighted by comparison with contemporary China. Here the pictographic character of the language presumably fostered the idea of mapping from an early date.[4] Long-developed expertise based on survey work is reflected in the earliest two surviving Chinese maps, painted on silk and unearthed during the 1970s from a tomb of the early second century BCE—one (96 cm square) incorporating both the physical and human features of what is today southern Hunan and parts of Guangxi and Guangdong at a scale of about 1:180,000, the other (98 x 78 cm) a section of this area at about 1:100,000, drawn up for defense purposes and buried with a commander, who had perhaps even commissioned it.[5] References in Chinese authors have been dated as far back as the seventh century: they frequently draw attention to the value of maps in warfare[6] and in administration. It was with a deliberate eye to the establishment of the future Han dynasty that, during the capture of Xianyang (the former Qin capital) in 207 BCE, Xiao He ignored the temptation to grab loot indiscriminately, and instead made his way to the secretariat, took all the ordinances, reports, and maps, and stored them.[7] As the story implies, Chinese imperial officials needed such documents to fulfill responsibilities that Roman governors could mostly leave to local people, who would naturally be familiar with their own area.

Dilke does explain briefly (180) when and why the methods used by seafarers in the West eventually changed with the gradual compilation of accurate charts, aided by the development of the magnetic compass from around the twelfth century. He does not add, however, that progress on land lagged behind, and that surveying techniques were slow to improve. It was not until the sixteenth century that the military needs of leading states (Spain's campaigns in Flanders, for example) really encouraged terrestrial mapping in Europe. And it was only Jean-Baptiste Colbert's concern to develop France during the 1660s that at last impelled Giovanni Cassini to find a means of determining precise longitude and to initiate the first definitive mapping of an entire country.

[4] Needham (1959), 498; cf. (1954).
[5] See Gutkind Bulling (1978), and essay 2 below.
[6] Compare the solitary and limited prescription of Vegetius 3.6, quoted in essay 9 below.
[7] Needham (1959), 535.

Dilke's book now opens the way for a fundamental reexamination of how most Greeks and Romans—not to mention many other peoples with comparable lack of "map consciousness"—perceived their environment and oriented themselves when they traveled far from home. Many, of course, never traveled far, while others who did (like airline passengers today) no doubt hardly gave the matter a thought. Among Roman officials, when the need arose to leave the highway and penetrate unfamiliar countryside, it must have been common enough practice just to collar a peasant guide,[8] who would reckon not by simple distance, but in terms of something like the *parasang* or *schoenus*, the amount of ground which could reasonably be covered in an hour, taking all the circumstances of the journey into account.[9]

But such blinkered vision could hardly serve the ruler of an empire, or a historian like Thucydides or Ammianus; a proper understanding of the outlook of such individuals as these is vital. The topic impinges upon a multitude of other issues: determination of the character and extent of a "literate mentality" in the ancient world, for example,[10] and of how the Romans perceived "frontiers";[11] more generally, the whole conduct and recording of any war waged far from home. Every kind of evidence merits reexamination for clues. Such unlikely items as directions to letter carriers[12] and the routes of processions[13] may help to elucidate inhabitants' perception of their cities, for instance. In general, studies such as R. M. Downs and D. Stea, *Maps in Minds: Reflections on Cognitive Mapping* (1977) and P. Gould and R. White, *Mental Maps* (ed. 2, 1986) are suggestive. Even though their concern appears to be exclusively contemporary, the attempt to apply some of their approaches to the Greek and Roman worlds ought not only to be illuminating, but also to assist in opening up a new dimension in the study of ancient society and thought. Chapter 6 in Gould and White, with its section "Mental Maps of the Military," can readily be related to investigation of the foreign policies of ancient states.[14]

[8] Note *SEG* 17.755 lines 27–30 for Domitian's disapproval, reinforced by Hadrian in 129 CE: see Hauken and Malay (2009, lines 22–24).
[9] See Mitchell (1976), 121–22.
[10] Reynolds (1986), 142–43.
[11] Lintott (1981) and essays 6 and 7 below.
[12] Note *P.Oxy.* 2719.
[13] Price (1984), 111.
[14] Compare Millar (1982), 15–20, and essay 3 below.

To be sure, these thoughts stimulated by Dilke's book range beyond the scope of his study. Within it he has rendered great service in producing what should now become a standard survey of ancient maps and mapping. The hope must be that it will also serve as the starting point for wider investigations which draw upon expertise from the full range of relevant disciplines.

2
China and Rome
The Awareness of Space

No doubt it was because I had produced a modest textbook *Atlas of Classical History* (1985), that Professor John North approached me to write a review of Oswald Dilke's 1985 monograph *Greek and Roman Maps* for the *Journal of Roman Studies*.[1] Since (to be frank) I had never given ancient maps much thought, this seemed a challenge, but one not too daunting insofar as he asked for only 350 words. I kept the book to read over the Christmas break, and was duly impressed by its masterly exposition of the material.

But once I had finished the book, it struck me that there was one fundamental point unconsciously conveyed by it which Dilke never even articulated, let alone discussed—namely, that it showed ancient Greece and Rome in general to be *not* map conscious societies. Dilke (understandably enough) had maintained such a close focus on his theme that he had never questioned whether Greeks and Romans made maps and used them for the kinds of purposes which educated Westerners of the late twentieth century take for granted.

To be sure, there were scientists or thinkers (or whatever we care to term them) who very ably developed techniques and gathered data from which maps either were, or could be, made. The long tradition of their work culminates in Ptolemy, the second-century CE figure whose achievement was not adequately appreciated until with his help Europe finally became map conscious in the fifteenth and sixteenth centuries. In ancient times, however, these scientific discoveries characteristically made no impact upon society as a whole, not even upon those rulers or military commanders or tax assessors who acquired responsibilities that took them beyond the already familiar surroundings of "home" and who would benefit tremendously (we might feel) from the information that a map could convey to them. Now given, on the one hand, that this "cognitive gap" did persist, and on the other

[1] Essay 1 above.

that Greeks and Romans cheerfully sought to expand their control well beyond "home," often with staggering success, it should surely be of fundamental concern to inquire how they did in fact visualize their surroundings and organize space and spatial relationships for themselves.

In pursuing this topic for my review of Dilke's book, I was pleased to discover that Pietro Janni had already started to address it; happily, the debate has since been joined by Tønnes Bekker-Nielsen, Claude Nicolet, Nicholas Purcell, and others. At the time, however, I was so surprised and baffled by this new perspective on Greek and Roman worldview that I thought it might help to acquire a comparative dimension by setting it against the outlook of another ancient society at least outwardly similar in some other respects. I opted for setting imperial Rome against ancient China, and found the comparison instructive, even though time along with difficulty of access to the material severely limited my pursuit of it.

Recently, however, circumstances have favorably conspired to prompt a return to this effort. As just mentioned, the debate among classicists has advanced; and my own expanding interest in the mapping of the classical world (in past, present, and future) has brought me into rewarding contact with the great *History of Cartography* project. First conceived by David Woodward and the late Brian Harley, this project is now directed by David alone from Madison, Wisconsin, and is published by the University of Chicago Press.[2] The first volume, covering Europe and the Near East from prehistoric times to around 1500 CE, appeared in 1987. The first part of the second volume, covering traditional Islamic, Ottoman, and Indian cartography, appeared last year [1992]; and its second part, including chapters on traditional Chinese cartography, is due for publication soon [1994]. I have been privileged to see proofs of three of those chapters, all written by Cordell Yee. I should underline at once that most of what follows here about the ancient Chinese outlook as I perceive it derives from his contribution, one which certainly transforms my own previous shaky grasp. That said, Yee is not to be considered in any way responsible for mistakes or shortcomings in my account.

To attempt a comparison between the ancient Chinese and Roman outlooks repays the effort, if nothing else for the new light in which it places Rome. On the Chinese side the comparison can only be made for the ruling intellectual elite. Evidently, the potential sources for an investigation into the use of maps publicly or privately at regional and local levels, as well as lower

[2] See further essay 18 below.

down the social scale, do exist, but have yet to be tapped. Equally, it has to be acknowledged that the number of Chinese maps actually surviving from the period of the Roman empire is pitifully few. Yet they are also a tantalizing few—all of them relatively recent discoveries in fact—and (despite difficulties of interpretation) sufficient to provide general confirmation for the larger body of literary references to maps and their use during the Eastern Zhou, Qin, and Han periods (early eighth century BCE to early third century CE).

Let me illustrate from the two major sets of finds. The first comprises seven maps drawn in ink on four wooden boards; three of these boards have a map on either side. They were all discovered in 1986 at Fangmatan Forestry Station near Tianshui in Gansu province. They were buried in the tomb (no. 1) of an officer in the Qin army who died in 239 BCE (or just possibly earlier in the same century). The boards are about one centimeter thick, about 26 cm wide, and 15 to 18 cm in height. Although puzzles remain about how the maps relate to one another, there is no doubt that all seven—drawn at approximately 1:300,000 scale—portray parts of the valley of the Wei river and its tributaries which cut through the Qinling mountains. The grave site is within this region, which was strategically important as a defensible pass for eastbound traffic into the Qin heartland. On the maps, black lines represent rivers and tributaries. Gullies, passes, transport checkpoints, and stands of at least four different types of trees are all labeled; settlement names appear in boxes. Some distances are given, although the relevant terminal points are not clarified. South seems to be the typical orientation (figs. 2.1 and 2.2).[3]

The second set of finds comprises three maps drawn in vegetable colors on silk and unearthed in 1973 in the tomb (no. 3) of a high official at Mawangdui just outside Changsha in Hunan province. The burial occurred in 168 BCE. All three of these maps are thought to portray parts of the Changsha state, which at the time included modern Hunan as well as adjacent portions of Guangdong and Guangxi. Not much can be said about the first map because it emerged so badly tattered, although the lower part does seem to represent a city with inner and outer walls. Fortunately, the other two maps were in far better condition. South is at the top in both. One (96 cm square) represents the southern part of the Changsha state, and may fairly be considered a topographic map insofar as it highlights mountainous areas and rivers. Rivers

[3] It had been thought that one of the maps (1B) carried a label signifying a north orientation, but infrared photography has shown this to be a misreading: see Venture (2016), 261–62.

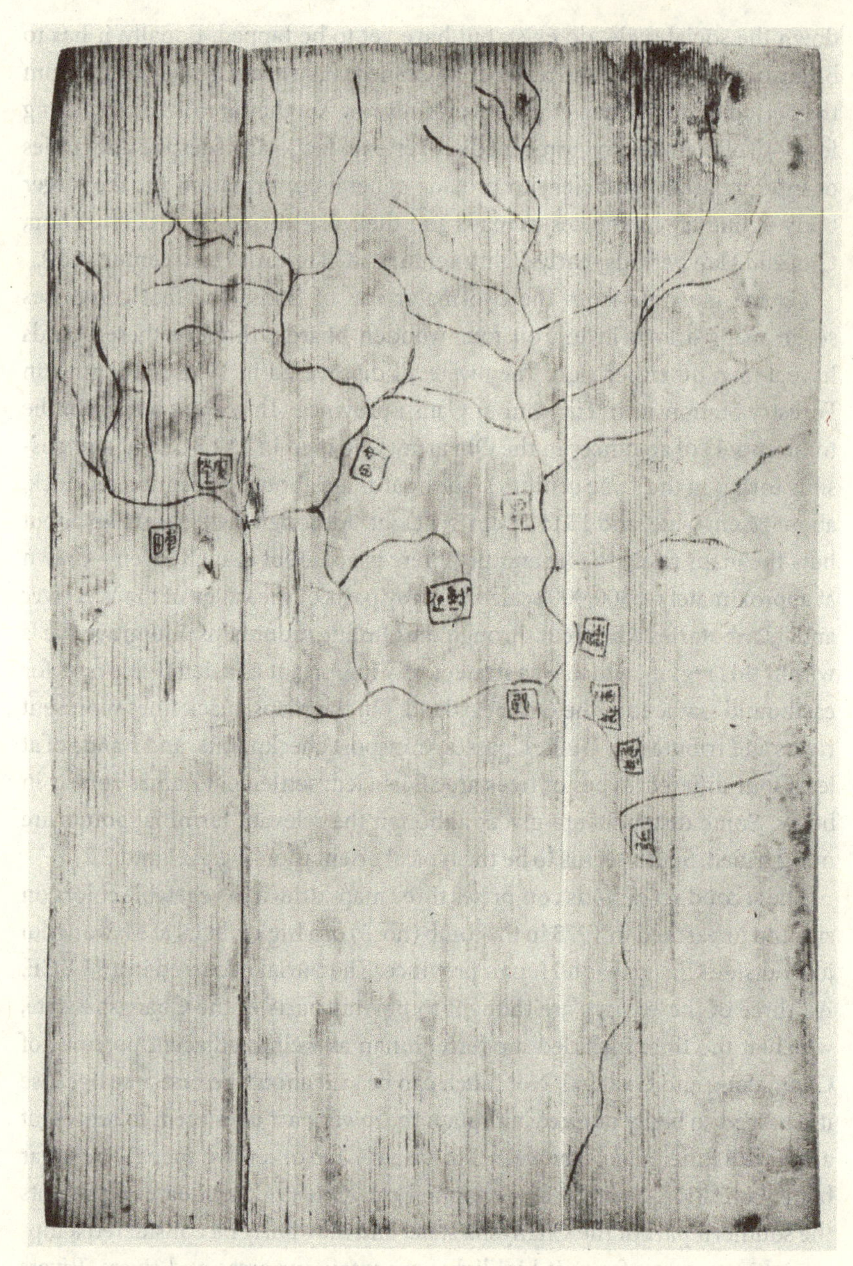

Fig. 2.1 Fangmatan map 1A. Photo: Wei (2014), vol. 7, 347.

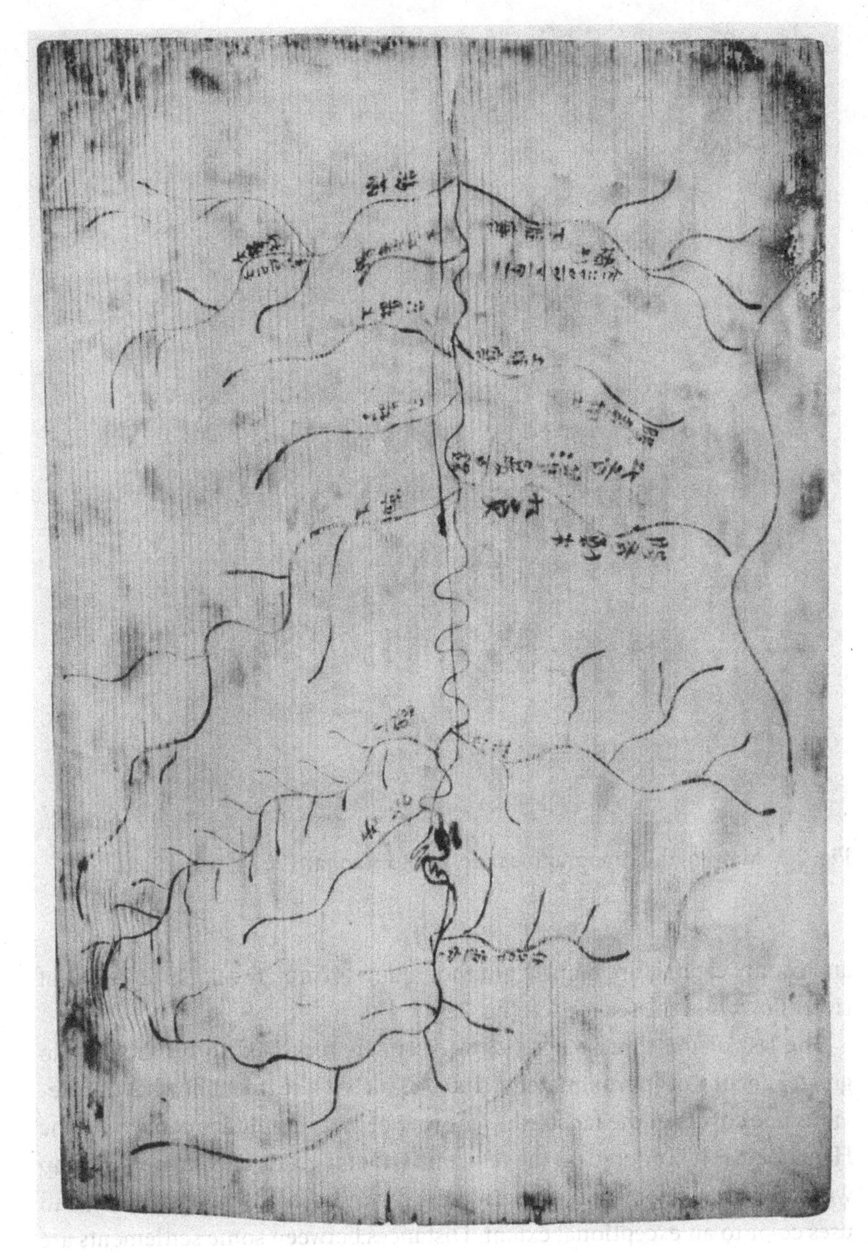

Fig. 2.2 Fangmatan map 3A. Photo: Wei (2014), vol. 7, 350.

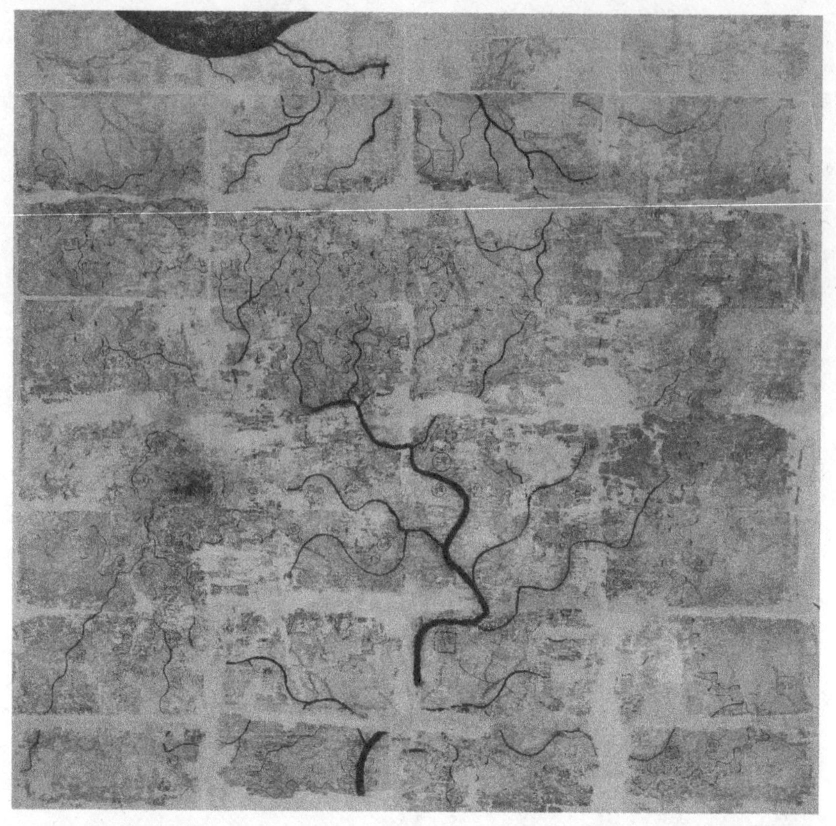

Fig. 2.3 Mawangdui topographic map. Photo: Yanjiao (2014), vol. 2, 152.

and county capitals are named, but mountains are not; the upper network of rivers flows into the sea top left (fig. 2.3).

The last of the three maps (98 × 78 cm) is reckoned to be showing in greater detail—at approximately 1:100,000 scale—a segment of what the previous one covers in the far south. This was an area of military concern to the Han, because it bordered on the reluctant tributary state of Nanyue. The map was surely drawn for military use; it marks army posts and headquarters, and uses color to an exceptional extent. Distances between some settlements are given, and in certain instances the number of households is also specified (fig. 2.4).

In general, all these finds reflect remarkable sophistication in the compilation and interpretation of maps; in no sense are they beginners' efforts. It follows therefore that the emergence of map consciousness on the part of

Fig. 2.4 Mawangdui military map. Photo: Yanjiao (2014), vol. 2, 159.

the ruling elite should go back at least to the period of intellectual ferment that marked the fifth to third centuries BCE in China, if not even earlier. To us, the maps are immediately striking for their "modernity"—with their standardized symbols, for example, and their sense of scale. The scale turns out to be by no means always accurate or uniform throughout a map, but still the sense of it is unarguable. Yee cautions against making modern Western "scientific" standards the main yardstick against which we assess any earlier civilization's cartographic achievement, and his warning particularly needs to be heeded in the case of ancient China just because the way in which its elite recorded spatial relationships is so close to ours.

Perhaps surprisingly, however, what Yee chooses not to pursue at all is the question of why Chinese civilization should have developed this method rather than some other of a distinctly different type. That is a question we raise much more readily in the case of ancient Rome, if only because the Roman outlook is so far removed from our own. But as Yee might agree, neither our own, nor that of ancient China, is necessarily "natural" or "inevitable." So is the character of its development in China to be attributed merely to chance? Or did Chinese culture already contain elements that can be identified as formative?

More specifically, the Chinese finds confirm a range of applications for maps quite unparalleled in the Greek or Roman world. Their importance to military commanders is especially clear, an impression reinforced, for example, by two passages from the third-century BCE *Guanzi* or *Book of Master Guan*:[4]

> In military affairs it is the duty of military commanders to examine maps, consult with the court astronomer, estimate accumulated stores, organize the brave warriors, acquire a broad knowledge of the realm, and determine strategy.

A chapter entitled "Maps" elaborates further:

> All military commanders must first examine and come to know maps. They must know thoroughly the location of winding mountain passes, streams that may inundate their chariots, famous mountains, passable valleys, arterial rivers, highlands and hills, the places where grasses, trees and rushes grow, the distances of roads, the size of city and suburban walls, famous cities and deserted ones, and barren and cultivated lands. They should completely store up [in their minds] the ways in and out of, and the contrasts, in the terrain; afterwards they can move their troops and raid towns. In the disposition [of troops] they will know what lies ahead and behind, and will not fail to take advantage of the terrain. This is the constant value of maps.

[4] Translation of both passages by Yee (1994), 73.

Both these passages, as well as the surviving maps themselves, reflect a comprehensive vision of landscape seemingly not matched in the Romans' way of organizing space. The one form of map that attained any wide usage among them (their own invention, it must be said) is the *itinerarium* or road map. From a cartographic perspective, it is a map stripped down to its absolute bare essentials. It simply shows travelers their route with the names of the towns they must pass through, and (if more details are offered) the distances between those places, and amenities like lodgings which are likely to be found in them. What such an itinerary does not furnish is any real insight into the landscape through which travelers will pass. This, to be sure, is information which strictly speaking they do not need to have, assuming that they can rely upon the road taken.

The old and well-developed Roman practice of *limitatio* likewise confines itself to no more than a partial vision of landscape: the practice, that is, of dividing up a community's land according to a grid pattern commonly termed "centuriation." Although substantial areas all over the Roman world came to be surveyed and divided thus, the network created took in only cultivable land, and was characteristically imposed with scant regard for variations in elevation. Whether terrain was flat or rolling, the grid looked the same (fig. 2.5a, b).

Romans seem unconcerned by their apparent inability to "organize" landscape conceptually wherever it lacked the vital markers that gave it shape for them—the markers, most notably, of towns, centuriation, and paved roads, some if not all of these indeed being features which Romans themselves introduced to the landscape. Given this perspective, it is no surprise that Strabo's description of the Iberian peninsula, for example, in his *Geography* book 3 concentrates upon the stable (and preferably romanized) communities, their cultivable land and navigable rivers. By contrast, the rest of the peninsula is pitiful and repellent to him. Its great tracts of mountain and forest breed dozens of lawless, unproductive tribes whose very names it is painful to record (names and groupings, it may be added, liable to reflect a degree of Roman fabrication or simplification).

It is surely this selective vision that we find reflected in what Artemidorus, interpreter of dreams, has to say in the second century CE on the instructive topic of understanding dreams in which you imagine flying along above the earth. Predictably enough, these indicate that you will be making a journey. Artemidorus continues:

> It is possible to ascertain from what is seen on the earth the kind of events that will be encountered during the trip abroad. For example, if one sees plains,

(a)

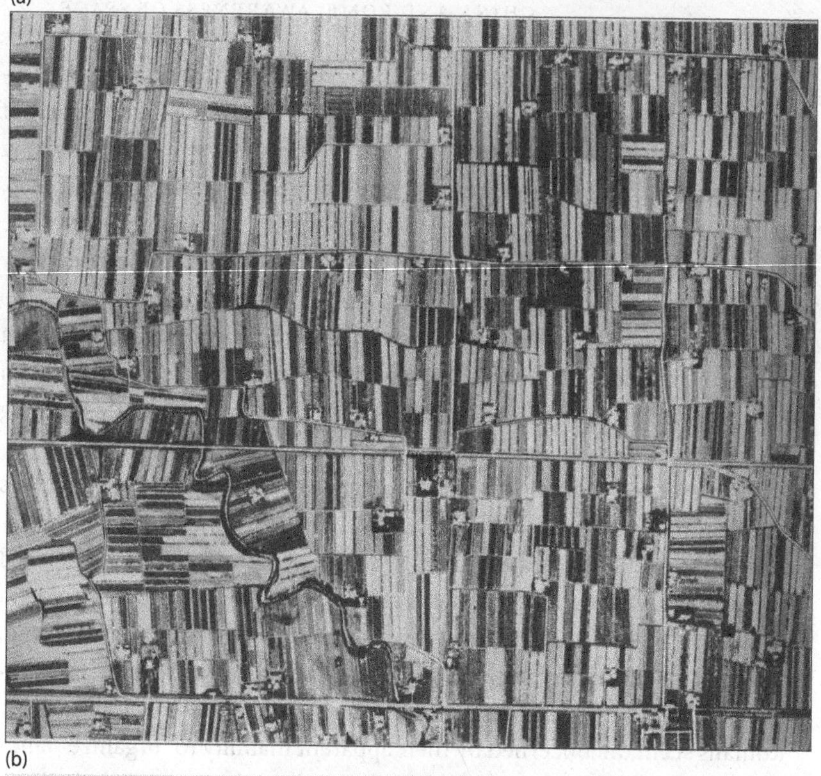

(b)

Fig. 2.5a, b (a) The pattern of Roman centuriation (here probably dating to the second century BCE) as still preserved in fields of the Po valley plain near Forum Cornelii (modern Imola, Italy), seen from the air. Photo: British School at Rome. (b) Fragments from an official plan inscribed on marble to record the centuriated subdivisions of part of the territory of Arausio (modern Orange, France). This plan was made around 100 CE at a scale of approximately 1:6,000. See fig. 18.1 for a drawing of all that can be reconstructed of it. Photo: Piganiol (1962), planche XXI.

> grain-lands, cities, villages, fields, all kinds of human activity, beautiful rivers, marshes, a calm sea, harbors, or ships that are sailing with a fair wind, all this foretells a good trip. On the other hand, valleys, ravines, wooded glens, rocks, wild animals, river torrents, mountains, and steep cliffs signify that nothing but misfortunes will occur on the trip. (*Oneirocritica* 2.68)

In the same vein he explains elsewhere that if you dream of "mountains, glens, valleys, chasms, and woods," these:

> signify sorrows, fears, disturbances, and unemployment for all. For slaves and criminals, they mean torture and thrashings. They portend harm for the rich because in these places something is always being chopped into pieces and thrown away. It is always better to pass through these regions, find the roads in them, descend from these places onto the plains, and to awaken from one's sleep when one no longer remains in them. (*Oneirocritica* 2.28)

It is clear that by contrast the Chinese vision of the landscape was much more comprehensive, one that consistently embraced countryside as well as town, upland as well as lowland, streams, forests, and other features. It is no surprise, therefore, that the Chinese elite made and used several different types of map for a wide variety of purposes. At one end of the spectrum there were maps of whole empires and their neighbors for political and strategic purposes (though the earliest surviving example only dates to the twelfth century); at the other end it was taken for granted that local registers of population, land ownership, livestock, and agricultural production would all be accompanied by maps. So deeply ingrained is the use of the map that it is taken as a key symbol of sovereignty. In particular, acknowledgment of submission or defeat is symbolized by the handover of a map of the state to an overlord or conqueror. As the *Han Feizi*, a philosophical text of the third century BCE, puts it:[5]

> To serve a great power always requires substantial concessions, wherefore one must offer up one's map and submit, and put the state seal in pawn for military aid. If the map is offered up, the territory will be cut up; if the state seal is handed over, prestige will be diminished.

[5] Translation by Yee (1994), 73.

It must likewise be because maps symbolize power that they are buried with officials or commanders. Such maps will illustrate the authority these individuals once exercised, and perhaps assist them in gaining a place of respect in the spirit world.[6]

There is an obvious temptation to draw the conclusion that the Chinese and Roman mindsets are far apart, but that may be too extreme. Romans, after all, were very much concerned to "organize" those elements in their surroundings of primary importance to them, and their surveyors developed excellent techniques with which to do this. Yet how Romans conceptualized the remoter areas beyond those surveyed remains more of a puzzle, although in general for long periods such remoter ones seldom posed the frequent menace that Chinese states habitually faced from their neighbors. Roman officials, too, gathered the kind of census data that their Chinese counterparts did. As we know from the jurist Ulpian, the owner of land had to set down

> the name of each property, the community and the *pagus* to which it belongs, its nearest two neighbors; then, how many *iugera* of land have been sown for the last ten years, how many vines vineyards have, how many *iugera* are olive plantations and with how many trees [and so on]. (*Dig.* 50.15.4 pr.)

What Romans hardly did, however, was to make that decisive mental leap which would enable them to grasp how valuable it could prove to record spatial relationships in cartographic form as opposed to just in writing. Even in the Chinese case, it is important to remember, maps never took precedence over, let alone superseded, the written word. China's culture, like Rome's, remained first and foremost a literary one. But the Chinese elite, unlike Rome's, did somehow discern, and very thoroughly exploit, the potential of recording spatial data cartographically as well. Everything seems poised for the Romans to have had the same realization too, but they never did. Their surveyors covered vast space, and they even sent copies of their local maps to Rome, but evidently no one ever took the initiative to collate all that material and to draw upon it for state business. Why none of this ever occurred in the Roman case must remain matter for further discussion.

[6] For the artistic and ritual significance of the Fangmatan and Mawangdui maps, see now Hsu (2010), 44–52.

In conclusion, the comparison with China does impress upon the Roman historian once again how underadministered the Roman empire seems, and how inadequate a grasp its rulers appear to have of it—at any rate, by the standards which modern Westerners share with the ancient Chinese. In relation to the total population, China's administrators were few in number; but in the case of the Roman empire the difference was even more extreme. China's rulers, high and low, made and consulted maps constantly as a matter of routine. How the Roman emperor and his officials comprehended from day to day their immense, sprawling territory, let alone the lands which lay immediately beyond, remains one of the many mysteries surrounding the functioning of Roman government. Even so, if their grasp was slight, what difference did it make?[7]

[7] See further now Wang (2015).

3
Grasp of Geography in Caesar's War Narratives

Today's readers typically struggle to comprehend the nature of Caesar's geographical grasp during his campaigns. So much that we take for granted must be set aside: in particular, Caesar gains no assistance from technology, digital or otherwise. Moreover, for him the several categories into which we routinely divide the acquisition and processing of geographical information—such as cartography of various types and scales, ethnography, and intelligence-gathering—remain undifferentiated. What we are prone to underestimate above all today is the depth of his geographical ignorance when he proceeds into Gaul and from there to Britain. It seems beyond belief that any leader in command of thousands of men could have been so rash and irresponsible. The fact is, however, that Caesar's plunge into vast territories of which he knew next to nothing was regular behavior on the part of Roman commanders—and indeed of their successors across much of the globe to the end of the nineteenth century CE. They had no practical alternative.[1]

Gaul—the landmass beyond "the province" that Rome had annexed in the late second century BCE—was not totally unknown to Romans. Envoys from such Gallic peoples as the Aequi and Allobroges had visited Rome during the years immediately prior to the start of Caesar's governorship in 58, and it must have been possible to glean geographical impressions from them. Caesar also evidently consulted some (now lost) Greek ethnographic writing, because in his *Gallic War* he once cites the third-century polymath Eratosthenes (6.24). The likelihood is that he had also read the much more recent work of Posidonius. Caesar shows keen awareness, too, of the notorious defeats inflicted upon Roman armies in Gaul near the end of the second century by migrant peoples, the fearsome, long-remembered Cimbri and Teutones.[2]

[1] See further Rambaud (1974); Austin and Rankov (1995); Bertrand (1997); Riggsby (2006), 21–45.
[2] Note especially 1.12.

It is natural enough for Caesar to begin his *Gallic War* by formulating a geographic and ethnographic overview of the whole of Gaul. In the absence of cities and highways (characteristically Roman features), the two main markers used for the purpose are ones that Caesar in fact continues to employ throughout the work, namely what he terms *civitates*—peoples, that is, whose territories he juxtaposes one to the other—in relation especially to principal rivers that act as boundaries, for example the Garumna, Matrona, Sequana, and Rhenus.[3] Thus he immediately identifies three main peoples—Belgae, Aquitani, Celts/Gauls—and further among the latter the Sequani and Helvetii; the neighboring Germans are also mentioned. Rivers aside, the elements of physical landscape that Caesar references in this opening description as (presumably) familiar to his readers are the (Atlantic) Ocean, Spain, and the Pyrenees mountains. The Roman province (Gallia Transalpina) is mentioned, too, although strikingly its relation neither to Italy nor to Gaul is clarified beyond repeated indication that Gaul lies to its north.[4]

To draw a reliable map on the basis of Caesar's overview—as we might wish to do nowadays—would hardly be feasible, but this limitation is made irrelevant by the fact that Caesar never indicates awareness of maps, and in all likelihood there hardly were any. His readers' outlook we can reckon to have been the same. Generally speaking, the outlook reflected even in such handbooks on generalship as survive from classical antiquity is no different. It is true that we learn of maps displaying the Roman world or regions of it, and a medieval copy of one survives (the so-called Peutinger map); but all are of later date than Caesar's time in the 50s BCE, when the regions controlled by Rome were still very scattered. It is also true that Romans made accurate maps at large scales of their cultivable land, although in this familiar territory the purpose was only to create a legal record of who owned what.

So Caesar's readers, having digested his opening overview, were then content—we may imagine—with the supplementary information on the local level that he continues to furnish as his narrative unfolds. Accordingly, for example, when the Raurici, Tulingi, Latovici, and Boii are first mentioned (1.5), their location in relation to the Helvetii is explained. The nature of the physical landscape in which notable events take place may be sketched—the mountains, rivers, and lake that hem in the Helvetii, for example (1.2), or the contrasting character of the only two routes by which they might migrate

[3] Modern Garonne, Marne, Seine, and Rhine, respectively; cf. 4.10.

[4] 1.16; cf. 4.20.

(1.6); equally, the border between the Sequani and Helvetii running along the Jura mountains (1.8), and the flow of the Arar (modern Saône) river (1.12). A forest of immense size, the Bacenis, extended far into the territory of the Suebi "and formed a natural barrier preventing the Cherusci and Suebi from raiding and inflicting damage on one another" (6.10).

The random inclusion of figures for area and distance creates an air of precision and geographical mastery. Caesar is somehow able to state, for example, that the territory of the Helvetii extended 220 miles from north to south and 165 from east to west (1.2), that Britain is almost 2,000 miles in circumference (5.13), and that the Ardennes forest stretches for more than 460 miles (6.29). His record may specify how far he found himself at this juncture or that—from the enemy's forces, for example ("just over seven miles," 1.21), or from an important location such as Bibracte ("no more than sixteen and a half miles," 1.23). Time may be substituted for distance in instances where the latter's precision is unattainable. Thus the remote Hercynian forest in central Europe, which Caesar read of in Eratosthenes, is said to take nine days for a man traveling light to traverse, and its full length would not be reached even after a sixty-day journey through it. Its size, Caesar adds, "cannot be described more accurately, for the Germans have no means of measuring units of distance" (6.25).

Throughout the Gallic and British campaigns Caesar is tireless in seeking to acquire reliable local geographical knowledge by sending out his own scouts as well as by interrogating whatever local individuals, envoys, traders, deserters, or captives he may encounter. As a result, the Aeduan chief Diviciacus tells Caesar of an advantageous route (1.41), and envoys from the Ubii advise him on how best to approach the territory of the Suebi (6.9). He sends Gaius Volusenus Quadratus to reconnoiter Britain from the sea, while at the same time summoning traders from all over Gaul to tell him what they know (4.20–21):

> . . . at that point he was unable to ascertain either the size of the island, the nature and numbers of the peoples living there, their skill in warfare, their established customs, or which harbors were suitable for a fleet of fairly large ships.

Caesar is aware of comparable zeal to acquire intelligence of every kind on the part of the Gauls, although in his opinion their leaders are prone to evaluate what they learn with insufficient caution.[5] He at least is shrewd enough

[5] 4.5; cf. 7.20–21.

to be more critical: thus in the light of reports that he has received from elsewhere, he already knows that German envoys have failed to furnish him with fully accurate information (4.9).

Caesar has the self-assurance not to panic at the unexpected, as when he has been advancing with his army for three days into the territory of the Nervii, only then to learn that the enemy's forces are massed in wait across the Sabis river a mere nine miles or so away (2.16). He is alert, too, to the risks run by Roman forces when they penetrate country that is rugged, wooded, and marshy, where hostile local peoples can exploit the environment in ways quite beyond the capacity of newcomers (6.34). As Caesar realizes, the worst predicament for any commander in such an isolated and vulnerable situation is for his men to dwell upon the potential consequences of such ignorance and thereby lose their nerve. This type of crisis—fed by rumor and panic—he must overcome at an early stage when he wishes to advance against Ariovistus. The men, he records, "declared that it was not the enemy they feared, but the restricted, narrow route of the march, the depths of forest between themselves and Ariovistus, or the arrangement of satisfactory transport for the grain supply" (1.39). In reaction, Caesar severely reprimanded the centurions "primarily for thinking that it was their business to inquire or think about either the direction or the strategy of the march" (1.40). Ironically, in the case of Britain, Caesar is proud to present Romans' total ignorance—shared by almost all Gauls, he adds (4.20)—as sound justification for his landing there, and thus gaining glory as the first Roman to report on its geography and ethnography. He evaluates his crossing of the Rhine river in the same terms (4.19).

Altogether, successful campaigning in largely unknown territory demanded a mix of bravado, circumspection, and quick thinking, with constant improvisation, risk-taking, and spur-of-the-moment decisions. Disaster could occur all too quickly, but time and again Caesar displayed the skill and energy to avoid it. The siege of Uxellodunum—a crucially important episode narrated by Hirtius (8.33–43)—is a characteristic instance of how Caesar excelled in this environment. His very arrival here is unexpected. He finds the Romans confronted by a well fortified and supplied town in an impregnable position. So he determines to cut off its water supply, and does so by using archers to prevent access to the river, while commissioning extraordinary siege-works that prevent the townspeople from accessing their spring below the walls; the Romans even tunnel under the spring to divert its channels.

Caesar's *Civil War* presents a marked contrast to the *Gallic War* narrative in the degree of geographical knowledge expected of readers. In this work the campaigns are fought for the most part within Roman territory. Hence the brief remarks made about, say, the geography of Pelusium and the Nile delta, or the mountain range linking Pontus with Lesser Armenia, are the exception.[6] A concise review of omens favoring Caesar's victory at Pharsalus refers matter-of-factly to happenings at Ephesus, Elis, Antioch in Syria, Ptolemais, Pergamum, and Tralles (3.105). To be sure, their locations are not integral to a grasp of campaigns in this instance, but that awareness is called for earlier in the same book when the run-up to the battle of Pharsalus is narrated. Consider the accounts of how Caesar's generals obstruct Scipio in Macedonia and Thessaly,[7] or of the movements of both Caesar and Pompey immediately prior to the battle (3.78–80). These demand an informed grasp of the geography of Greece. Such explanatory asides as "the Haliacmon river which divides Macedonia from Thessaly" (3.36), or "Gomphi, which is the first town in Thessaly as you come from Epirus" (3.80) are conspicuously rare. It is no less exceptional for two brief reflections on landscape to be offered in a chapter of the *Spanish War* (8):

> Throughout almost the whole of Further Spain the fertility of the soil and the equally plentiful supplies of water blunt the effect of sieges and make them difficult.
>
> The majority of the towns of the province are given a fair degree of protection by hills and are built on naturally elevated sites, so that it is not easy to approach or climb up to them.

Overall, there is no question that the *Civil War* narratives are written by authors who command an impressive grasp of the geography of Rome's territories and assume the same of their readers. Where and how in the course of their upbringing Romans acquired and retained such insight remains a puzzle, however.[8] Again, as with the *Gallic War*, the use of maps is not even

[6] *Alexandrian War* 26, 27, 35.
[7] Especially 3.34–36.
[8] See further essay 18 below.

hinted at. Meantime at the local level a strong chance of going astray in the course of a journey persisted. By altogether omitting from the *Civil War* the famous episode of his crossing the Rubicon in January 49, Caesar need not mention how he lost his way for hours (granted, during the night) in trying to reach the river by back roads.[9]

[9] Suetonius, *Divus Julius* 31; Whittaker (2002), 81.

4
Trevor Murphy's *Pliny the Elder's Natural History: The Empire in the Encyclopedia* (2004)

Without question, in their own quiet way the newly attested Dii Itinerarii[1] must have viewed the writing of this fine book with favor. It stands as further confirmation of how rewarding the current revival of interest in Pliny the Elder continues to prove, as demonstrated most recently by Sorcha Carey's *Pliny's Catalogue of Culture: Art and Empire in the Natural History* (2003). Murphy differs from most of his immediate predecessors, however, by focusing attention less on the author of the *Natural History* and more on the work itself, on the encyclopedia "as a cultural artefact," as he phrases it (11), and hence "as a source for ancient Roman culture" (2). This is a creative approach, undertaken in an introduction, five chapters, and conclusion. Murphy appreciates that Pliny—despite his active administrative career in several provinces, and his death caused by the desire to observe the eruption of Vesuvius at close range—gathered most of his learning from books. This was no cause for embarrassment on Pliny's part. On the contrary, it was only because of Roman conquest that detailed knowledge of the world had been unlocked, and only thereby did the opportunity arise to encapsulate it all proudly in an encyclopedia. Greeks had not written this type of work, and such earlier Romans (perhaps Cato the Elder, certainly Varro and Celsus) with the courage to try had plainly lacked the range and depth of far-reaching knowledge that Pliny could command by the Flavian period. He presented it not so much to instruct like a textbook, but rather to offer a work of reference, to classify and validate knowledge (be it familiar, novel, or marvelous), and hence above all to celebrate the Roman power which underpinned this authoritative grasp of the world. To Murphy, the *Natural History* can fairly

[1] Or Itineris: *AE* 2000.1191, from Pannonia Superior.

be compared to an ancient map, as well as to a Roman triumphal procession where newly discovered peoples, lands, plants, animals (or their images) are all paraded for public viewing and instruction.[2]

Murphy's first two chapters reflect upon the shape of the *Natural History* as a whole from the reader's perspective, and the logic informing it. He discerns that within the ostensibly tidy sectioning of the work readers soon encounter a diffuse, breathless style, a devotion to antitheses, and an idiosyncratic linking of ideas which leads repeatedly to digression. This last, disorienting characteristic he sees as no mere accident, but as a deliberate choice on Pliny's part, both artistic and aesthetic, and a justification for the provision (unusual in ancient works) of a table of contents that in fact fills the first book of the thirty-seven.

The encyclopedia's anchor throughout, Murphy goes on to propose in chapter 2, is Rome. In Pliny's eyes, Rome is the touchstone to which themes or foreign items are invariably related, or against which they are measured. Pliny not only boasts of the "20,000 things worthy of consideration" that he has included (54), but he also (again unusually for ancient works) takes pains to cite his sources. His purpose is in part to demonstrate his engagement with other writers past and present, but he is eager, too, to disclose his debt to intellectually active social superiors among his contemporaries, recalling the Late Republican milieu where Cicero and his circle dedicated books to one another and exchanged stories. At the same time, just as Pliny deplores indulgence in luxury, so he deplores those who decline to share knowledge, especially at a time when he fears that the changes induced by Rome's very success are leading to many of the traditional Roman ways decaying unrecorded.

In chapters 3–5 Murphy taps the geographic and ethnographic sections of the *Natural History* as a fertile means of uncovering the quintessentially Roman character of the work's purpose. The long-established ethnographic tradition is introduced prior to a focus on Pliny's contribution to it, found in disjointed passages of his work with no immediately obvious connection to one another. Following the approach taken to Herodotus by François Hartog—who probed ethnographies for the light shed upon the ethnographer, not for truth-value—Murphy explores Pliny's treatment of luxury, of frankincense from Arabia, of Taprobane (its riches and society), and of the

[2] See further essay 18 below.

Seres, Essenes, and Hyperboreans. What unites these very diverse passages is seen to be a preoccupation with contemporary moral dilemmas, in particular how luxury is to be defined, and to what degree asceticism and suicide each have merit.

Chapter 4 interprets the geographical section of the *Natural History* (following a lead given by Nicholas Purcell) as the product variously of bird's-eye view, itinerary by land or sea, and narrative centered around the two mighty conflicting features that frame the physical landscape: rivers and mountains. The parade of fresh knowledge displayed in triumphal processions at Rome is identified as a favored setting and organizing principle for Pliny's absorption of the novel and unfamiliar.

Pliny's fullest description of a primitive people is the centerpiece of chapter 5. No matter that modern scholarship can demonstrate the inaccuracy of his account of the Chauci (16.2–4). Rather, Murphy seeks to account for their place in Pliny's worldview, and finds it in their unstable liminal environment, caught between ordered Nature and Roman power in one direction and the primeval, uncontrollable disorder of Ocean on the other. Their plight acts to remind Romans that there is a point beyond which the natural order cannot be maintained, so that Roman ambitions should extend no further. Moreover, the sheer insecurity of the Chauci illustrates the fragility of civilization. It, too, may eventually be overwhelmed by floods (evoked most vividly by Seneca), and Rome itself may sink down into the Cloaca Maxima sewer (described with special pride by Pliny 36.104–108) which underpins it.

The place that Pliny claims for himself in his encyclopedia is the focus of the conclusion. In imperial Rome knowledge reflected power wielded by emperors, and should properly be validated and controlled by them. The bold author whose work sought to encompass all knowledge must therefore not seem to be rising above his emperor, and Pliny's preface in particular was crafted with care to dispel any such impression. In the longer term, his encyclopedia enjoyed undreamt-of success. It continued to be copied, and the sources it drew upon suffered neglect. For over a millennium it defined knowledge; in the Renaissance it was first a model and then a challenge to supersede.

Murphy's ambitious reading of the *Natural History* and of the thinking behind it is a persuasive one that coheres. Its thrust could be variously reinforced and extended. As a further small illustration of the emperor as the authority to whom all curious knowledge must be presented, I did miss the story which for long eluded Fergus Millar, of the mother who worried that she would lose her son to the emperor because of his unique ability to understand the language of birds.[3] The likeness to be seen between ethnographic images displayed in Roman triumphs and the "mock villages" that became such an indispensable feature of late nineteenth- and early twentieth-century World's Fairs has been commented upon by Mary Beard,[4] but perhaps too recently to be taken into account. Limitation of the imaginative modes that dominated the conceptual geography of the Romans to five (131, following Purcell) overlooks a sixth in constant use by Pliny, namely the division of the Roman world into its provinces. Purcell drew attention to the empire's "cellular" nature, "a great mass of individual units whose only common matrix was relationship to Rome."[5] Yes, to be sure. But it is no less striking conceptually that as soon as Rome established two contiguous provinces (Hispania Ulterior and Citerior in the 190s BCE), the senate at once ordered the land boundary between them to be defined.[6] From Augustus' Principate onward we may fairly imagine widespread awareness of the principal cells' spatial relationship to one another. Monuments large or small marked not only the termini of great roads and the edges of the empire (cf. 175) but also the boundaries within the quilt of provinces. Pliny assumes this conceptual grasp of his readers, as does Ptolemy in his *Geography*, among other authors, and it is no less fundamental to an understanding of the Antonine and Bordeaux itineraries.[7]

Murphy is right to be struck (129) by the enumeration in the table of contents (book 1) of the total numbers of towns, races, rivers, mountains, and other features that Pliny's readers will find recorded in each of the geographical books (3–6). This is none other than the format expected of the well-prepared Roman itinerary,[8] as seen for example on the Vicarello cups or in the Bordeaux itinerary (fig. 4.1a, b).[9]

Murphy's larger vision of the *Natural History* as "like an ancient map of the world" with Rome as its center (20) is especially exhilarating, although for us the cartographic image which it recalls cannot be the lost map of Agrippa (157) but surely can be the later, unmentioned Peutinger map. Here Rome dominates at the center, seas and continents are remolded (cf. 46–47), and

[3] Porphyry, *On Abstinence from Living Things* 3.3.7, cited in Millar (1992), 637–38.
[4] Beard (2003).
[5] Purcell (1990), 8.
[6] Livy 32.28.11 with Richardson (1986), 77.
[7] See further essay 7 below.
[8] See further essay 9 below.
[9] Note that this itinerary totals not only distances in miles (*milia*) but also the lesser stopping-points (*mutationes*) as well as the larger ones (*mansiones*).

(a)

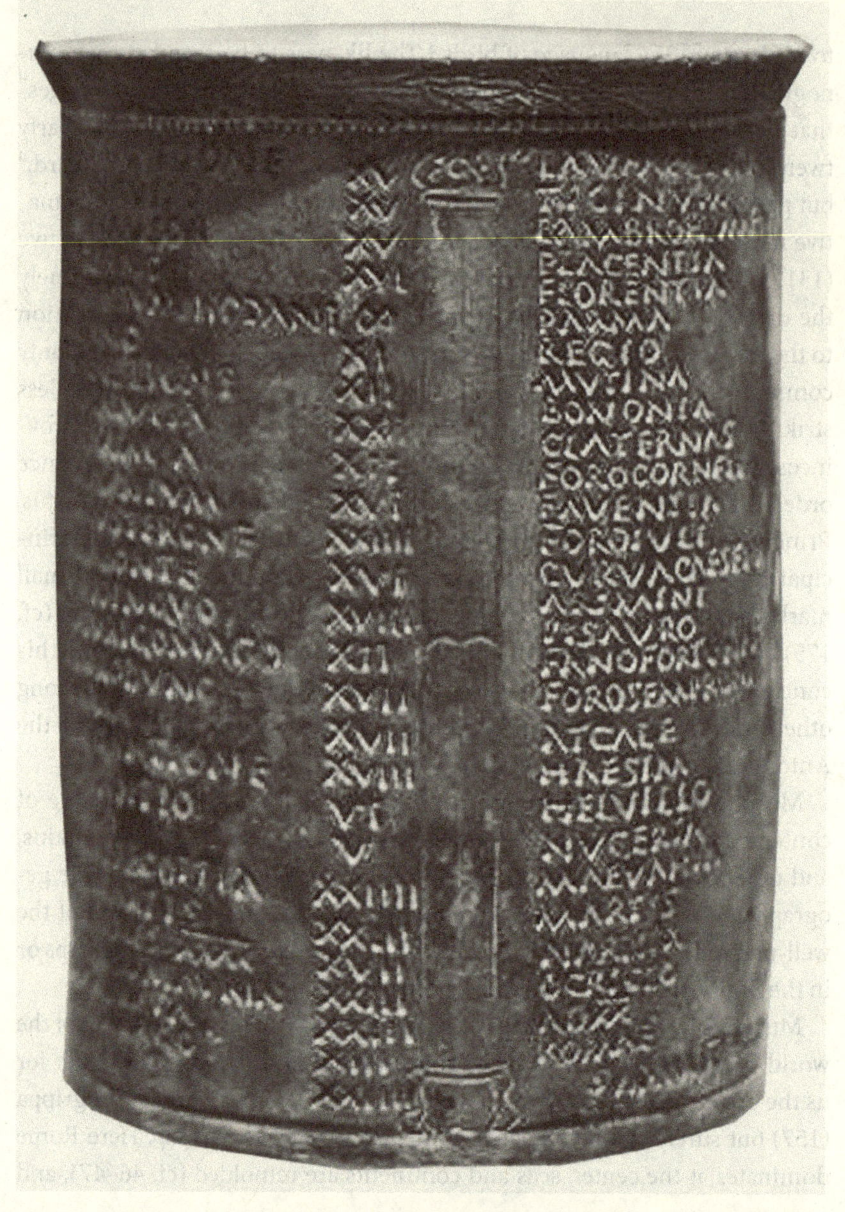

Fig. 4.1a, b Four inscribed miniature silver beakers, probably produced during the first century CE, were found in a sacred spring at Vicarello, northwest of Rome. This one (*CIL* XI.3284) is the shortest in height (9.5 cm). Transcribed below is its list (b) in four columns of over 100 stages on the long land journey from Gades (modern Cádiz, Spain) to Rome, specifying the number of Roman miles from one to the next. The prominent column in (a) here is the fourth, with Laumellum in the Po valley at the top (*BAtlas* 39D3) and then the stages—via Mutina, Ariminum, and through the Apennines—to journey's end in Rome, with the total distance, 1,835 miles, recorded below. See Cassibry (2021), 17–62. Photo: Jesús Rodríguez Morales.

(b)

A G A D I B V S ROMA
ADPORTV XXIIII
HASTA XVI
VGIAE XXVII
5 ORIPPO XXIIII
HISPALI VIIII
ABHISPALICORDVBAE
CARMONE XXII
OBVCLAE XX
10 ASTIGI XV
CORDVBAE XXXV
ABCORDVBATARRACONE
ADDECVMVM X
EPORA XVIII
15 VCIENSE XVIII
ADNOVOLAS XIII
ADARAS XXIIII
ADMORVM XVIIII
ADDVOSOLARIA XVIIII
20 MARIANA XX
MENTESA XX
LIBISOSA XXIIII
PARIETINIS XXII
SALTIGI XVI
25 ADPALAE XXXII
TVRRES SAETAB XXV
SAETABI XXV
SVCRONE XVI

VALENTIA XX
SAGVNTO XVI
ADNOVA XXIIII
ILDV XXII
INTIBILI XXIIII
DERTOSA XXVII
SVBSALTV XXXVII
TARRACONE XXV
ATARRACONENARBONE
PALFVRIANA XVI
ANTESTIANA XIII
ADFINES XVII
sic ABRAGONE XX
ADPRAETORIVM XVII
BAETERRAS XVI
AQVASVOCONIAS XV
GERVNDA XII
CINNIANA X
IVNCARIA XII
SVMMOPYRENAE XVI
RVSCINONE XXV
ADCOMMVSTA VI
NARBONE XXXIIII
ANARBONETAVRINOS
BAETERRAS XVI
sic CESSIRONE XII
FR∘NTIANA X
FOR∘DOMITI VIII

S E X T A N T I O N E XV
AMBRVSIO XV
NEMAVSO XV
VGERNO XVI
5 TRAIECTVM RHODANI ∞
GLANO XI
CABELLIONE XII
APTAIVLIA XXII
CATVIACIA XII
10 ALAVNIVM XVI
SEGVSTERONE XXIIII
ALABONTE XVI
VAPPINQVO XVIII
CATVRIGOMAGO XII
15 EBORODVNO XVII
RAMA XVII
BRIGANTIONE XVIII
DRVANTIO VI
TYRIO V
20 INALPECOTTIA XXIIII·
ADMARTIS XXIII
ADFINES $\overline{XXXX}$ XVII
AVGVSTATAVRIN XXIII
QVADRATA XXIII
25 RIGOMAGO XIII
CVTTIAS XXIIII

LAVMELLVM XII
TICINVM XXI
LAMBROFLVMEN XX
PLACENTIA XVII
5 FLORENTIA XV
PARMA XV
REGIO XVIII
MVTINA XVII
BONONIA XXV
10 CLATERNAS XI
FOROCORNELI XIII
FAVENTIA X
sic FOROIVLI X
CVRVACAESENA XIII
15 ARIMINI XX
PISAVRO XXIIII
FANOFORTVNAE VIII
FOROSEMPRONI XVI
ATCALE XVIII
20 HAESIM XIIII
HELVILLO X
NVCERIA XV
MAEVANIA XVIIII
MARTIS XVI
25 NARNIA XVIII
OCRICLO XII
AD $\overline{XX}$ XXIIII
ROMAE XX
SVMMA [illegible] [illegible]CCCXXXV

Fig. 4.1a, b Continued

the display of routes, principal settlements, road stations, and distances is unashamedly both triumphalist and encyclopedic. At the same time, mountain chains, and even more so the freely flowing rivers—the great ones with their multiple tributaries and branches—stand out as prominent, formative components of the landscape, just as Pliny regarded them.[10]

It is a tribute to Pliny's skill and energy that he was able to complete an encyclopedia which reorients an entire Greek and Roman intellectual universe by placing Rome firmly at its center, a work that no ancient successor sought to improve upon, one that continued to be copied in full despite its length. Meantime its sources faded away, like Ptolemy's predecessor Marinus or the jurists excerpted for the *Digest*. In the fifteenth century it was from Pliny that

[10] See further essays 13 and 14 below. For a view of the world with Rome its center-point as long established Roman practice, note now Gargola (2017).

the naturalists of the Renaissance took their start, just as its mapmakers and explorers were inspired by Ptolemy. Murphy convincingly succeeds in setting Pliny's outlook and methods within the political, social, and cultural context of his own time and of Roman tradition. His perceptive investigation, while pitched for a specialist audience, is concise, well documented, and a pleasure to read. It also underlines, incidentally, what valuable service a new, sound English translation of books 3–6, if not more, could perform in opening the *Natural History* to a wider range of modern readers.[11]

[11] See further essay 5 below.

5

An English Translation of Pliny's Geographical Books for the Twenty-First Century

The project in progress discussed here stems from a perception that Pliny the Elder's *Naturalis Historia* has untapped potential to advance understanding of ancient geographical knowledge and thought. An English translation could usefully assist that purpose. The project's initial impetus and its design relate primarily, therefore, to the geography in the work's opening books rather than to a preoccupation with its author himself.

For many years past there has been widespread recognition that the sole usable English translation of Pliny's geographical books—made for the Loeb Classical Library over eighty years ago (see further below)—is long overdue for replacement.[1] My former pupil and Ancient World Mapping Center director Brian Turner, now associate professor of history at Portland State University, Oregon, first resolved to meet this need. At the outset he consulted me, and I agreed to help. But as this informal help continued, he then insisted that I become his full partner. I consented, on the two conditions that he should have the final say on any disputed issues, and that his name should precede mine on the title page of the book we now have a contract to publish with Cambridge University Press.[2]

Devoted admirers of Pliny might interject that Turner and I should be reprimanded for separating the geographical books from the rest of the *Natural History* and then limiting our translation to them. To be sure, the objection has merit. Even so, it may be considered utopian in character as well as impractical, given the exceptional length of the work and the great diversity of its aspects.[3] Moreover, we readily acknowledge the need to keep in mind

[1] The point is made at the end of essay 4 above.

[2] My thanks to Brian Turner for commenting on this essay in draft and confirming that it reflects our joint thinking to date.

[3] The choice to focus on just one aspect of Pliny's work is of course far from novel: note the length of the section "Subject Areas in the *Natural History*" in Doody (2015).

Pliny's own conception of the work as a single whole, as well as the important fact that its different parts do interrelate rather than remaining merely self-contained. This is particularly the case with the geographical books, because they come first. They lay the foundation for the treatments of peoples, animals, plants, and minerals that will follow. Notably, for example, in book 7 on the human animal (immediately following the geographical books) Pliny expands on the character and lifestyle of Scythian cannibals, northern Arimaspi, and numerous extraordinary peoples on all three continents first mentioned in books 3–6.

What books should our translation cover? Books 3–6 without question, but we think it essential to include book 2 also (on the universe and its elements). Throughout, at the appropriate points, we shall insert the subject headings listed in book 1, and we shall also reproduce its listing of each book's sources. In a substantial appendix we mean to translate notable "geographical" passages that occur in later books. The draft list of such passages appended here is no more than a provisional selection, but it serves to demonstrate our thinking as well as to underscore our concern to recognize that books 2–6 form an integral part of a much larger whole in which geography continues to feature constantly (fig. 5.1).

Which editor's Latin text should we translate, or should we prepare our own edition? To address the second question first, without hesitation we resolved not to prepare our own edition. This would be an unavoidably herculean task, given (on the one hand) the mass of manuscripts and their difficulties, not to mention our lack of qualifications for editing texts, and (on the other hand) the fact that satisfactory editions are already available. Admittedly, no single editor since Carl Mayhoff for Teubner in 1906 has published a text of books 2–6 in their entirety. But today there are volumes in the (French) Budé series for all but the latter two-thirds of book 5 and the first and third quarters of book 6.[4] Our decision, therefore, has been to translate the Budé text where there is one—all of it published since 1980, except for book 2, which dates to 1950. Otherwise we turn to the (German) Sammlung Tusculum text; everything we need from this series was published in the 1990s.[5] By this means we leave decisions about textual matters to experts alone.[6] We also

[4] Book 2: Beaujeu (1950); book 3: Zehnacker (2004); book 4: Zehnacker and Silberman (2015); book 5, sects. 1–46: Desanges (1980); book 6, sects. 46–106: André and Filliozat (1980); book 6, sects. 163–220: Desanges (2008).

[5] Book 5: Winkler and König (1993); book 6: Brodersen (1996). In addition, the *Gesamtregister* compiled by Bayer and Brodersen (2004) is invaluable.

[6] All Budé texts are presented with an apparatus criticus. Sammlung Tusculum volumes record editors' variant readings in an extensive appendix.

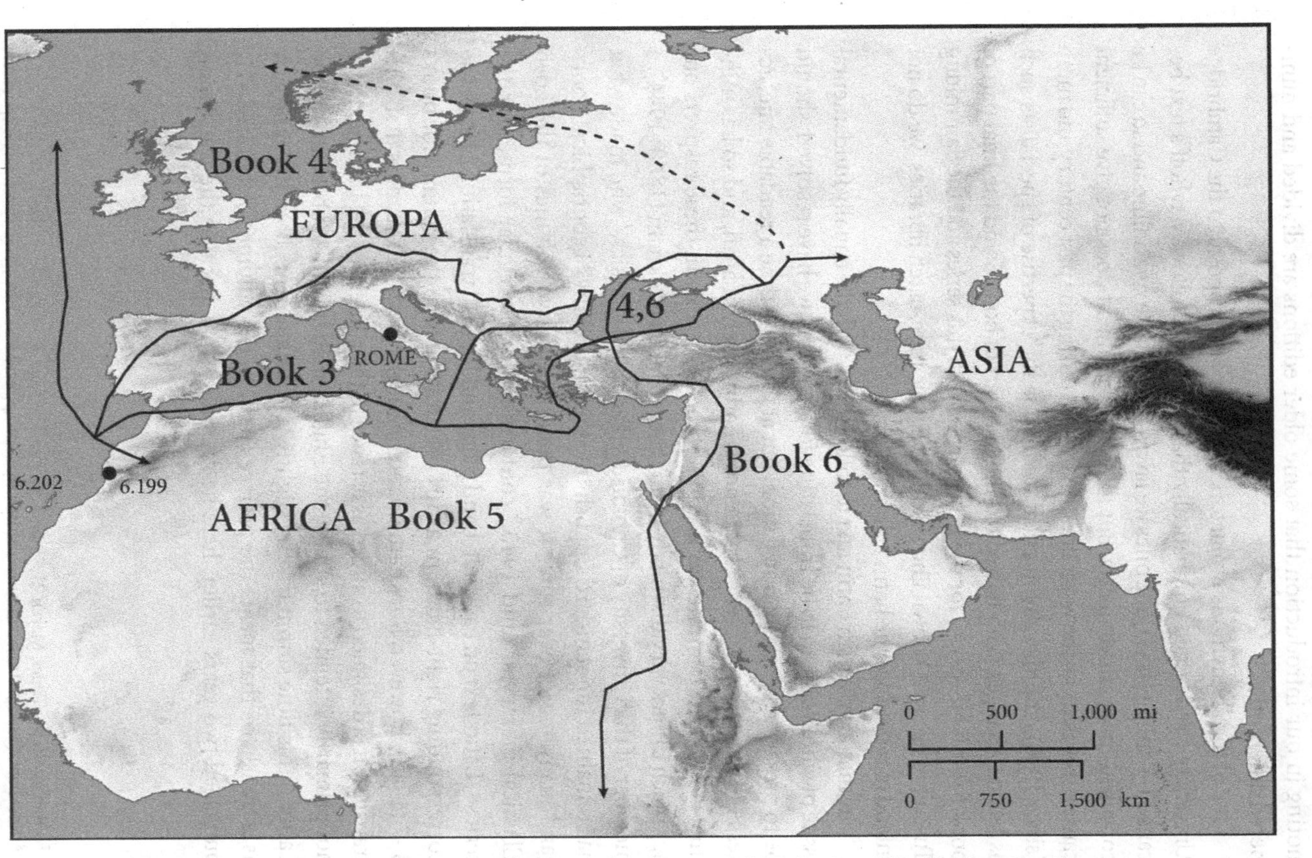

Fig. 5.1 Pliny's coverage by continent in *Natural History,* books 3–6. Map: Ancient World Mapping Center.

divide and number the sections of text as in Budé and Tusculum only, while noting in our introduction that some older editions are divided and numbered differently.

One of the anonymous evaluators of our project proposal to the Cambridge University Press urged emphatically that we should follow Mayhoff's text because it—unlike (as yet) the ones we favor—is available free online and so can be consulted more conveniently. But to us, this argument does not give sufficient cause to set aside all the work done on the text during the past century and more (since 1906), and thereby in effect to devalue the expertise of specialists such as Jehan Desanges, who has edited the first part of book 5 and the final part of book 6 (both on Africa) for the Budé series. Both this series and the Sammlung Tusculum are well known and readily available in research libraries; we do not intend to reproduce the Latin texts of either.

The matter of which Latin text to follow links with the equally fundamental issue of who we mean to be translating for. Our aim, we have resolved, should be to present a scholarly translation, but at the same time a readable one, accessible to a broad audience that knows Latin barely, if at all, and so has minimal concern to examine any edition of the Latin text. In these respects our approach differs very distinctly from that of Harris Rackham (1868–1944), a member of the University of Cambridge Faculty of Classics,[7] who made what has remained the standard English translation of Pliny's geographical books (and many more) ever since it was published in two volumes of the Loeb Classical Library around 1940,[8] conforming to the regular format for the series—Latin text on each left-hand page, and English translation of it on the corresponding right-hand. In a curt prefatory note to the volume for books 3–7 (1942), however, Rackham unapologetically warns his readers that "this translation is designed to afford assistance to the student of the Latin text; it is not primarily intended to supply the English reader with a substitute for the Latin."[9] This note summarizes a slightly fuller statement about his approach as a translator that forms the preface (p. v) to the volume for books 8–11, published two years earlier: here he maintains that, because the translation

[7] Obituary in the *Times*, March 21, 1944.

[8] Rackham (1938, 1942). Another volume by him appeared in 1940, and three more (posthumously) in 1945, 1950, and 1952.

[9] A more restrictive purpose in fact than that articulated by James Loeb when he initiated the series in 1912. This note has been dropped from reprints issued after 1969.

is to be printed facing the original text, its purpose should be "to assist the reader of the original to understand its meaning."[10]

Rackham's translation of books 2–6 has proven long-lasting, but really by default, because a replacement in English has yet to appear. The lack of one for this important text has become all the more frustrating when such a wealth of new English translations of Greek and Latin geographical works has appeared during the past twenty years. In the circumstances it is a further disappointment that Rackham evidently showed minimal concern for how his translation of Pliny's geographical books might be framed to last. I say this with particular reference to his practice in one prominent respect vital for these books: the rendering of place-names. If Rackham did in fact formulate methodical principles by which to render the names of places and also of peoples (as any translator of Pliny should), he never explains them; instead he proceeds with a surprising and unfortunate degree of inconsistency. His practice seems more puzzling still in the light of his warning that he did *not* intend to supply a substitute for Pliny's Latin.

In this event, his logical course would surely be to reproduce a name just as Pliny records it in Latin, even when the place or feature is identifiable and its current equivalent modern name readily established. In fact Rackham duly does just this in the case of Lixus and the others underlined in the following specimen passage (5.9, as presented in both Latin and English by him):

> Ad flumen Anatim CCCCXCVI, ab eo Lixum CCV Agrippa, Lixum a Gaditano freto CXII abesse; inde sinum qui vocetur Sagigi, oppidum in promunturio Mulelacha, flumina Sububum et Salat, portum Rutubis a Lixo CCXXIV, inde promunturium Solis, portum Rhysaddir, Gaetulos Autoteles, flumen Quosenum, gentes Velatitos et Masatos, flumen Masathat, flumen Darat, in quo crocodilos gigni.

> Agrippa says that to the river Anatis is a distance of 496 miles, and from the Anatis to Lixus 205 miles; that Lixus is 112 miles from the **Straits of Gibraltar** and that then come the gulf called Sagigi Bay, the town on Cape Mulelacha, the rivers **Sebou** and **Sallee**, the port of **Mazagan** 224 miles from Lixus, then **Capo Blanco**, the port of **Safi**, the Gaetulian *Free State*,

[10] Rackham's 1940 preface is reprinted in the volume for books 12–16 (1945), vi. The first volume to appear (1938) begins with a concise introduction, but this makes no reference to the character of the translation.

> the river **Tensift**, the Velatiti and Masati tribes, the river **Mogador**, and the river **Sous**, in which crocodiles are found.

This prudent choice of retaining what is in the Latin eliminates the risk to be incurred by substituting the modern current name instead; sooner or later there may be some change to it, rendering the translation increasingly out of date and hard to comprehend. Even so, in other instances (set in bold type here) Rackham opts for exactly that alternative: he dispenses with Pliny's name and substitutes a contemporary one of the post–World War I period. By now, up to a century later, predictably enough this is often a name superseded decades ago, so that today's readers are liable to find its use variously disorienting, quaint, or offensive for its association with a rejected colonial past. To create further bafflement, the choices that Rackham makes between use of a name's ancient form or its equivalent in his day seem merely random. Thus he could just as well have substituted a modern name for ancient Lixus (Larache, say) but for whatever reason he did not, and equally he could have retained ancient Rhysaddir instead of substituting modern Safi.

In other instances still—again at random, it seems—Rackham chooses to set aside Pliny's form of the name and to use instead an English translation of its meaning in Greek or Latin. Here he could have done this for promunturium Solis ("Sun's cape"), though he preferred to substitute its modern name Capo Blanco. But the Gaetuli Autoteles are rendered as the "Gaetulian Free State" (set in italic type here),[11] and elsewhere Cynossema is rendered as Bitch's Tomb (4.49), Hierasycaminos as Holy Mulberry (6.184), Zeugma as Bridgetown,[12] and so on.

To add to the sense of confusion, Rackham acts to mislead his readers by the way in which he translates a sentence eight sections later (5.17), where Pliny names the Autoteles again and then refers back to them with the pronoun *horum*:

> Gaetulae nunc tenent gentes, Baniurae multoque validissimi Autoteles et horum pars quondam Nesimi, qui avolsi his propriam fecere gentem versi ad Aethiopas.

[11] It may be relevant to recall that the Republic of Ireland was officially named the Irish Free State between 1922 and 1937.

[12] 5.67, 86, 90; 6.119, 120, 126.

Rackham translates:

> The country is now occupied by the Gaetulian tribes, the Baniurae and the Free State, by far the most powerful of them all, and the Nesimi, who were formerly a section of the Autoteles, but have split off from them and formed a separate tribe of their own in the direction of the Aethiopians.

So here Rackham in the first instance renders Autoteles (consistently) as the Free State. But then at once, when he considers it necessary to specify the name of this people again in order to clarify *horum* in Pliny's compressed Latin, he opts instead for Pliny's own Autoteles. He gives no indication that these are in fact the Free State just mentioned rather than an entirely different people, as readers of this English translation would reasonably infer; only a painstaking check of the Latin could correct that impression.[13]

The manifestly unsatisfactory nature of Rackham's practice has convinced us that, so far as seems practical, our translation should methodically retain the form of each name in Pliny's Latin. We do so even when Pliny names an *oppidum* or the like by its ethnic adjective rather than by its toponym: so in 5.29, for example, we translate Simittuensian and Uchitan rather than converting to Simittu and Uchi.[14] If within a list Pliny should vary his usage between ethnic adjective and toponym—as in 5.29 with Canophicum—we reflect the variation. By the qualification "practical" I mean that we do not abandon our concern for readability and for modern English usage, although we still expect readers to recognize that Pliny's world in the distant past must inevitably appear unfamiliar in multiple respects. Accordingly, for a limited number of frequently used names we have opted to use the common anglicized forms rather than Pliny's Latin ones: for example, Rome instead of Roma, Egypt instead of Aegyptus, Nile instead of Nilus. Pliny's Aethiopia, however, we retain as a repeated reminder to readers that the part of Africa Pliny has in mind when he uses this name is by no means equivalent to modern Ethiopia.

In plenty of other instances, however, where the Latin name-form may not sound instantly familiar but still can hardly be a source of confusion,

[13] A lesser instance of such inconsistency is Rackham's retention of Pliny's Zeugma—rather than substitution of Bridgetown, as before—in his translation of 34.150.

[14] However, if the location of the community is known, it is to be named with its toponym on the map in preparation (see further below), not its ethnic adjective.

we have resisted the temptation to anglicize. Hence we keep, for example, Alpes, Danuvius, Europa, Gallia, Hispania, Italia, Euxinus, and Pontus, although for this last pair we do add in square brackets after the name's first appearance: "[= Black sea]." Where the name of a physical feature is a noun, such as *mare* or *oceanus*, and adjective, we favor using the latter's common anglicized form if there is one: so, for example, Atlantic ocean instead of Atlanticus, Caspian sea instead of Caspium, Aegean instead of Aegaeum, Egyptian instead of Aegyptiacum, Indian instead of Indicum. By the same token we readily anglicize Latin ethnic forms if there is a common English equivalent: so, for example, Amazons instead of Amazones, Greeks instead of Graeci, Indians instead of Indi, Macedonians instead of Macedones, Numidians instead of Numidae, Romans instead of Romani. We acknowledge that in all our choices of whether to anglicize a name or not there is an element of subjectivity, but this is unavoidable; rather, the goal should be to strike a practical balance between retaining Pliny's usage and making the translation readable.

We strive to avoid repeating a name where Pliny uses only, say, a pronoun or adjective to refer to it again. In some instances, of course, the name simply has to be repeated for the sake of achieving clarity in English, and we then enclose it in square brackets to signify that it does not recur at that point in the Latin. So we handle *horum* as follows in the compressed sentence quoted earlier (5.17):

> Gaetulae nunc tenent gentes, Baniurae multoque validissimi Autoteles et horum pars quondam Nesimi, qui avolsi his propriam fecere gentem versi ad Aethiopas.

> Nowadays Gaetulan peoples dominate, Baniurae and Autoteles (by far the most powerful) and Nesimi, who were once a sub-group of theirs [Autoteles] but split from them to become a people in their own right located in the Aethiopians' direction.

Similarly, we use square brackets to identify any other words that must be introduced to make Pliny's Latin meaningful (beyond routine minimal additions that a readable translation into English calls for).

Another practice we have adopted stems from concern to present most effectively the long lists of features, peoples, and places that Pliny offers. Rackham, with ample justification, presents them for the most part in a normal way for English prose—using "the" frequently, and also "of," as in "Straits of . . . ," "port of . . .":

> Agrippa says that to the river Anatis is a distance of 496 miles, and from the Anatis to Lixus 205 miles; that Lixus is 112 miles from the Straits of Gibraltar and that then come the gulf called Sagigi Bay, the town on Cape Mulelacha, the rivers Sebou and Sallee, the port of Mazagan 224 miles from Lixus, then Capo Blanco, the port of Safi, the Gaetulian Free State, the river Tensift, the Velatiti and Masati tribes, the river Mogador, and the river Sous, in which crocodiles are found.

However, we have concluded that such repeated use of "the," "of," and the like is unnecessary in this context, and that their omission (for the most part) permits better replication of Pliny's staccato shorthand with its frequent lack of a verb. So we translate this passage in leaner fashion as follows (from the Budé Latin, which marks lacunas at the start):

> . . . as far as Anatis river < . . . > 496 < . . . >. Agrippa says that Lixus is 205 from it, and that Lixus is 112 from the Gaditan strait. Then the bay called Sagigi, a town on cape Mulelacha, Sububa and Sala rivers, Rutubis harbor 224 from Lixus, then Sol's cape, Rhysaddir harbor, Gaetulan Autoteles, Quosenum river, Selatiti and Masathi peoples, Masath river, Darat river in which crocodiles are born.

In Rackham's translation "miles" occurs four times, because he automatically follows each distance figure with the unit; but in no instance here does Pliny's Latin state it, and we see no merit in adding it in instances like these where the unit is beyond all doubt. We never convert Roman miles, or any of the other units of measurement stated by Pliny, into a modern equivalent; instead, our book's front matter includes an explanatory note about such conversions and the difficulties they pose. Likewise, we just translate Pliny's directional indicators—*adversus*, *ante*, *contra*, *infra*, *sub*, *super*, and so on—literally, without substituting a compass direction.

Observe that we do not capitalize the initial letter of physical-feature nouns such as "cape," "strait," and so on. This practice is indeed contrary to regular English usage and may act to disconcert some readers at first, especially in the cases of, say, an ocean, sea, or mountain. Even so, no obstacle is introduced to comprehension of Pliny's material, and the practice curbs the number of capitalized words in a text already teeming with them. Observe, too, that we do not convert Pliny's long list beginning "Then the bay called Sagigi . . ." into a normal sentence by introducing a main verb, as Rackham does unnecessarily; we simply leave it as a list. Nor do we insert "and" before the mention of the last name, as Rackham does in accordance with regular English usage, but again unnecessarily in our view. We also do not retain Rackham's repetition of the river-name Anatis, but refer back to it only with a pronoun, as Pliny does; what "it" refers to here in our translation can hardly give readers any doubt.

There can be no avoiding the delicate perennial problem for translators of whether a Latin word used by their author should always be rendered as the same English word. In tackling Pliny, we have again sought to strike a balance. In some instances we find no sound reason to maintain consistency: for example, *nobilis* can as well be translated as, say, "outstanding" or "splendid" (depending on the context), *clarissimus* as "very well known" or "famous." This said, we are concerned not to translate such adjectives either in a distinctly more muted way than Pliny does, or in an even more effusive way.

More generally, too, we reject going to the extreme adopted by Tony Woodman in his painstaking, but provocative, translation of the *Annals* of Tacitus—a great stylist of course, quite unlike Pliny—where Woodman was determined to translate a Latin word with always the same English word. This was a heroic effort on his part that produced some curious choices, such as "dene" for *saltus*. To be sure, he admitted defeat in some instances, and we share his conviction with reference to readers: "it is positively valuable to be reminded constantly that ancient Rome was an alien world."[15] Even so, we heed the verdict of Barbara Levick in her thoughtful review, which declared Woodman's approach to be unduly rigid; hence her summing-up: "this is a book that is useful to have and an irritation to read."[16]

Common ethnic and geographical terms used throughout by Pliny pose a problem for his translators in that for the most part there seems no knowing how far each represents a deliberately precise choice on his part. For certain, he does evidently treat *desertum/deserta* ("desert") and *solitudo/solitudines* ("wilderness") as the same.[17] However, is he consciously differentiating when,

[15] Woodman (2004), xxxiv (defeat), xxxvi.
[16] Levick (2005).
[17] Cf. 6.73, 77.

for example, he uses *litus* in some contexts and *ora* in others, or may these two Latin terms be also considered interchangeable? Turner and I cannot say, but since a pair of alternates in common English use is ready to hand, our view is that we may as well differentiate between *litus* always to be translated "shore," and *ora* "coast" likewise. By contrast, when the same issue arises with regard to *amnis, flumen, fluvius*, again any differentiation that Pliny may have in mind eludes us, as do also three acceptable alternate words in English; so without more ado we just translate all three as "river." Terms that Pliny does seem to use with at least some degree of deliberation (even if the basis for his choices remains obscure), and that we are able to translate consistently with the same word, include *castellum* "fortress," *civitas* "state," *gens* "people," *iugum* "range," *locus* "place," *oppidum* "town," *populus* "community," *portus* "harbor," *sinus* "bay," *urbs* "city." This said, allowance must also be made for the fact that general Latin usage for a term may itself rule out such consistency: *mundus*, for example, in Pliny as in other Latin authors can signify either "universe" or "world."

At the same time, even when consistent translation of a term can be achieved, we recognize that occasional exceptions should be accommodated nonetheless. In the specific case of *Persicus sinus*, for example, "bay" jars in modern English, because "Persian gulf" for this well-known body of open water is standard usage. In a different instance, when *gens/gentes* is translated as "people," this choice creates difficulty if Pliny also speaks of *natio/nationes* in the same context (as he does occasionally), because the translation "nation" may well encourage readers to conceive that term in misleadingly modern terms (just as "race" for *gens* is to be avoided for the same reason). When this problem arises in 6.14–15, for instance, we have chosen "tribes" for *nationes*, albeit with reluctance, because this word too has its undesirable connotations:[18]

deinde multis nominibus Heniochorum *gentes*.

V regio Colica et *gentes*, Achaeorum *gentes*, ceterae eodem tractu *gentes*

15. Subicitur Ponti regio Colica, in qua iuga Caucasi ad Ripaeos montes torquentur, ut dictum est, altero latere in Euxinum et Maeotium devexa, altero in Caspium et Hyrcanium mare. reliqua litora ferae *nationes* tenent Melanchlaeni, Coraxi. . . .

[18] Sammlung Tusculum Latin text, with my italics.

Then Heniochan peoples [*gentes*] with their many names.

V Colica region and peoples [*gentes*]. Achaean peoples [*gentes*]. Other peoples[*gentes*] in the same region

15. Colica region lies below Pontus where the Caucasus range veers towards the Ripaean mountains (as already mentioned [5.98]), one side sloping toward Euxinus [sea] and Maeotis [lake], the other to the Caspian and Hyrcanian seas. The rest of the shoreline is occupied by the savage Melanchlaeni and Coraxi tribes [*nationes*]. . . .

In showing respect for Pliny's prose, we try to preserve his sentences where possible. We do this even at the risk of taxing our readers, who will often have to remain alert to the punctuation if they are to comprehend the annotated lists satisfactorily, as in this typical instance:[19]

In ora regio Sordonum intusque Consuaranorum, flumina Tecum, Vernodubrum, oppida Illiberis, magnae quondam urbis tenue vestigium, Ruscino Latinorum, flumen Atax e Pyrenaeo Rubrensem permeans lacum, Narbo Martius Decumanorum colonia XII p. a mari distans, flumina Araris, Liria.

On the coast the Sordones' region and inland that of the Consuarani, Tecum and Vernodubrum rivers, towns Illiberis (mere vestige of a once-great city), Ruscino Latinorum, Atax river flowing from Pyrenaei through lake Rubrensis, Narbo Martius a colony of the Tenth legion 12 miles from the sea, Arar and Liria rivers.

However, there are passages where Pliny rambles so breathlessly that, for comprehensible translation into English, we find it essential to introduce a break or two into an interminable sentence. In this challenging instance in book 4, where we introduce two breaks, observe that even Budé's Latin places a semicolon after *Rhoxolani* in 80:

80. Ab eo in plenum quidem omnes Scytharum sunt gentes, variae tamen litori adposita tenuere, alias Getae, Daci Romanis dicti, alias Sarmatae,

[19] 3.32, Budé Latin text.

Graecis Sauromatae, eorumque Hamaxobii aut Aorsi, alias Scythae degeneres et a servis orti aut Trogodytae, mox Alani et Rhoxolani; superiora autem inter Danuvium et Hercynium saltum usque ad Pannonica hiberna Carnunti Germanorumque ibi confinium, campos et plana Iazyges Sarmatae, montes vero et saltus pulsi ab iis Daci ad Pathissum amnem, 81. a Maro, sive Duria est a Suebis regnoque Vanniano dirimens eos, aversa Basternae tenent aliique inde Germani.

80. From this point all the peoples are generally Scythian, although different groups have occupied coastal areas, in one instance Getae (called Daci by Romans), in another Sarmatians (called Sauromatae by Greeks) and those of them called Hamaxobii or Aorsi, in another Lower Scythians (including ones of slave origin) or Trogodytae, and then Alani and Rhoxolani. In the uplands between the Danuvius and Hercynian forest as far as the Pannonian winter-quarters at Carnuntum and the German borderland there, Sarmatian Iazyges occupy the plains and lowland, while Daci—expelled from there by them—occupy the mountains and forests up to Pathissus river.

81. From the Marus (or if it is the Duria separating [the Daci] from the Suebi and Vannius' kingdom) Basternae hold the opposite side and thereafter other Germans.

Observe here our deliberate placement of phrases within parentheses and dashes to encapsulate what would otherwise be separated by commas. Also note our separation of one section from the next with a line left blank, a layout that we see as invaluable assistance to the reader's comprehension. By its very nature Pliny's text is dense, and the Loeb Classical Library's space-saving format does little to relieve its rebarbative character. In its volumes only the start of some sections is accompanied by a slight line-indentation, and the marking of section numbers beside the Latin text is not repeated beside the translation.

*****

A discussion of plans for a translation of Pliny's geographical books would be incomplete without consideration of two aids that must surely accompany it: notes and maps. Turner's initial intention was to provide concise footnotes on the modest scale that Frank Romer, for example, does in his translation of Pomponius Mela's *Description of the World*. As Romer explains:[20]

[20] Romer (1998), ix.

The notes to the translation are motivated by items that stand out in Mela's narrative. These notes include cross-references within Mela's text, references to other ancient writers, selected bibliography, and other useful information that suits his miscellany.

However, in due course we learned that Duane Roller was investigating prospects for achieving with Pliny's geographical books what he already had with Strabo's *Geography*: a complete translation of its seventeen books published by Cambridge University Press in 2014, followed in 2018 by a historical and topographical *Guide* over 1,000 pages in length from the same press. It emerged from discussions that Roller would welcome the opportunity to prepare a similar *Guide* to Pliny's geographical books on the basis of the translation that Turner and I were making, without needing to make his own. Turner and I were agreeable, and Cambridge University Press has committed to publishing first our translation and then later, when ready, Roller's *Guide*.

For our translation, the notable consequence is that the already modest scale of notes envisaged becomes further reduced. Certainly, we still mean to supply cross-references to other passages in the *Natural History*, as well as the years of consular dates, brief explanation of technical terms like *conventus*, and so forth. Everything else in need of comment, however, we now leave to Roller's *Guide*—with regret, naturally, that inquiring readers will be temporarily ill-served during the unavoidable interval between the appearance of the translation and that of the *Guide*.

As to a map or maps, with twenty-first-century digital cartography at our disposal and data from the *Barrington Atlas* and its *Map-by-Map Directory*, today we are empowered to transform the quality of what was offered when printed paper was the only possible format and cartography for antiquity remained in disarray. Notably, of the latest three Budé volumes, the fourth quarter of book 6 (2008) does include a grayscale foldout map with inset, but neither book 3 (2004) nor book 4 (2015) has any map. Sammlung Tusculum offers some (small, grayscale) for books 3, 4, and 5, though none for book 6. For us, by contrast, there is already an instructive model in the color digital map made by the Ancient World Mapping Center to accompany Roller's translation of Strabo.[21]

[21] Visit awmc.unc.edu/applications/Strabo/.

Consequently, the Ancient World Mapping Center is preparing a matching digital map for Pliny's geographical books. Like the one for Strabo, it extends seamlessly all the way from the British Isles to India, using the Center's *Map Tiles* as its base and conforming in style to the Center's mapmaking tool *Antiquity-A-La-Carte*. The map is georeferenced, and the modern physical landscape is returned—so far as can be achieved—to how it is likely to have been in antiquity. Users can pan and zoom as they wish. The scale at which the map may be displayed is of course variable, with a zoom possible up to about 1:50,000. All features, peoples, and places mentioned by Pliny that can be located are to be marked, following his form and spelling of the name.

The marking of each name on the map offers a link to the relevant Pleiades project entry,[22] where fuller information—including the modern equivalent name, if any, and bibliography—may be found. Where applicable, the map is also to indicate a community's official Roman status as specified by Pliny (colony, with Latin rights, ally, tax-exempt, free, tribute-paying), as well as to outline the approximate extent of Roman provinces, *conventus* districts, and the Augustan regions of Italy. Each type of data is entered on a separate layer which users of the map can introduce or remove as they wish. Distances may also be calculated on the map.

*****

The focus here has been on aspects of translating Pliny that are especially important for achieving a sound, lasting English version of the geographical books for the twenty-first century, one that strives to balance accuracy with readability. Even so, our translation cannot always flow smoothly. Yet should readers express dissatisfaction on that score, the appropriate response must be that this limitation merely reflects Pliny's material and his manner of presenting it; our translation has not rendered these books harder to comprehend. The temptation to adapt and supplement the translation further is one that we have resisted. Fortunately, Pliny is not an author who requires these efforts in the way that another (anonymous) one I have translated for a forthcoming volume does.[23] This mid-fourth-century work, *Expositio Totius Mundi et Gentium*,[24] mercifully short, is written in Latin so execrable, with syntax so opaque (no doubt worsened by copyists' slips), that the need repeatedly arises for the translation to become a conjectural reconstruction of what may be meant; otherwise the translator will deliver unsatisfying

[22] pleiades.stoa.org.
[23] Graham Shipley, ed., *Geographers of the Ancient Greek World*.
[24] Attributed to Iunior Philosophus by *FGrHist* V.2023.

gibberish. As a result, readers of a translation are offered prose more meaningful and more readable than the Latin original. Pliny is often hard to comprehend, but he is seldom so mystifying that his translators are called upon to resort to such uncomfortable speculation.

Geographical Passages in Books 7–37 for Inclusion in the Translation (Draft List)

7.95–99 Extent of Pompey's conquests; 191–206 Inventions: Who made them, and where
8.225–29 Localization of animal species
9.44–46 and 49–53 Tunny-fish
10.74–79 Migration and localization of birds; 132–35 Unusual birds
12.51–57 and 63–65 Frankincense and its export to the Mediterranean; 82–84 Arabia and the value of Rome's eastern imports from beyond the empire; 107–109 Gums and mosses
14.59–76 Vineyards and the quality of their wines
16.2–6 Germania, with and without woodland; 159–62 Reeds and bamboo for arrows; 238–40 Trees of great age in Graecia and Asia Minor
18.210–17 Difficulties of forecasting the seasons
19.2–15 Flax: Importance, cultivation, processing
27.1–3 Roman peace permits worldwide transport and use of plants
31.5–6 and 9–30 Whereabouts and nature of waters, healing and deadly
32.15–19 and 21 Remarkable fish; coral
37.30–46 Amber; 201–205 Conclusion, with highest praise for Italia

6

Boundaries within the Roman Empire

It would be difficult to overstate the degree to which territory within the Roman empire was demarcated. Quite literally, there were (in Latin) *fines, limites, termini,* (in Greek) *horoi* everywhere, boundaries dividing provinces, communities, tax districts, landholdings, and more. Demarcation of territory was a hallmark of Roman civilization, and throughout the Roman world a traveler would encounter boundaries of one type or another at every turn. Plenty of these did indeed predate Roman annexation of a region, but there is no question that the development of Rome's rule and its foundation of new communities led to a marked increase in the creation of boundaries. In recent scholarship generally, it is the empire's external boundaries—their nature, significance, and permeability—which have attracted the most attention and debate. Meantime fortunately, over the past decade and more, a small number of scholars has nonetheless sought to penetrate many of the complexities associated with the empire's internal boundaries, and they have made remarkable progress. Accordingly, this contribution ventures some conclusions and questions that arise from their findings. It also examines a recent discovery as an instructive illustration of the obstacles commonly encountered when considering internal boundaries.

The first conclusion is the large one that from the time of Augustus onward, if not earlier, the creation and maintenance of internal boundaries has to be recognized as a constant, major preoccupation of the authorities at every level throughout the empire. This is activity that demands greater prominence in modern appreciation of the work of emperor or governor or city council than it has typically been accorded to date. Understandably enough, what has most engaged scholars' attention are those instances where some dispute over a boundary involving a community arises, which then prompts the involvement of the Roman authorities.[1] We should always bear in mind that the power to adjudicate such disputes is one which the Romans

[1] See most recently Burton (2000) and (2002), 115–17; Campbell (2000), especially 452–71; Elliott (2004).

reserve strictly for themselves. These matters are not referred to a provincial *concilium* or *koinon*, nor is the Greek practice of arbitration by a third community continued. Emperors are naturally reluctant to adjudicate boundary disputes themselves, but they are insistent that governors become personally involved. The volume of work thereby created for governors could prove substantial. It would seem that, during the second half of the first century CE at least, the city of Histria made a point of petitioning each successive governor of Moesia to confirm its right to tax the revenue from fishing at one of the Danube mouths. When the extent of the relevant area was challenged around the end of the century, Histria won its case.[2] Its repeated concern to uphold its rights, without diminution of the area to which they applied, paid off. The lesson was doubtless not lost on other communities: maintaining recognition for boundaries was only a prudent precaution on the part of a community.

One practical justification for boundary disputes involving a community to be decided by the Roman authorities exclusively was that Rome alone maintained an archival record of all its surveys. To be sure, this point is already well appreciated. However, some other features of the archive have been less recognized. One is that it evidently preserved maps or plans not only of regular, centuriated land division, but also of other, less regular types of boundary demarcation. Another feature is that the archive was without question sufficiently organized to permit searches to be made, and thus for requested items to be located and copied. In the late 60s CE, the Galillenses in Sardinia, locked in dispute with Roman colonists known as the Patulcenses Campani,[3] avowed that they would present from the *tabularium principis* what is referred to interchangeably as both *tabula* and *forma*. Accordingly, the proconsul allowed them three months to do so, on condition that if this version of the map was not forthcoming by then, he would instead follow the *forma* already in the province.[4] Why the Galillenses failed, even after a further two months' grace, is left unexplained; but it is their expectation of being able to deliver a copy from the *tabularium principis* which matters for present purposes.

Most remarkable in this connection are the inscriptions (two examples, neither in good condition) from central Tunisia of the early second century which specifically cite the reestablishment of a boundary between

[2] See *IScM* 1 (1983), nos. 67–68 = Burton (2000), no. 21.
[3] *BAtlas* 48B3.
[4] *ILS* 5947.

two peoples on the basis of a *forma* sent by the emperor Trajan. The natural assumption is that this *forma* recorded the original delimitation of the boundary made when a large area was centuriated in 29 or 30 CE, seventy years or so earlier:[5]

> Ex auctoritate / Tac(apitanos) / BAVIB+ISATV / [---]DIA[---] // Imp(eratoris) Nervae Tr / aiani Caes(aris) Aug(usti) / [secun]dum formam m[i]/[s]sam sibi ab eod/[em---?]AECNMEO posita/ est NF MIN / SUMUM venire / non potuit // Term(inus) inte[r] Tac(apitanos) et N[ybgenios] / N[yb]g(enios) (?).

We may imagine that the unassuming *forma* mentioned here (in the singular) must in reality comprise copies of a whole set of many documents, because the area involved was extensive.

Altogether, inscriptions preserve at least four other instances where use of an earlier *forma* is explicitly referred to: in Dalmatia and Africa with the same phrasing used in both instances, *[s]ecundum formam Dolabellianam*[6] and *[secun]dum formam [Mar?]tianam*;[7] at Canusium in Apulia, where Vespasian restored *fines agrorum public(orum) . . . ex formis publicis*;[8] and in Campania, where the same emperor restored the boundaries of the lands consecrated to Diana Tifatina by Sulla *ex forma divi Augusti*.[9] In none of these instances do we know where the earlier *forma* was obtained, but the comprehensive archive in Rome must be among the possibilities.

I am among the scholars who have dismissed Oswald Dilke's claim that there was a "civil service maps and plans department" in Rome, made on the basis of inference from the Late Roman title *comes formarum*. On this basis the claim is assuredly wrong, because *forma* as used in this title has a quite different connotation.[10] But it has become clear that epigraphic testimony does give the claim some validity. I now believe that there was a *tabularium* for *formae* in Rome (at a location still to be identified), where survey results were stored systematically, and material could be retrieved and copied in the event of a dispute or when a border was in need of restoration.[11]

[5] *AE* 1910.20, with Trousset (1978), 134–36.
[6] Wilkes (1974), 268 no. 26.
[7] *CIL* 8.23910.
[8] *AE* 1945.85.
[9] *ILS* 251, 3240; *AE* 1971.80.
[10] See Dilke (1985), 167, with essay 1 above; Dilke (1987a), 244, repeats the claim.
[11] Note the valuable discussion by Moatti (1993), 63–97.

Preoccupation with those boundary disputes which came to involve the Roman authorities should not lead us to overlook the point that most fixing of boundaries never proved a matter of sufficient contention for the Roman authorities to be drawn in. Moreover, even when we can be confident that there was such involvement, it was not invariably because a dispute had occurred. Burton's analysis is a notable recent instance of such misjudgment.[12] In his text he duly observes that testimony which beyond all doubt reflects the occurrence of a serious dispute over fixing a boundary must be distinguished from testimony where no dispute is apparent. However, in the appendix summarizing all the relevant epigraphic and literary texts known to him, he has scant regard for this important distinction. This is unfortunate. Instead, greater caution and discrimination are called for in classifying this material. Granted, there will always remain the possibility that a serious dispute did arise over the fixing of a particular boundary, even though the surviving testimony happens to preserve no reference to it. Nonetheless, until fresh testimony to the contrary emerges, we can only assume that where no reference is made to a dispute, there was none. For purposes of classification, it is vital that appropriate working criteria be established and then applied consistently.

In a case like the following from Dalmatia, where the key verb is *renovo*, it does seem likely enough that bad maintenance or flood is the cause of intervention to renew a bridge and boundary-markers, rather than any dispute:[13]

> [L(ucius) F]unisulanus Vet/[to]nianus leg(atus) pr(o) pr(aetore) / [po]ntem et terminos / [re]novari ius(s)it per Cas(s)ium Fron(t)one(m) / o(ptionem) leg(ionis) IIII F(laviae) f(elicis) in / [fun]do Ves<i>o C/SCDLV.

In the many documents where the verb chosen is *restituo*, by contrast, there is room for ambiguity: would a Roman authority have recorded the "restoration" of a boundary unless some kind of dispute had prompted this action? I hesitate to give an unequivocal affirmation or denial. The same hesitation is appropriate in those instances where we have a record of a Roman official (sometimes on the emperor's authority) making what Burton terms an "authoritative demarcation." The following two examples may suffice.

[12] Burton (2000), discussed by Elliott (2004), 3–5.
[13] Wilkes (1974), 266–67 no. 21.

From Macedonia, 119/120 CE:

. . . Terentio G[en]tiano leg(ato) A[ug(usti)]/ pro pr(aetore) termin[i]/ positi per Cl(audium) A[---]/num Maaxim[um(!) / (centurionem)] / leg(ionis) I Minerv(i)ae [in]/ter Geneata[s et ---]/ xinos.[14]

From Bulgaria, 135 CE:

Antius Rufinus in/ter Moesos et Thra/ces fines posuit.[15]

At the least, it is vital for us to stay alert to the various different types of boundary-fixing documents that survive, and to recognize the doubts that stem from their incomplete records.[16]

*****

A recent discovery in Slovenia offers apt illustration of several of the difficulties that testimony about internal boundaries within the Roman empire typically presents. It is a boundary stone discovered in the bed of the Ljubljanica river, at the mouth of a supposed drainage channel and close to the confluence of the river and one of its tributaries (fig. 6.1a, b). Marjeta Šašel Kos emphasizes in her publication that the stone "was found more or less in situ."[17] In other words, she is convinced that it has barely moved, if at all, from where it was originally placed. It is moreover certain that it was meant to be dug deeply into place. Only an upper section about 30 cm high was smoothly worked, while the rest of the stone (stretching on down for about 100 cm) was left unworked. On one side the upper section is inscribed AQUILEIEN / SIUM, and on the other EMONEN / SIUM. The narrow panel on the top of the stone carries the word FINIS facing AQUILEIEN / SIUM below. The only dating criteria are the material of the stone itself, and the lettering style, which together point to the period from Augustus to Claudius in Šašel Kos' view.

If this boundary demarcation stemmed from a dispute, we are offered no clue to it by this stone, nor to any alternative reason for why the demarcation

[14] *AE* 1924.57.

[15] *ILBulg* 184, 357, 358, 386, 390, 429.

[16] Note further discussion in Campbell (2005). The demarcation of woodland areas in Syria Phoenice by Hadrian, and again by Caracalla, is remarkable: see *BAtlas* 68, 69, with *AE* 2001.1966; 2002.1524–26.

[17] Šašel Kos (2002), 373.

Fig. 6.1a, b Stone marking the boundary between the territories of Aquileia and Emona, retrieved from the bed of the Ljubljanica river (Slovenia): (a) shows its two upright sides as each would be seen on approach from Emona's territory (left) and from Aquileia's (right); (b) enlargement of the top side, along which FINIS is inscribed and only visible from above, as here. The word is more conveniently legible along with AQUILEIENSIUM on approach from Emona; on approach from Aquileia, it appears in reverse upside down. Photo: Marjeta Šašel Kos.

was made when it was. Instances are known of demarcations where only a limited proportion of the marker stones carried information beyond the barest essentials, as here.[18] If there were such "fuller" stones in this instance, clearly they are still missing. Needless to add, the one surviving stone makes no mention of any involvement by Roman authorities. Comparable stones are known elsewhere—concise markers of the boundaries fixed, apparently without special contentiousness, between two communities or peoples.[19]

The findspot of the Slovenian stone, however, proves puzzling, to say the least. As Šašel Kos explains,[20] to date there has been no precise notion of where the boundary between the territories of Aquileia and Emona ran, although the reasonable assumption had been that it was at the road station In Alpe Iulia/Ad Pirum or thereabouts (fig. 6.2). Now, however, the new discovery points to a location much further east, indeed within as few as thirteen km of Emona, and a remarkable 100 km or so by main road from

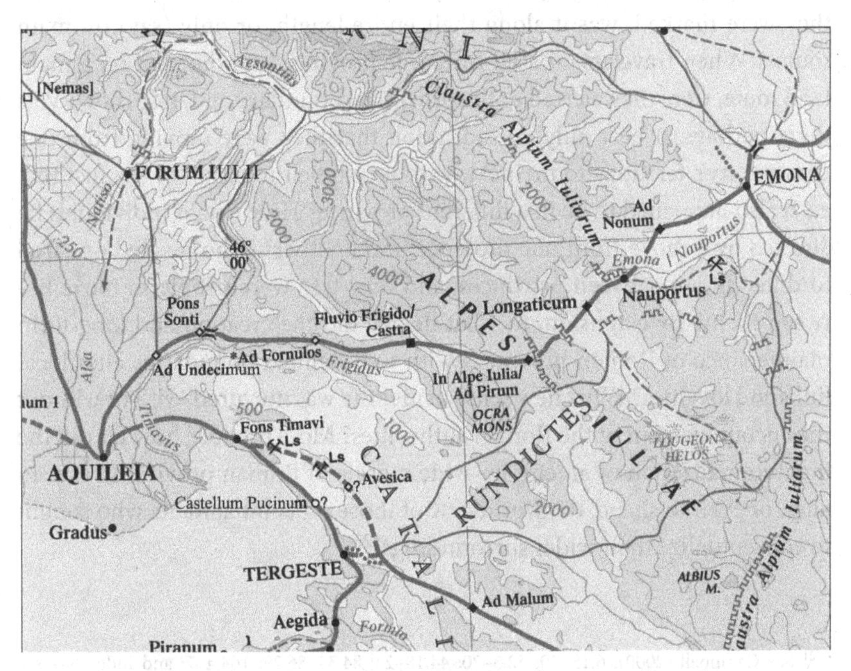

Fig. 6.2 Aquileia–Emona region at 1:1,000,000 scale, oriented north. Map: *BAtlas* 19–20.

[18] See, for example, Trousset (1978).
[19] See the examples mentioned by Šašel Kos (2002), 376.
[20] Ibid., 377–78.

Aquileia. The latter location (accepted by Šašel Kos) has the merit of situating Nauportus within the territory of Aquileia, reinforcing literary testimony which points to commercial links between the two centers. On the other hand, as she recognizes, this location does leave Emona with an unexpectedly stunted section of territory to the west of the city. Due consideration should surely be accorded to an alternative possibility (not pursued by Šašel Kos) that the newly discovered stone was marking an Aquileian enclave within Emona's territory. In this event, the belief that Emona's territory did extend further west for some distance would still hold. There are repeated references (both specific and general) to enclaves in the writings of the *agrimensores*.[21] This said, how and exactly when such an Aquileian enclave might have come to be created in the vicinity of Nauportus remains an open question.[22]

To conclude, let me draw attention to one category of internal boundary overdue for further investigation, namely those between Roman provinces. To what degree, for example, were provincial boundaries marked? Where they were marked, was it along their entire length, or only (say) on main routes? When travelers crossed from one province into the next on a main land route, were they alerted to the provincial boundary and, if so, how? Was there perhaps at best only an indication that they were moving from one community's territory to the next, without explicit mention of the fact that each also happened to be in a different province? How would individuals be affected by a move from one province into another? Who needed or cared to differentiate between provinces, and how free were people to move between them? By extension, to what degree did the creation and long-term maintenance of Roman provinces with well-defined boundaries alter local links and loyalties? What new sense of identity was nurtured, with or without the encouragement of the Roman authorities? More broadly, how far did the institutionalization of an empire-wide mosaic of Roman provinces serve to alter, or even to shape, the worldview of those of its inhabitants who sought to conceptualize their wider surroundings?[23]

[21] See Campbell (2000), 6.15–20; 42.3–20; 44.18–24; 84.34–86.29; 104.4–7; and index 547 s.v. "praefectura." Note the tale in Philostratus (*Vit. Apoll.* 1.38) where the Roman governor of Syria sends an embassy to the king in Babylon to assert Rome's right to a couple of villages near Zeugma (thus within the Roman province) which pay tax to him as if they were still his.

[22] For some enclaves elsewhere, note the discussion by Veyne (1959), with particular reference to *ILS* 6488, a remarkable dedication of the Severan period. For the territory of Emerita and its *praefecturae*, see Canto (1989), 175–90, 198. My thanks to Brian Campbell for these references; he shares my impression that the topic of enclaves would repay further study.

[23] For some preliminary answers to these questions, see essay 7 below.

In the very opening sentence of his *Roman Histories*, Appian bluntly states that he considers it necessary to begin by setting out the boundaries (*horoi*) of the *ethne* ruled by the Romans. From the description that follows there can be little question that he is thinking in terms of Roman provinces, and he makes repeated reference to the *horoi* of each or of a group of them. He then sums up his account by stating how the Romans "establish a ring of great garrisons around the empire and guard all this land and sea just as one would an estate [*chorion*]."[24] This vision of the entire empire as a well demarcated entity, clearly subdivided into individual province components, lays a robust foundation meriting our closer attention both in outline and in detail.

[24] Praef. 7.

7

Rome's Provinces as Framework for Worldview

This year [2004] is the twentieth anniversary of the publication of Pietro Janni's seminal book *La Mappa e il Periplo: Cartografia Antica e Spazio Odologico*. Its main thesis has convinced many scholars who seek to understand how Romans visualized their wider surroundings beyond the immediate vicinity of home, myself included. Among the strongest and most recent affirmations of support must surely be that of C. R. Whittaker (2002). Four brief passages may serve as illustration: "when it came down to mapping on the ground . . . Romans viewed their localities and environment . . . as 'hodological space,' the term adopted by Janni" (102); "space itself was defined by itineraries, since it was through itineraries that Romans actually experienced space, that is, by lines and not by shapes" (102); "I believe . . . itineraries dominated and infiltrated all the other categories of ancient representations and perceptions of space" (83); "a Roman's sense of space and visual perspectives were shaped by the horizontal, linear movement of itineraries over land and sea" (87). Finally, Whittaker goes on to urge (98):

> The conversion of Constantine to Christianity . . . radically transformed the world view of Romans. Travel made the world a smaller place. There was a new emphasis on the heroic journeys, both physical and spiritual, which were fused into one by the pilgrimage to Jerusalem.

I do not quote Whittaker to dismiss his, and Janni's, view. I believe that it contains much truth: the use of itineraries *was* unquestionably *one* means by which Romans organized space in their minds. Rather, my concern here is to argue that Whittaker makes "hodological space," or the "itinerary model," too comprehensive and too exclusive an explanation. One way or other, all else is subsumed to it. Thus, in particular, Ptolemy's work is set aside as esoteric and unnoticed by Romans, and the Peutinger map is regarded—in the traditional way—as no more than a set of route itineraries in diagrammatic, pictorial form.[1]

My view, however, is that both Ptolemy and the maker of the Peutinger map tapped the same Hellenistic cartographic tradition to create the base elements for their maps, namely the shorelines, principal rivers, and principal mountain ranges. These elements by definition create shapes; they are not merely lines as itineraries are. It is a serious misconception to see the Peutinger map as first and foremost a set of itinerary lines to which all other landscape elements have subsequently been added as no more than superfluous "decoration." To create such a map in this way is a virtual impossibility, and any attempt would be most unlikely to result in the well-known cities of the empire appearing in correct relation to the shorelines and principal rivers. In fact, from a cartographic perspective, one of the Peutinger map's most impressive features is that the placement of principal cities does cohere with the physical landscape, distorted though it is. This landscape, which underpins the entire map, could not have been derived just from itinerary data.[2]

A view of the Peutinger map as a work in which outstanding features of the physical landscape are important, even fundamental, inevitably casts doubt upon the validity of the "itinerary model" as a fully satisfying explanation for how Romans visualized their wider surroundings. Itineraries alone can hardly create much sense of spatial relativity. In order to conceptualize their world (however imperfectly), most Romans would surely need some set of images (however sketchy) for the purpose, beyond the type of information that one-dimensional lists could supply. The rough equations of landmasses with well-known shapes could conceivably have been of assistance: Italy like an oak-leaf, Britain like a shield, and so forth.[3] Even so, individual shapes still need placement relative to one another. Whatever representations were made on globes could perhaps foster a grasp of relativity but, as Whittaker himself observes,[4] any such "cosmic maps" were likely to be too small to show much in the way of physical or topographic features.

No, if there was some set of images that were commonly related to one another to create a vision of the Roman world, in all likelihood we have to look for it elsewhere. The set proposed here first caught my attention when

[1] Whittaker (2002), 92, 83, 93.
[2] See further essays 13 and 14 below.
[3] Whittaker (2002), 84.
[4] Ibid., 83.

reading the *Antonine Itinerary*[5]—not the standard place-by-place, distance-by-distance pattern of the presentation, but headings such as:

> Item de Pannoniis in Gallias per mediterranea loca, id est a Sirmi per Sopianas Treveros usque (231.8–10)
>
> Iter quod ducit a Durrachio per Macedoniam et Trachiam Bizantium usque (317.3–4)
>
> Inde per loca maritima in Epirum et Thessaliam et in Macedoniam, sic (324.1–2)

References of this type to provinces or principal regions (including the Alps) can only be meaningful to readers who have some vision of the placement of such entities relative to one another, however hazy it may be. Otherwise, a formulation like *de Pannoniis in Gallias* will merely be redundant. Conceivably, some or all the headings in the *Antonine Itinerary* collection as we have it are additions made by a post-Roman editor.[6] Most, however, seem an integral part of the work,[7] together with the summary total of the distance for the entire journey that follows in each instance. It is no doubt these headings that gave rise to the title bestowed (at whatever stage) on the land part of the work: *Itinerarium Provinciarum*.

The notion of provinces as ready-made, well-established components for creating a vision of the Roman world hardly seems likely to predate Augustus' rule.[8] Only from that date is the empire a single cohesive entity from the Iberian peninsula to Egypt, subdivided into provinces that are each defined units adjoining one another (or large islands) (fig. 7.1).[9] From then onward, however, a comprehensive framework is established, and a web of main land

[5] Discussed further in essay 9 below.

[6] For the manuscript tradition, note the observations by Salway (2004), 68–69, and see further essay 9 n. 4 below.

[7] The repeated concern to clarify the province within which a landing-place was situated that characterizes the beginning of the *Itinerarium Maritimum* likewise seems an integral, and thus original, feature of that work. I am at a loss to account either for the writer's purpose in offering this information up to 493.1, or for the abrupt exclusion of it thereafter.

[8] Compare, however, the earlier attempt by Eratosthenes to divide his map into "seals" (*sphragides*, regions marked by distinctive lines and landmarks), so that its representation of the *oikoumene* should be "readily drawn, copied, and memorized"; see Geus (2004), 20–26.

[9] Profs. W. Eck, H. Meyer, and H. von Hesberg have all kindly drawn my attention to the volume edited by the latter *Was ist eigentlich Provinz? Zur Beschreibung eines Bewusstseins* (1995): its stimulating contributions range widely (over administration, art, language, religion, etc.) but do not touch on the vision advanced here. The same is true of G. I. Luzzatto's entry "Provincia. Diritto Romano" in *Novissimo Digesto Italiano* (1967).

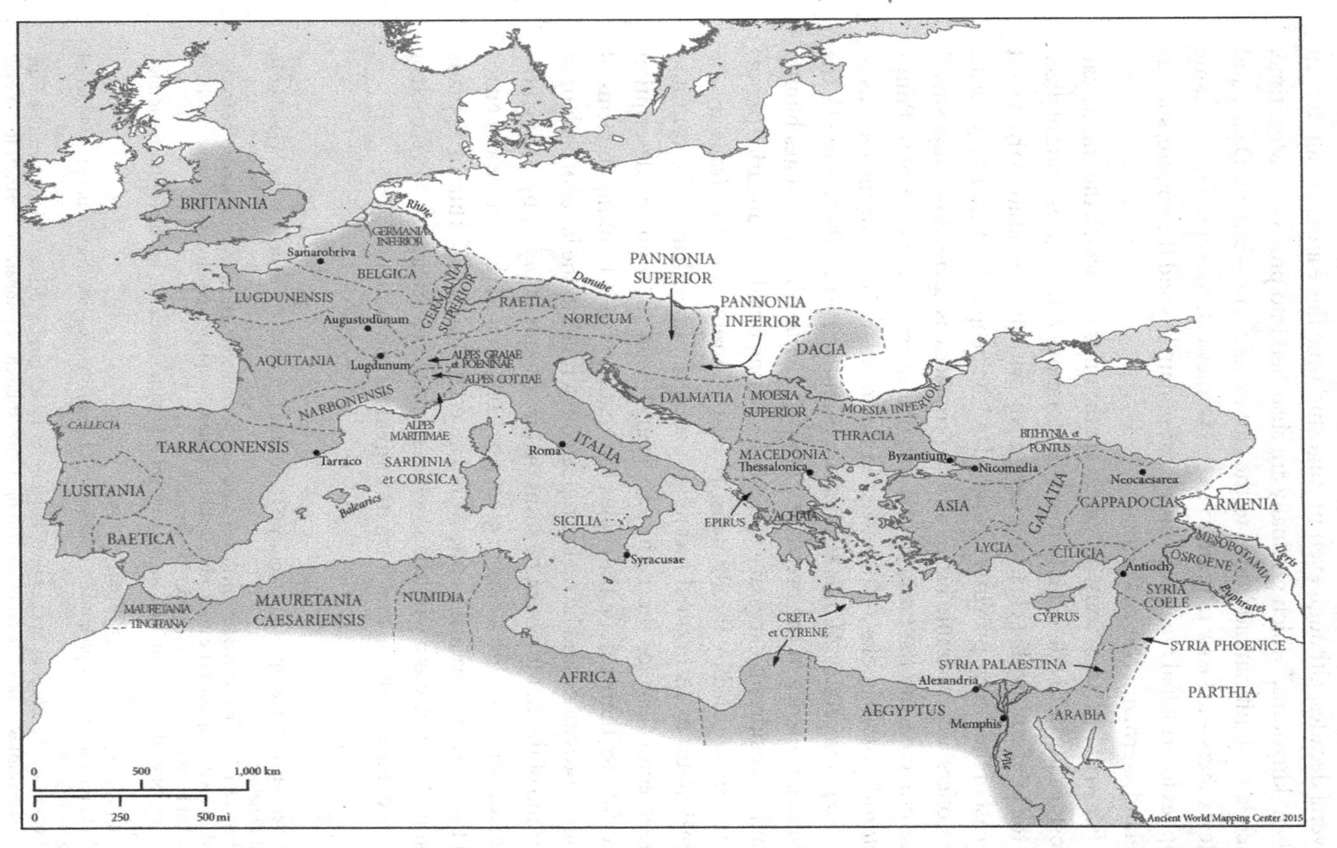

Fig. 7.1 Provinces of the Roman empire around 200 CE outlined on a modern map base. Map: Ancient World Mapping Center.

routes develops.[10] Strictly speaking, "holes" persist within the framework for several decades, although even in Augustus' time these areas are all Roman "client kingdoms," which gradually are absorbed into provinces. Meantime, such later significant additions to the empire as occur—Britain, Dacia, and a few others—are easy to graft on to this framework conceptually. The same applies to principal areas or peoples that remain beyond the empire, such as Ireland, Germany, Nubia, Parthia.

For written descriptions, provinces are the units into which the empire can most readily be divided. It is to them that Strabo (17.3.25) gives pride of place in the brief outline that closes his *Geography*, with its enumeration from west to east of the twelve provinces assigned to "the people" by Augustus. Cassius Dio offers a comparable list of all the provinces as divided by Augustus.[11] The provinces are the basis for the detailed record of the empire in Pliny's *Natural History* books 3–6; likewise in the fourth century for the records of the *Expositio Totius Mundi et Gentium*,[12] and of Festus, *Breviarium*.[13] The opening sentence of the preface to Appian's *Roman Histories* states bluntly that he considers it necessary to begin by setting out the boundaries (*horoi*) of the *ethne* ruled by the Romans. Here it is surely correct to translate the Greek noun by its original meaning "peoples," rather than as the common Greek equivalent for the Latin *provincia*.[14] Even so, from the description that follows there can be little question that Appian is thinking in terms of Roman provinces, and he makes repeated reference to the *horoi* of each or of a group of them. Moreover, he sums up his account (*praef.* 7) by stating how the Romans "establish a ring of great garrisons around the empire and guard all this land and sea just as one would an estate [*chorion*]." Thus in Appian's vision the empire as a whole is a well demarcated entity, and so too is each of the provinces within it that form its individual components.[15]

[10] See *BAtlas* 100 (early second century CE), 101 (early fourth). On routes, see further essays 16 and 17 below.

[11] 53.12.4–7. As his remarks immediately following demonstrate, Dio does not consider the framework of his description to be undermined either by the subsequent division of some of the provinces listed or by Rome's annexation of further territory.

[12] See especially 21 to the end in Rougé (1966); English translation in Shipley (forthcoming), no. 31.

[13] 4.1–14.5 in Arnaud-Lindet (1994).

[14] On the use of Greek *ethnos* and *eparcheia* for Latin *provincia*, note the comments of Mitchell (2000), 125–26.

[15] Compare the allusions in Claudius' "Lugdunum Table" speech (delivered to the senate during his censorship) which, while no more than figurative, still reflect comparable alertness to the *fines* or *termini* between provinces (*ILS* 212 col. 2 lines 21–22, 26–28, 30–31, discussed further in essay 12 below). A passage preserved from Ulpian, *De Officio Consulis*, indicates that there were circumstances

The great increase in the number of provinces from the time of Diocletian's reorganization onward hardly acts to cloud the type of vision reflected by Appian. The increase stemmed, after all, from a splitting of the long-established principal components, and for broad descriptive purposes those larger components remained clearly in focus. This is certainly the vision maintained by the composer of the list of maximum fees for cargoes on forty-nine and more voyages in the Tetrarchs' Price Edict. The list is headed *ex quibus locis ad quas provincias quantum nav<a>li excedere minime sit licitum*, but it gives no hint of the new division of provinces. Rather, the list (with its chosen starting-points for trans-Mediterranean voyages primarily situated in the East) is entirely comprehensible to anyone who can distinguish east from west, and is familiar with the empire's main territorial components as well as with some of its most notable ports.[16] To name only two late fourth-century authors, Festus[17] and Optatus of Milevis both demonstrate awareness of the increased number of provinces in their own day, but for clarity—or out of pride or nostalgia, too—both still outline the empire to their readers in larger, traditional units.[18] Hence the Catholic bishop Optatus objects to the Donatists' claim that the church exists only among them:[19]

> So in order that it can be among you in a scrap of Africa, in a corner of a small region, shall it not be with us in another part of Africa? Shall it not be in the Spains, in Gaul, in Italy, where you are not? If you want it to be among yourselves alone, is it not to be in the three Pannonias, in Dacia, Moesia,

in which a magistrate would need to be aware of which provinces adjoined Italy: "'Continentes provincias' accipere debemus eas, quae Italiae iunctae sunt, ut puta Galliam: sed et provinciam Siciliam magis inter continentes accipere nos oportet, quae modico freto Italiae dividitur" ("'Mainland provinces' we must regard as those contiguous to Italy, such as Gaul: but on balance we should regard Sicily too as among 'mainland' ones, separated from Italy though it is by a narrow strait"; *Dig.* 50.16.99; cf. 5.1.9.) In the same vein, note the distinctions made by an unidentified emperor's edict (preserved on papyrus, third century?) between Italy and its *provinciae transalpinae* and *provinciae transmarinae* (*FIRA*[2] I.91 col. 1 line 10, col. 2, lines 4–5).

[16] See Crawford and Reynolds (1979), 184–86.

[17] Note 4.6, 5.3, 6.3, for example.

[18] Compare Aurelius Gaius' commemoration of his travels (essay 9 below with fig. 9.1).

[19] *Traité contre les Donatistes* 2.1.3–4 (Labrousse [1995]): "Ergo ut in particula Africae, in angulo parvae regionis apud vos esse possit [sc. ecclesia], apud nos in alia parte Africae non erit? In Hispaniis, in Gallia, in Italia, ubi vos non estis, non erit? Si apud vos tantummodo esse vultis, in tribus Pannoniis, in Dacia, Mysia, Thracia, Achaia, Macedonia et in tota Graecia, ubi vos non estis, non erit? Ut apud vos esse possit, in Ponto, Galatia, Cappadocia, Pamphilia, Phrygia, Cilicia et in tribus Syriis et in duabus Armeniis et in tota Aegypto et in Mesopotamia, ubi non estis, non erit? Et per tot innumerabiles insulas et ceteras provincias quae numerari vix possunt, ubi vos non estis, non erit?"

> Thrace, Achaia, Macedonia and throughout Greece, where you are not? So that it can be among you, shall it not be in Pontus, Galatia, Cappadocia, Pamphylia, Phrygia, Cilicia and in the three Syrias and in the two Armenias and throughout Egypt and in Mesopotamia, where you are not? And shall it not be across so many innumerable islands and other provinces almost impossible to count, where you are not?

By the first century CE, if not earlier in many instances, I imagine that the point at which travelers by a main land route crossed from one province to another would normally be marked. Recent scholarship has chosen to focus so much on the varied character and significance of the empire's external frontiers (with all the associated difficulties of defining those satisfactorily) that the abundance of internal frontiers has attracted little attention by comparison.[20] But there can be no question that these internal ones existed, and that they were lines dividing the territories of neighboring communities, and by extension of provinces. With his Greek perspective, Strabo described the erection of conspicuous markers (*horoi*) as "ancient custom" (*ethos palaion*). Thus, for example, Alexander was credited with having built altars at the furthest point of his expedition, and Theseus with having erected a pillar on the border between Megarian and Athenian territory, inscribed on one side "This is not the Peloponnese, but Ionia," and the reverse on the other.[21] Just such pillars were erected by Romans, among them those inscribed on one side *F(ines) terr(ae) Thrac(iae)* and on the other *F(ines) terr(ae) Odess(itanorum)* that marked the boundary between the province of Thrace and the city of Odessus in Moesia Inferior.[22] Moreover, even under Roman rule boundary disputes between communities remain common occurrences, and frequently long-lasting. Indeed, one possible means for a community to attract the attention of the imperial authorities to itself was to engage in a fierce boundary dispute with a neighbor.

The itinerary with the fullest record of where a traveler crosses from one province to another is that of the Bordeaux pilgrim dated to 333.[23] In a recent article on this work, Jás Elsner draws attention to the pilgrim author's "acute awareness of provincial boundaries" and comments: "this care both to notice

[20] See essay 6 above.

[21] Strabo 3.5.5, with 9.1.6 and Plutarch, *Theseus* 25, for the pillar. Alexander: Diodorus 17.95.1; Curtius 9.3.19; Plutarch, *Alexander* 62; Arrian 5.29.1; Peutinger map 11A3; cf. Lucian, *Vera Historia* 1.6.

[22] Gerov (1979), 226 (with discussion of date) and Tafel II.3.

[23] Geyer and Cuntz (1965), 1–26, with analysis in Calzolari (1997).

and to delineate boundaries is more than a taxonomic fetish. It shows implicit awareness of administrative, ethnic, even cultural differences across the terrain which the linear thrust of the text so relentlessly traverses."[24] Quite so; and, as Elsner himself proceeds to stress, this awareness reflects a sense of administrative and spatial geography which is more than merely linear. At the same time the references to crossing from one province to the next helpfully subdivide for the reader what might otherwise seem a long, disorienting succession of quite unfamiliar stopping-place names.

Whether this means of subdividing the stages of a lengthy journey was an exceptional choice on the Bordeaux pilgrim's part seems open to question. To be sure, among our few surviving materials of comparably detailed type, this record of provincial border-points is an unusually full one, but the same may be said of the entire itinerary. Egeria, who probably traveled during the 380s, is likewise alert to provincial border-points.[25] Altogether, however, the Bordeaux pilgrim stands out for considerateness to the traveler who seeks to reduce risks and surprises to a minimum. In this important respect, the *Antonine Itinerary* and above all the Peutinger map both prove cavalier and inconsiderate. They are often content with intervals of thirty or forty miles, or even more, between stopping-places, in other words well beyond the distance that the typical traveler will be able to cover in a day.[26] The Bordeaux pilgrim, by contrast, offers reassurance by consistently taking pains to mention a *mansio* or *mutatio* every few miles.

Under normal circumstances most free individuals,[27] it would seem, could move about within the Roman empire, as well as in and out of it, just as they wished.[28] If they chose to cross a provincial border, that was a routine

[24] Elsner (2000), 187–88.

[25] See Hunt (2004), 99–102, 106. Note also how Egeria summarizes her return journey through Asia Minor to Constantinople by reference to provinces: "and then the next day I ascended mount Taurus and made a now familiar journey through successive provinces I had traversed outbound—namely, Cappadocia, Galatia, and Bithynia—to arrive at Chalcedon" ("Et inde alia die subiens montem Taurum et faciens iter iam notum per singulas provincias, quas eundo transiveram, id est Cappadociam, Galatiam et Bithiniam, perveni Calcedona"; 23.7); and how she remarks in the course of describing the Christian year at Jerusalem: "and since in this province some of the people know both Greek and Syriac..." ("Et quoniam in ea provincia pars populi et grece et siriste novit..."; 47.3).

[26] See Salway (2001), 32, and essay 13 below.

[27] But not necessarily senators and exiles, for instance, nor the dangerously insane (cf. *Dig.* 1.18.13–14), nor everyone within and from Egypt; and certainly not slaves. For restrictions in Egypt, see Adams (2001), 157–58. For governors, note Marcianus' observation "ne qui provinciam regit fines eius excedat nisi voti solvendi causa, dum tamen abnoctare ei non liceat" ("that someone who governs a province may not go beyond its borders except to discharge a vow, and even then he may not lawfully be away overnight"; *Dig.* 1.18.15.)

[28] Payment of *portoria* might impose a check (hence stations named Ad Publicanos); see further below.

matter, seldom of concern to a traveler or to the authorities, therefore.[29] Consequently, we do not hear of efforts by the latter to mark or control provincial boundaries for policing or security purposes, and it is understandable that the land part of the *Antonine Itinerary* does not go to the trouble of indicating where the traveler will cross from one province to the next.[30] Yet in at least one instance—whatever the reasons—Rome did evidently feel the need to mark an entire border along its length, as documented by the work of M. Antius Rufinus late in Hadrian's reign, who "inter Moesos et Thraces fines posuit."[31]

Even if it remained exceptional to mark an entire provincial border in this way, on main land routes a variety of means is attested that served to alert travelers to the point at which they crossed from one province to another. Although Ammianus Marcellinus (21.15.2) does not mention how he knew Mobsucrenae[32] to be the last *statio* in Cilicia for travelers proceeding north from Tarsus, it is striking that this is how he identifies the place where Constantius II died. Equally, he identifies Dadastana,[33] where Jovian died, as "qui locus Bithyniam distinguit et Galatas."[34] The itineraries preserve plenty of names in the style of *Fines, Ad Finem, Ad Fines*: often these will signify only the boundary between communities or great estates, but in some cases it is unquestionably one between provinces.[35] Explicit attention is not drawn to that point when Ad Fines makes its surprising appearance in the *Antonine Itinerary*'s summary of the main stages of the long journey Sirmium–Lauriacum–Augusta Vindelicum–Ad Fines–Treveri, but the distance totals elucidate the reason for its inclusion: hereafter to Treveri, *leugas, non m.p.*

[29] Pliny's argument (*Ep.* 10.77) that Iuliopolis (*BAtlas* 86B3) needs the assistance of a centurion in part because of its location *in capite Bithyniae* hardly seems compelling. In principle, a border location ought not to have placed it under greater strain from passing travelers than any other town along a main route: this indeed seems to be Trajan's reaction in response (10.78).

[30] The papyrus record of Theophanes' return journey from Hermopolis in Egypt to Antioch in Syria, probably during the early 320s, does not record such crossings either: see *P.Ryl.* IV.627–28, 638, and essay 10 below. Why the first part (only) of the *Itinerarium Maritimum* should trouble to do so is puzzling, as observed above (n. 7).

[31] See *ILBulg* 184, 357, 358, 386, 390 and 429, with Gerov (1979); for Antius, *PIR*² A784 with https://pir.bbaw.de/id/1513

[32] *BAtlas* 66F2 Ma(m)psoukrenai.

[33] *BAtlas* 86B3.

[34] 25.10.12, phrased notably in this order (rather than *Galatas et Bithyniam*) even though Jovian, too, had been traveling westward. Compare Ammianus' identification of the Cilician Gates (Pylae) as "qui locus Cappadocas discernit et Cilicas" (22.9.13).

[35] For example, the *mutatio* Fines (*BAtlas* 86B3) at the border of Bithynia and Galatia in *ItBurd* 574.3–4 (strictly speaking, no doubt more accurate than Ammianus' Dadastana above). Ad Fines on the route westward from Augusta Taurinorum marked the border between Italy and Cottius' kingdom: see *BAtlas* 39A3 with France (2001), 326.

(232.3). This Ad Fines[36] is therefore the boundary-point between Raetia and Germania Superior.

Herodian's mention[37] of the *methorioi bomoi* at the border of Pannonia and Italy—where the emperor Maximinus sacrificed in 238 as he marched south to overcome rivals—seems to be unique; but from its sheer casualness we might infer that other borders, too, were marked by altars.[38] It is natural to infer that the altars erected near the Rhine between Rigomagus and Antunnacum, with a dedication "finibus et genio loci et IOM," marked the boundary between Upper and Lower Germany.[39]

On roads where a provincial boundary was stated by milestones to be the endpoint from which distances were measured, a traveler proceeding in the appropriate direction as far as that boundary would readily recognize when it was reached. It is clear enough that a provincial boundary was not a common choice for endpoint when milestones were erected. But it is also quite evident that there was no consistent code of practice for making these choices,[40] and we do find several instances where a provincial boundary was settled upon. Most notable perhaps are the milestones marking Trajan's work after his establishment of the new province of Arabia in 106: "redacta in formam provinciae Arabia viam novam a finibus Syriae usque ad mare Rubrum aperuit et stravit..."[41] It is likewise a Trajanic milestone that records repairs to the Via Egnatia "a Dyrrac(hio) usq(ue) Acontisma per provinciam Macedoniam."[42] Two Hadrianic milestones have been found southwest of Amaseia—at that date in the province of Cappadocia—on a route "ab Amaseia ad fines Galatorum."[43] At the end of the second century Septimius Severus and his sons used milestones to commemorate roadwork in their new province of Osrhoene: "viam ab Euphrate usque ad fines regni Sept(imi) Ab(g)ari a novo munierunt per L(ucium) Aelium Ianuarium proc(uratorem)

[36] *BAtlas* 19A2. On Gallic leagues and their use, see now Rathmann (2003), 115–20.

[37] 7.12.8; cf. 8.1.1.

[38] A feature of this particular border which by contrast seems exceptional was the *praetentura Italiae et Alpium* maintained under M. Aurelius. What form the *praetentura* took (military zone?) is uncertain: see *ILS* 8977 with *BAtlas* 20B3 and Šašel (1974).

[39] *BAtlas* 11H2; *CIL* 13/2.7732.

[40] See Rathmann (2003), 112–15.

[41] *ILS* 5834.

[42] *AE* 1936.51; for Acontisma/Hercontroma (*BAtlas* 51D3) as border-station, see *ItBurd* 603.8. Another milestone (*AE* 1936.52) from the Via Egnatia has identical wording, except that it defines the extent of Trajan's repair work less precisely as "a Dyrrac(hio) usq(ue) Neapoli per provinciam Macedoniam."

[43] *BAtlas* 87A4; French (1988), nos. 060, 339.

Aug(usti) prov(inciae) Osrhoenam (!) . . ."[44] Stone plaques have been found—inscribed in Greek and erected in 233—that served as milestones on a route measured from the boundary between Syria Palaestina and Egypt.[45] Maximinus and his son recorded the scope of their repair work to a road in Africa specifically as "viam a Karthagine usque ad fines Numidiae provinciae longa incuria corruptam adque dilapsam restituerunt."[46]

At some border-points, we may be sure, monuments were erected which left travelers in no doubt that they were crossing from one province to another. That was plainly the intention of an arch, the Ianus Augustus, erected in Augustus' time at the Baetis river (perhaps in connection with a bridge) to mark Baetica's eastern boundary. Milestones measured distances from the arch, and its function was evidently remembered long after the province's eastern boundary had in fact been shifted elsewhere (fig. 7.2).[47] On a desert

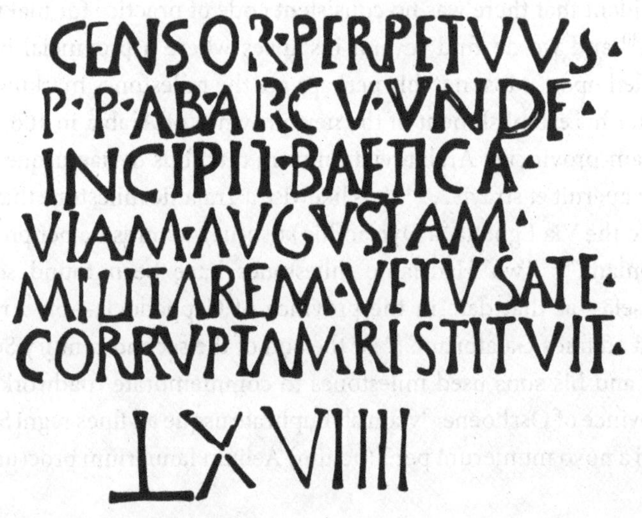

Fig. 7.2 Seven lines of the Latin inscription on a Roman milestone erected by the emperor Domitian in 90 CE, found six miles west of Corduba, Spain. He proclaims his restoration of the *Via Augusta militaris* "from the arch where Baetica begins." Figure: Ancient World Mapping Center after Sillières (1990), 102 = *Hispania Epigraphica* 24858.

[44] *AE* 1984.920; for the installation of boundary markers "inter provinciam Osrhoenam et regnum Abgari fines," see *AE* 1984.919.

[45] *AE* 1973.559, 559 bis.

[46] *ILS* 488; Rathmann (2003), 207.

[47] *RE* Suppl. 6 s.v. Ianus Augustus; Pekáry (1968), 107–108; Haley (2003), 34–35; *CIL* 2.4697 = *CIL*[2] 2/5.1280; and now Bellón Ruiz (2021).

route about 70 km northwest of Palmyra a single great column was erected—probably around 100 CE—to show quite unmistakably where Palmyrene territory began.[48] At the traditional frontier point of Arae Philaenorum, we know that around the end of the third century four freestanding columns were erected, surmounted by statues of the four Tetrarchs, to mark the border between the new provinces of Tripolitana and Libya Superior and also the new dioceses of Africa and Oriens.[49]

It remains unclear whether the statues of the emperor in a four-horse chariot, which Claudius permitted the Alexandrians to set up to him in 41, were ever erected. The request, at least, was to place these at three "entrances" to Egypt (*eisbolai tes choras*), at Taposiris, near the Pharos in Alexandria, and at Pelusium.[50] An associated puzzle is whether the statues were intended to carry any statement to the effect that they marked an "entrance" to Egypt. There must be the same uncertainty about the bridge across the Dravus river at Poetovio. When crossing it,[51] the Bordeaux pilgrim was aware of leaving Noricum and entering Pannonia Inferior: what, if anything, on the bridge itself drew attention to the existence of the provincial boundary?

In certain instances the presence of a *statio* for the payment of *portoria* might alert travelers to the fact that they were crossing a border. There was evidently such a *statio*, for example, on the border between Italy and Cottius' kingdom (later the province of Alpes Cottiae).[52] It must be recognized, however, that the "zones" into which the empire was divided for payment of *portoria* were large—often spanning several provinces—and few; consequently, at most provincial borders such dues were not levied.[53] Even where they were payable, the surviving evidence is too random and meager to claim that a *statio* was situated, say, on every main route or at every landing-place. A *statio* might indeed be situated at such locations but also, it is clear, in a major center such as Lugdunum Convenarum;[54] equally, it seems that payment of dues might be made at Ostia.[55] It is quite understandable that the

48 *BAtlas* 68E4 *Khirbet el-Bilaas*, with publications cited s.v. in *Map-by-Map Directory* 1045. A border dispute was ongoing, and apparently continued nonetheless; it must have been with either Apamea or Emesa, or conceivably even both.

49 *BAtlas* 37D2 Arae Philaenorum and Fines Africae et Cyrenensium, with publications cited in *Map-by-Map Directory* ss.vv., 553–54; *BAtlas* 101M5.

50 *P.Lond.* 1912 col. 3 lines 44–48, with *BAtlas* 74B3, B2, H2.

51 *ItBurd* 561.5, with *BAtlas* 20C3.

52 See n. 35 above. Compare the incident where Apollonius of Tyana in Philostratus' *Life* (1.20) is questioned by a customs collector as he leaves Zeugma (*BAtlas* 67F2) for Mesopotamia—crossing an external border of the empire, therefore.

53 See *BNP*, s.v. Toll IV.

54 *BAtlas* 25F2; France (2001), 65–68, 316.

55 France (2001), 135–38, 322–23.

so-called customs law of 62 CE from Ephesus retains the clause,[56] dating back to 75 BCE, which requires anyone entering the Asian customs zone at a location without a *statio* to proceed promptly to the nearest one in order to pay the expected dues.

For Ptolemy's mapmaking, Rome's provinces (and, further east, Parthia's satrapies) are entities of fundamental importance to his procedure and purpose. As he explains in 1.19:[57]

> We have written down for all the provinces [*eparchiai*] the details of their boundaries [*perigraphai*]—that is, their positions in longitude and latitude—the relative situations of the more important peoples in them, and the accurate locations of the more noteworthy cities, rivers, bays, mountains, and other things that ought to be in a map [*pinax*] of the *oikoumene*. . . . In this way we will be able to establish the position of each place, and through accuracy in particulars we will be able to establish the positions of the provinces themselves with respect to each other and to the whole *oikoumene*.

He continues in the same vein in 2.1.7–9:

> We will keep to the same principles also in each continent with respect to its parts as we do for the whole world and the entire *oikoumene* with respect to the continents, that is, we will again begin by recording the more northern and western countries and the adjacent seas and islands and the more noteworthy things of each kind. We will distinguish these parts of the continents by the boundaries of the satrapies or provinces, making the guide, as we originally promised, only as detailed as will be useful for recognizing and including places on the map, while leaving out the great mass of reports about the characteristics of the peoples (unless perhaps some bit of current knowledge calls for a brief and worthwhile note).
>
> Moreover this method of exposition will also make it possible, for anyone who wishes, to draw the parts of the *oikoumene* on planar surfaces, individually, or in groups of provinces or satrapies, in whatever way they might fit the proportions of the maps. The localities contained by each chart will then be inscribed at the appropriate scale and relative placement.

[56] *SEG* 1989.1180 lines 40–42.

[57] Here, and further below, I draw upon Berggren and Jones (2000), an invaluable translation.

Not only does Ptolemy then proceed to present his data, as promised, by province or satrapy or region, but in his last book (8) he also provides captions for twenty-six regional maps. A great advantage of making regional maps, he explains, is that the scales can be varied according to the extent of territory and density of data to be marked; each such map does not need to be at the same scale. His twenty-six comprise ten of Europe, four of "Libya" (Africa), and twelve of Asia. Predictably enough, provincial boundaries are one of the main determinants for settling what the scope of each of these regional maps shall be, and some of the captions state this explicitly—those for Spain and Gaul, for example. The results are rendered most effectively in the expert reconstructions by Carl Müller, published posthumously.[58]

It is seemingly an interest in the dimensions of provinces or regions which is the distinctive purpose behind the compilation of the two Late Antique Latin lists that are preserved with the titles *Dimensuratio Provinciarum* and *Divisio Orbis Terrarum*.[59] Apart from a broad definition of the location of each province or region by reference to major physical features (typically mountains, rivers, seas), the only further information offered is figures for the length and breadth of each in Roman miles.

It is true that neither the Peutinger map nor the Madaba mosaic map marks any boundary lines, but their absence need not mean that the makers in either case were ignorant or unconcerned about boundaries.[60] Both makers are to be seen as "professionals," who (like Ptolemy quoted above) were deliberately selective in determining what they wanted to show and how their map might most effectively convey their aims. Since the Peutinger mapmaker gives special prominence to land routes, he might reasonably prefer not to introduce a large amount of further linework of a different type, especially when the extreme format already so distorts the landscape. In all likelihood, too, the inclusion of borders would act directly counter to the vivid demonstration that Rome's sway stretches unbroken across the entire *oikoumene* from west to east, allowing ease of communication everywhere. This said, the Peutinger mapmaker does not hesitate to name provinces, regions, and peoples very conspicuously; their placement confirms a sound sense of their spatial relationships.

The Madaba mapmaker chooses to mark neither borders nor any routes linking places,[61] but the wording of this map's didactic legends testifies time

[58] Müller (1901); and see now Stückelberger and Grasshoff (2006), 766–907.

[59] *GLM* 9–19.

[60] For the Madaba map, see Piccirillo and Alliata (1999) with *OCD*[5] s.v.

[61] That routes are shown, according to Whittaker (2002), 87, seems a misperception on his part.

and again to awareness of boundaries, both traditional ones that hark back to the biblical world which is so central to the map's purpose, and ones in the contemporary world. Thus attention is drawn to the *horoi* of Egypt and Palestine,[62] and to the eastern boundary (*horion*) of Judaea (sect. 32). To Beersheba, "today Berossaba," not only is the information appended that it marks the southern boundary of Judaea, but the chance is also taken to repeat from *Second Samuel* 24:2: "the boundary of Judaea to the south reaches down to it from Dan near Paneas, which marks the northern boundary" (sect. 104). Asemona is described as a "city by the desert bordering [*diorizousa*] Egypt and the crossing to the sea" (sect. 106), while the legend for Gerara reads: "once a royal city of the Philistines and boundary [*horion*] of the Canaanites to the south, where is the Saltos Geraritikos" (sect. 107) (fig. 7.3).

The Madaba map could fairly be cited in support of Whittaker's traditional claim that Constantine's conversion "radically transformed the world view of the Romans."[63] One manifestation of this shift was Christian pilgrimage: it may well have boosted the number of long-distance travelers, and they certainly went in a changed spirit to novel destinations. At the same time, however, we should not omit to appreciate how one fundamental way in which non-Christians had been visualizing the world remained unaffected by the triumph of Christianity; rather, it was reinforced by it. I refer to the notion of Roman provinces as an organizational framework.

As the Christian church expanded, it evidently founded its organization as a matter of course on cities and provinces.[64] The result of this slow, shadowy growth at last becomes clearly visible under Constantine, when the Canons of the Councils of Nicaea in 325, and of Antioch two years later, formalize previous haphazard practice. It is now established that a new bishop should if possible be consecrated by all the existing bishops of cities in the province, once his appointment has been confirmed by the "metropolitan," the bishop of the chief city of the province. Moreover, in order to address various concerns, each metropolitan should summon and preside over a synod of all his province's bishops twice yearly.[65] The substantial authority placed in the hands of a metropolitan was justified at Antioch: "the bishop

[62] Sect. 129 in the presentation of Piccirillo and Alliata (1999).
[63] Whittaker (2002), 98.
[64] My sketch of this large theme derives from Hunt (1998), 240–50.
[65] *Concilium Nicaenum*, Canon 5, in Jonkers (1954), 41; ibid., *Concilium Antiochenum*, Canon 20, p. 54.

Fig. 7.3 The notices quoted about boundaries relating to Beersheba, Gerara, and Asemona appear on this segment of the Madaba Map (the first two middle and lower left, the third middle right). Map: Piccirillo and Alliata (1999), 85.

in the metropolis undertakes responsibility for the whole province, because it is in the metropolis that all those with business to settle assemble from everywhere."[66]

Evidently it was not at all the intention of ecclesiastical authority to position itself in an alternative location of its own choice, away from the center of secular authority. In the East, the church came to match the pattern of secular authority even more closely by 381, when a council at Constantinople confirmed that if a dispute could not be resolved within a province of one of the five secular dioceses (Egypt, Oriens, Asiana, Pontica, Thrace), then the

[66] *Concilium Antiochenum* Canon 9, in Jonkers (1954), 50.

matter might be referred to other provinces of that diocese.[67] Meantime, predictably, alterations by the secular power to the boundaries of its provinces for whatever reason were liable to have serious repercussions for the authority of individual bishops—fuel for any number of disputes and rivalries. Some church leaders eventually concluded that there was no value in always seeking to keep in step with the secular authorities. As bishop Innocent of Rome declared early in the fifth century: "it is not right that the church of God should be changed to suit the flexibility of worldly requirements, nor should it be subject to the promotions and divisions which the emperor may presume to make for his own reasons."[68]

Broadly speaking, the fact is that in territorial terms the organization of the Christian church within the Roman empire replicated that of the existing secular administration. The church, too, was organized on a provincial basis; to conceive its sway spatially would be to think of it encompassing a set of provinces.[69] In this important respect, therefore, the otherwise radically different outlooks of Christian and non-Christian coincided and reinforced one another. Later, in the West, Rome's provinces evidently remained the standard framework by which educated people organized their view of the world far into the Middle Ages.[70]

In short, then, I suggest we would be right to perceive a sense of the empire's provinces as spatial entities, and of the geographical relationship between them, developing from the early first century CE. This sense—*alongside* the linear sense gained from itineraries, together with representations in a variety of artforms[71]—becomes a further recognized means by which Romans envision their wider surroundings (fig. 7.4a, b). By chance, in time it turns out to be one notably reinforced by Christianity. It did not need to be at all a sophisticated perspective, and it was well able to accommodate boundary

[67] *Concilium Constantinopolitanum* Canons 2 and 6, in Jonkers (1954), 107, 110.
[68] *Epistulae* 24.2 (*PL* 20.548–49), quoted by Hunt (1998), 244.
[69] Compare the passage from Optatus of Milevis quoted in n. 19 above.
[70] Note the recent generalization by Gautier Dalché (2001), 370, in discussing a work where, exceptionally, this framework would seem to have been absent: "alors que les provinces romaines furent, durant presque tout le Moyen Âge, le cadre le plus fréquent dans lequel la géographie, même d'inspiration contemporaine et moderne, se plut à classer et à situer les êtres et les phénomènes, alors même que les mappemondes de grande taille montrent parfois des lignes qui semblent être des restes des limites tracées sur les modèles antiques, l'*Expositio* ne s'intéresse que de façon sporadique aux provinces . . ."
[71] Forms very familiar to us through the magnificent sculpture from the mid-first-century Sebasteion at Aphrodisias, and through Hadrian's coin series, to cite only two examples. On the series, see further now Dueck (2021), 174–75.

Fig. 7.4a, b Coins of Hadrian with regions or provinces personified on the reverse: (a) *as*, Mauretania; (b) *denarius*, Hispania. Note that in both these instances large regions divided into two or three provinces are featured. Photos: American Numismatic Society (1995.11.212 and 1944.100.45550).

shifts, additions to the empire, and the subdivision of large provinces into smaller units.[72]

Once this is said, however, one limitation to such a sense of the empire's provinces demands to be recognized: typically it was, so to speak, an outsider's perspective. It reflected the viewpoint of the Roman authorities, or of individuals seeking to grasp their surroundings well beyond home. It was seldom a vision that altered those individuals' self-identity. Rather, this remained rooted in their *origo*, the community of their birth and family.[73] In some cases, it is true, a region demarcated by Rome as a province already had an ethnic identity that predated the annexation. Galatia, Lycia, and Judaea seem indisputable instances,[74] and it was no doubt precisely in order to counter Jewish identity that Hadrian renamed the latter province Syria Palaestina after the suppression of the Bar Cochba revolt in the mid-130s.[75] At the opposite end of the range—to summarize a persuasive argument by Stephen Mitchell[76]—Pompey's annexation of the area newly termed

[72] Note the reliance upon it in commemorative inscriptions which summarize extensive travel experience, for example that of Aurelius Gaius quoted in essay 9 below (see also fig. 9.1); also in lists on certain portable sundials discussed in essay 18 below. Likewise striking is Ammianus' image of Rome's "baths built on the scale of provinces" (16.10.14 "lavacra in modum provinciarum exstructa").

[73] See further essay 8 below.

[74] For Galatia and Lycia, see Mitchell (2000), 122–24.

[75] See Smallwood (1976), 473.

[76] Mitchell (2002), 48–50.

"Pontus" by Rome bestowed upon its inhabitants a shared sense of identity as Pontici which they had not had previously, and which (over time) they adopted with pride. This set of circumstances in Pontus seems to find no parallel elsewhere.

More typically, the creation of a Roman province either divided existing ethnic identities or brought into one administrative framework various disparate communities which had never identified with one another; or it did both these things, all in the interests of Rome. As a result, the old ethnic identities might well disappear, but the preferred substitute was identity by city, not by province nor even a subdivision of one such as a *conventus*.[77] Rome on purpose offered most of its subjects little cause to identify closely with their province. The only body representing one or more whole provinces which Rome encouraged was the *concilium* or *koinon*, and its meetings involved no more than a very select group of top-class delegates, although its festivals did have broader appeal.[78] Even so, most provincials' lives remained centered around their own community, and their normal tendency was to regard neighboring communities as rivals. Only when seeking to grasp their world well beyond home were they likely to think in terms of their own province, the provinces contiguous to their own, and the others that comprised the whole empire.

[77] For Asia Minor, this shift is demonstrated by Mitchell (2000), 126–27.
[78] On these bodies, see Deininger (1964).

8
Worldview Reflected in Roman Military Diplomas

We hardly need to remind ourselves that the vast majority of people in antiquity never visualized their world with reference to any form of precise measurement. If they took stock of their wider surroundings at all, it was likely to be by means of a personal, unwritten "mental map" which we today have scant hope of recovering. Even so, the quest for some recovery may not be utterly hopeless in instances where a distinct, extensive category of relevant evidence exists, and where something of a shared vision—however rudimentary—can fairly be expected. A neglected instance of such a category, I believe, is the "diplomas" issued by the emperor to many men who served in the Roman military approximately between the mid-first century CE and the early third century. To be sure, the findings about geography and worldview to be teased out from these documents can only be modest at best, but as far as I am aware diplomas have never been investigated from this perspective, so a concise report on my cautious efforts to investigate it may have value.

Military diplomas—as modern scholarship terms them—are well known to students of Roman imperial history. Each comprised a pair of bronze tablets: on their outside face a legal document was inscribed, and it was inscribed again (more or less in duplicate) on the inside; the pair of tablets was then sealed. Broadly speaking, this design and the template for the document's wording were standardized. The recipients were for the most part noncitizen auxiliary troops. A diploma confirmed the rights and privileges that such a soldier received as a reward on a specified day, normally on his discharge after twenty-five years of service. He now gained Roman citizenship for himself (if he was not already a citizen, as typically auxiliaries were not) and was at last permitted to marry (serving soldiers could not be legally married), even receiving a grant (one only) of the legal status *conubium* for a wife, so that the children of this marriage would be Roman citizens. In addition, up until about 140, children born to such a soldier illegitimately during service could be legitimized and granted citizenship on their father's discharge—reflecting the fact that when serving plenty of men maintained a stable, long-term relationship which was marriage in all but name. As well as auxiliaries in the army, sailors in Rome's fleets received diplomas, and so did the men of the emperor's bodyguard, the Praetorian Guard. Guardsmen would all be Roman citizens already, but their diplomas still confirmed the privileges for a wife and children.[1]

Thanks to the mixed blessings that the use of metal detectors has brought, the number of known diplomas has dramatically increased during the past twenty years or so, more than doubling to around 1,000 currently, and sure to rise further.[2] Inevitably, within this total there is no lack of instances where what survives is only a fragment that happens not to preserve data relating to geography and worldview. Even so, such information is preserved in many hundreds of instances, although it has been all but ignored while other data yielded by diplomas has long received intense attention, stoking continual debate about deployment of units, consular dating, legal rights, and more. I thank Elizabeth Greene for alerting me to the potential of the overlooked information about geography and worldview.[3]

Perusal of the wording of a typical diploma allows us to establish the passages in the document where information relevant to geography and worldview is regularly to be found. Take this recently published instance, which is complete and very legible; its findspot is not known unfortunately (fig. 8.1):[4]

Imp(erator) Caesar, divi Traiani Parthici f(ilius), divi Nervae nepos, Traianus Hadrianus Aug(ustus), pont(ifex) max(imus), trib(unicia) potest(ate) XVI, co(n)s(ul) III, p(ater) p(atriae), proco(n)s(ul), equitib(us) et peditib(us) qui militaver(unt) in alis tribus et coh(ortibus) XII quae appell(antur): I Aug(usta) Gallor(um) Procul(eiana), et Aug(usta) Vocont(iorum), et Vetton(um) Hisp(anorum); et I Aug(usta) Nerv(iana) German(orum) (milliaria), et I Celtib(erorum), et I Thr(acum), et

[1] The bibliography on the nature and purpose of military diplomas is vast: for a synthesis of much current thinking, see Eck (2010). My brief summary ignores many variations and changes over time which do not affect the aspect of diplomas treated here.
[2] For this calculation, see Eck (2010), 34 n. 4.
[3] See in this connection Greene (2011), especially 194–235, 264–87; (2015) with appendix 1; (2017). Note in addition Cuff (2010).
[4] Eck, Holder, and Pangerl (2010), 197 (conflated text, with my bolding).

Fig. 8.1 Diploma issued to an auxiliary soldier in Britain, 132 CE; height 14.5 cm, width 11.5 cm. Photo: Werner Eck, Paul Holder, and Andreas Pangerl.

I Ling(onum), et I Hamior(um) sag(ittaria), et I Morin(orum), et I Sunuc(orum), et I, et II Dalm(atarum), et III Brac(araugustanorum), et IIII, et V Gallor(um) et sunt in Britannia sub Iulio Severo quinq(ue) et vigint(i) stipend(is) emeritis dimiss(is) honest(a) missione, quorum nomin(a) subscript(a) sunt ipsis liber(is) posterisq(ue) eorum civitat(em) dedit et conub(ium) cum uxorib(us) quas tunc habuissent cum est civitas iis data aut si qui caelib(es) essent cum iis quas postea duxiss(ent) dumtaxat singuli singulas.

a. d. V id. Dec(embres) C. Acilio Prisco, A. Cassio Arriano co(n)s(ulibus):

coh(ortis) I Hamior(um) sagitt(aria) cui praest M. Mussius Concessus, ex pedite Longino Sesti f(ilio) Moms, et Longino f(ilio) eius et Sestio f(ilio) eius et Sestiae f(iliae) eius.

Descriptum et recognitum ex tabula aenea quae fixa est Romae in muro post templum divi Aug(usti) ad Minervam.
Ti. Claudi Menandri; P. Atti Severi; T. Flavi Romuli; L. Pulli Daphni; P. Atti Festi; C. Iuli Silvani; C. Vetieni Hermetis.

In brief, this diploma issued by the emperor Hadrian is one for auxiliaries serving as cavalry (*equites*) and infantry (*pedites*) in three *alae* and twelve *cohortes* (all named) stationed *in Britannia*, who were discharged on December 9 of the year (132) in which Gaius Acilius Priscus and Aulus Cassius Arrianus were consuls. Specifically, this particular diploma was issued to an infantryman in a cohort of archers, the *Prima Hamiorum*, commanded by M. Mussius Concessus. The recipient's name is stated (Longinus), together with the name of his father (Sestius), and the recipient's *origo* (his place or people of origin) recorded just as MOMS, as well as the names of two sons and one daughter; no wife is mentioned in this instance, however. Finally, there is stated the location in Rome where an official permanent record of this discharge has been set up in public on bronze, followed by the names of seven witnesses.

The location in Rome where the discharge was publicly displayed—*in muro post templum divi Aug(usti) ad Minervam*—has obvious potential value for the topography of the city, and scholars have long taken note of the information. These references may be set aside for present purposes, except to underline the point that each diploma forges an explicit link with the city of Rome itself as the place where ultimate authority resides—a hazy vision of distant power for most recipients of diplomas, who were not discharged anywhere near Rome, and in all likelihood had never served there or even visited there.

It was not the practice for discharges requiring the issue of diplomas to be made piecemeal to this soldier or that on any random day. Rather, the imperial authorities maintained tight control of the entire process by enforcing a procedure whereby all the men who qualified across various units in a particular province were discharged together on the same day in one deliberately planned and coordinated step. Each auxiliary diploma most explicitly, indeed laboriously, reflects this approach. It names not just the unit of the individual recipient, but also all the units from which men were discharged on the same date, specifying both that date and the province: hence *in Britannia* above. While ceremonies at which the names of the units and the discharged men were solemnly read out might seem a possibility, the practice is not attested. But we may still wonder what geography or worldview, if any, a reading of these unit-names brought to the minds of men in possession of a diploma whenever they did read the names there or had the names read to them. For the most part the names stem from peoples and regions, and in certain instances offer extended geographical resonance which could well have been meaningful to long-serving soldiers, as in the following record (presented without abbreviations) of discharges in Germany on May 21, 74:[5]

> . . . equitibus et peditibus, qui militant in alis sex et cohortibus duodecim quae appellantur: I Flavia Gemina, et I Cannenefatium et II Flavia Gemina, et Picentiana, et Scubulorum, et Claudia nova; et I Thracum, et I Asturum, et I Aquitanorum veterana, et I Aquitanorum Biturigum, et II Augusta Cyrenaica, et III Gallorum, et III, et IIII Aquitanorum, et IIII Vindelicorum, et V Hispanorum, et V Dalmatarum, et VII Raetorum, et sunt in Germania . . .

No doubt because it was regular practice for a man to be discharged where his unit was based, a diploma did not normally record the place. Mention was made of the place, however, in the special case of a diploma issued on August 11, 106, to affirm that at Darnithithi (a location otherwise unknown) men who had distinguished themselves *in Dacia . . . expeditione Dacica* were rewarded with citizenship early by Trajan.[6] Also untypical is the inclusion in some diplomas of not just the name but also the *origo* of the discharged soldier's commanding officer; this occasional practice defies explanation.[7] More common initially was inclusion of the *origo* of each of the seven or more witnesses named for each diploma; but this practice faded out during the 70s.[8]

By contrast, the *origo* of the man receiving the diploma is invariably recorded, along with his name and his father's name.[9] For this unique document to be inscribed, the recipient himself must be consulted in advance; in

[5] *CIL* XVI.20 = *ILS* 1992; cf. among many comparable instances XVI.4 (July 2, 60, Illyricum, primarily Spanish unit-names) and 69 (July 17, 122, Britain). Also to be considered in this connection is the degree to which the order adopted for listing the names reflects a logical geographic sequence of the locations within a province where the units were stationed.

[6] *CIL* XVI Suppl. 160.

[7] Note *CIL* XVI Suppl. 179, 180 (both October 9, 148).

[8] Note most of *CIL* XVI.1–20.

[9] For discussion of the *origines* recorded on diplomas (with reference to perspectives that differ from mine here), see most recently Speidel (1986) and Mirković (2007).

particular, the information that he supplies about his dependents (or the lack of any) can only come from him. There is reason to suspect that the *origo* recorded for a man at his discharge was not invariably the same as that recorded at his enlistment (his *probatio*), although it is impossible to be sure.[10] It is quite evident that either men themselves or the engravers of their diplomas (or both) often reduced an *origo* to as few as four letters, so that uncertainty about what is meant may be embedded in the document, conceivably even by design.

Alternatively, uncertainty might arise more by accidental means from possible mis-hearing or mis-spelling on the part of the engraver. The diploma issued in Britain in 132, quoted above, offers a case in point: MOMS is not readily recognizable as the abbreviated name of any people known to us. The three scholars who published this diploma conjectured the town of Mom(o)-asson in Cappadocia on the road between Ancyra and Tarsus.[11] As they note, Anthony Birley—perhaps less plausibly—suggested to them Montana in Moesia Inferior.[12] My own inclination is to look beyond Tarsus to Mopsou(h)-estia,[13] a name known to have given rise to all manner of variants in Latin; on the Peutinger map it appears as Mompsistea.[14] For comparison, we may note that a soldier discharged a few years later, in 140, has ONIANDO EX LYCIA recorded as his *origo*, in other words Oenoanda.[15]

The single term *origo* which scholars use to refer to this item of data in all diplomas may seem inadequate when we consider that these *origines* vary greatly in scope, and that firmer geographical grasp on the recipients' part can be detected. Michael P. Speidel has urged that we should abandon the Latin *origo* for the looser English "home."[16] He also drafted a "rule" for discussion, namely that an auxiliary's native province would be given as his *origo* if he was sent elsewhere on recruitment, whereas a tribe or town would be given if he enlisted in a unit stationed or raised in his own province. The second part of the "rule" is understandable enough: it made sense for a man from Pannonia serving there to be more precise about his *origo* than merely stating "Pannonian"; accordingly, he might specify

[10] See Speidel (1986), appendix I.
[11] *BAtlas* 63E4.
[12] Eck, Holder, and Pangerl (2010), 195–96; *BAtlas* 21F6.
[13] *BAtlas* 67B3.
[14] 9B4.
[15] *CIL* XVI Suppl. 177; *BAtlas* 65C4.
[16] Speidel (1986), 473, where he notes Tacitus' choice of *patria* in recounting how Germanicus in 14 CE asked centurions for their *nomen, ordinem, patriam, numerum stipendiorum* . . . (*Ann.* 1.44.5).

Azalus, say, or *Eraviscus*, people-names that were unlikely to be well known elsewhere.

Speidel was surely right to conclude that the first part of his imagined "rule" was commonly ignored (that an auxiliary's native province would be given as his *origo* if he was sent elsewhere on recruitment). The distinction between what is to be understood as a province-name as opposed to a people-name is in any case blurred where instances such as Lusitanian, Pannonian, Thracian and Syrian are concerned.[17] Moreover, we find cases of men serving in Britain and Pannonia giving their *origo* as *Treverus* (from northern Gaul); likewise another man serving in Pannonia gives his as *Lucensis* (from Spain). There is an intriguing instance of a diploma issued in Dacia Porolissensis on July 2, 133, where an attempt has begun (on the outside only) to alter or "correct" the *origo* inscribed as PANNON to another name—perhaps the more localized CORNAC, presumably a reflection of the identity that the recipient in retrospect wished to claim (fig. 8.2a, b).[18] It is only natural to expect that men gave as their *origo* the place or people with which they most closely identified themselves and regarded as their "home." Many men from the East gave a city rather than a people.

It was not necessarily a concern that the *origo* recorded for a man would in all likelihood be unrecognized by anyone outside his own immediate circle in the province where he was serving. The *origo* DOBUNN recorded for a man discharged in Moesia Superior on January 12, 105, is a case in point; at least in this instance his unit name (*cohors prima Brittanica*) can help to situate the *origo* in the wider world.[19] Sailors—whose service was not linked to a province—routinely give as their *origo* communities or peoples that must have baffled most contemporaries, not to mention modern experts: for example, MAEZEIO,[20] GALLINARIA SARNIENSIS,[21] CARES(io).[22] Similarly, a Praetorian Guardsman just gives MOGIONIBUS.[23] By contrast, there is also no lack of cases where men demonstrate awareness and concern that they must gloss their *origo* if it is to be meaningful to others—either

[17] *Bessus*, too, even if it is used to signify *Thrax*, can hardly be regarded as a province-name; cf. Speidel (1986), 468.

[18] Daicoviciu and Protease (1961) = Roxan (1978), no. 35, with further comment on the *origo* intended to replace PANNON. For the Cornacates, see *BAtlas* 20F4.

[19] *CIL* XVI.49; *BAtlas* 8E2.

[20] From northern Dalmatia (*BAtlas* 20D4); *CIL* XVI.14.

[21] Uncertain location; *CIL* XVI.16 found in Corsica.

[22] From Sardinia (cf. Karensioi, *BAtlas* 48B2); *CIL* XVI.40.

[23] Uncertain location, possibly Pannonia; Roxan and Holder (2003), no. 303, found within Pannonia.

(a)

(b)

Fig. 8.2a, b Diploma (a) issued to an auxiliary soldier in Dacia Porolissensis, 133 CE, with detail (b) of his altered *origo*. Photos: Constantin Daicoviciu and Dumitru Protease.

because there are well-known cities elsewhere with the same name, or because the name on its own is considered too obscure to be meaningful. For the latter reason, presumably, the man mentioned above who gave his *origo* as ONIANDO EX LYCIA specified a province (Lycia) as well as a city. Comparable formulations are SELINUNT EX CILICIA[24] and CAECOM

[24] Roxan (1994), no. 171.

EX MOES(ia).[25] Concern to eliminate confusion with cities of the same name must account for the formulations NICOPOL EX BESSIA[26] and ANTIOCHIA EX SYRIA.[27] Placement of the "ex" formulation can equally be reversed, as in EX PAN(nonia) INFER(iore) IATUMENTIANIS.[28]

It is striking that many diplomas issued to sailors from around the late second century are laid out to present the name and *origo* of the recipient in larger lettering,[29] and that by the reign of Elagabalus in the second and third decades of the third century the *origo* is expanded to specify three levels—province, city, and village (not necessarily in that order).[30] Thus, for example, a sailor discharged from the Misenum fleet on November 29, 221, is recorded as: "C. Iulio Barhadati fil(io) Montano, Dolich(e) ex Syria, vico Araba, et Aureliae Bassae uxori eius civitat<e> s(upra) s(cripta) et [five children]."[31] Another sailor (fleet unknown), discharged in late 224, is recorded as: "M. Aurelio Spori fil(io) Victori cui et Drubio, Nicopoli ex Moesia, vico Dizerpera."[32] Presumably for some reason it had now become a matter of pride among sailors to demonstrate a more informed worldview in this way. Greater specificity in stating an *origo* can hardly have been a demand on the part of the authorities, because by the reign of Elagabalus diplomas had become no more than honorific, now that Septimius Severus had removed the ban on marriage during service, and Caracalla had extended Roman citizenship to almost all free inhabitants of the empire.

Of particular interest from a geographical perspective (as well as a "family" one) are those diplomas where the recipient includes a wife with her name and *origo*. When both have the same *origo* and he served far away from there—as in the case of Iulius Montanus immediately above—the question arises of whether they were a couple before he enlisted, or whether she somehow made the long journey from "home" to join him. In the latter case, we may imagine a sailor contriving to make the necessary arrangements for a partner once he knew that a ship was under orders to take the route required. By contrast, for an auxiliary soldier to arrange for a woman to join him when he could not leave his posting for a long period, and she would need to travel a long distance overland, must have presented a far sterner challenge—one

[25] Pferdehirt (2004), no. 31.
[26] Holder (2006), no. 392.
[27] Ibid., no. 454.
[28] Roxan and Holder (2003), no. 304.
[29] For example, ibid., pl. 57 (photograph of no. 308).
[30] On this dating, see the note by Holder (2006), 705.
[31] Roxan and Holder (2003), no. 307.
[32] Holder (2006), no. 463.

which raises questions of how he understood and visualized the route that would have to be taken.

When the recipient of a diploma and his wife have widely separated *origines*, the natural assumption must be that the man on service away from "home" meets, and forms a permanent relationship with, a woman from the area where he is serving. This is surely the case with the DOBUNN man from Britain—already mentioned—discharged in Moesia Superior in 105 from a unit that is known to have served previously in Pannonia. No doubt it was there that he met his wife whose *origo* is AZAL, from a Pannonian tribe in the region of Aquincum.[33] Given the fact that his diploma was found on the site of Brigetio not far west from there, it follows that after his discharge the couple went back, not to Britain, but to the wife's home area.

A comparable instance involving a sailor in the Misenum fleet is reflected in a diploma issued in 229.[34] This man, Titus Domitius Domitianus, gives his *origo* as CLAUDIOPOLI EX CILICIA VICO VINDEMI,[35] and names a wife whose *origo* is AFRAE (most probably signifying the province of Africa). Moreover, three sons are named: Diodotus; Caricus, that is, presumably, born in Caria (southwest Asia Minor), or perhaps born when his father was serving there; and Putiolanus (by the same token, a home-base birth at Puteoli close to Misenum). The exact findspot of the diploma is not recorded, but it is known to have been in the region of modern Konya in southern Turkey, so evidently after his discharge this sailor took his family back to his own "home" region of Asia Minor rather than to that of his wife in Africa. It is only natural to conjecture that they met during his service, rather than before his enlistment. However, the man's ship need not have been based for a period at an African port; for his wife, say, to belong to a family that had migrated from Africa to Italy and taken up residence in Misenum or Puteoli is also possible. Equally, as we know or can strongly infer from cases among auxiliaries, it happened that men formed relationships with the sisters or daughters of fellow soldiers.

The few items of personal information preserved by Domitius' diploma permit the most tentative reconstruction of the worldview that he and his wife may have formed as a result of his naval service and their marriage. This view was no doubt framed by the Mediterranean, with the city of Rome featuring prominently—ships from the Misenum fleet would routinely have visited Portus at the Tiber mouth—and visualized in relation at a minimum

[33] *BAtlas* 20F2; see further note 19 above.
[34] Roxan (1985), no. 133.
[35] For Claudiopolis, see *BAtlas* 66C3; the location of the *vicus* is unknown.

to Misenum/Puteoli, the Italian peninsula, Sicily, Africa and perhaps Malta, the Aegean and perhaps Crete, and the west and south of Asia Minor. By contrast, an auxiliary soldier's worldview is likely to have been rather different, with the city of Rome (as suggested above) remaining a remote, unvisited prospect. Even so, auxiliaries can be expected to have developed some "mental map," however rudimentary, of the geographical relationship of the province where they served to neighboring provinces.[36] Such a map in the heads of men serving far from their native provinces we can reckon to have been more complex than the worldview of men who only ever served close to their "homes." At least, the notion is reinforced that a mental sketch-map of a group of provinces, if not of the entire empire, was basic information valuable to every serving soldier. What remains a puzzle is how far, if at all, auxiliaries could, or would, relate a listing of unit-names to geography. Would men have been more readily stimulated to make such connections in instances where their *origo* happened to be the same as the name of the unit in which they served—an Ituraean in a *cohors Ituraeorum*, or an Emesene in an *ala Hemesenorum*? Possibly so, just as Domitius' idiosyncratic choice to name his second son Caricus and his third son Putiolanus argues for some grasp on his part of how these two different locations that he found so memorable related spatially to one another as well as to a wider world.

Plainly, the observations and questions sketched here with caution call for collation of further data and more systematic analysis. In addition, what may be gleaned about geography and worldview from diplomas needs to be matched with similar information from soldiers' funerary inscriptions in particular,[37] including the exceptional listing of men with their *origines* on a funerary altar at Adamclisi (Romania) erected around 100 CE.[38] Inevitably, the limitations of the standardized records that diplomas preserve preclude our gaining anything more than glimpses. But even glimpses are precious, especially when they serve to re-create the outlook of some of the empire's vast lower-class, non-Roman-citizen population—auxiliaries and sailors who evidently conceived the world more in terms of its peoples than of the cities with their greater appeal for both educated Romans and modern scholars.[39]

[36] For such a mental map, see further especially essay 7 above.
[37] Note Speidel (1986), 475–76.
[38] *CIL* III.14214, with discussion by Turner (2013).
[39] For the geographical consciousness of illiterate Greeks and Romans in general, see now Dueck (2021).

9
Author, Audience, and the Roman Empire in the *Antonine Itinerary*

From the Middle Ages onward, the anonymous collection misnamed the *Antonine Itinerary* (*Imperatoris Antonini Augusti Itineraria Duo, Provinciarum et Maritimum*) has consistently attracted attention. This contribution focuses on its substantial land part (1.1–486.17), which comprises seventy-five double-columned pages in the standard 1929 Teubner edition by Otto Cuntz.[1] It was eagerly exploited by the maker of the Hereford map around 1300, for example,[2] as well as by other medieval and Renaissance cartographers,[3] and it survives in as many as fifty or so manuscripts.[4] Today, as ever, it remains important to scholars with an interest in Roman travel, control of space, and related themes.[5]

It seems fair to claim that, after long debate, broad consensus has by now been reached about certain basic features of the collection—in particular, that it assembles individual itineraries of distinctly varied character and perhaps even date; that it was compiled around 300 CE; and that, despite the traditional title, there is no secure connection to travel by emperors. All these points have been convincingly established by Pascal Arnaud and others,[6] and I have no wish to dispute them, nor indeed to deny the outstanding value of the collection from many perspectives.

This said, in my view there are further fundamental questions relating to the collection which seem to have been explored much less, if at all. The

[1] For listing of the place-names (with etymological analysis of Celtic name-elements), note now Isaac (2002, 2004); Sims-Williams (2006). There is no knowing whether or not the land and much shorter sea (487.1–529.6) parts were assembled by the same individual. For discussion of the latter part, see Salway (2004), 68–85.

[2] See Westrem (2001), xxix–xxx, 429–31; and now Hiatt (2020), 35–36.

[3] Nicholas of Cusa (1401–1464), for example: see Friedman and Figg (2000), s.v.

[4] See Herzog (1989), 94–97 = (1993), 105–108. The earliest manuscript dates to the seventh century, the first of many printed editions to 1512 (Paris): for a review of these editions, see Leclercq (1927), cols. 1858–62.

[5] Several aspects are now addressed by Löhberg (2006).

[6] Arnaud (1993). Note especially Calzolari (1996), whose analysis, while focused on Italy, also ranges beyond; more briefly, Salway (2001), 23–25, 39–43.

attempt that follows may serve to sharpen our insight into the day-to-day functioning and administration of the Roman empire. As with my ongoing study of the Peutinger map, my approach here to the *Antonine Itinerary* consciously diverges from the prevailing fashion, which is to treat the collection as little more than a work of reference for issues of a very focused, specific type, such as the form of a place-name, or the distance given between two points, or the choice of route followed.

My present concern is not so much to examine individual components of the collection, but more to reflect upon the work as a whole. What can be determined, if anything, about the likely status of its author or "compiler"? Where and how did "he" find his material? To what extent did he edit it? Who did he envisage as his audience—and whether for extended reading, or for reference, or even both—assuming that he did intend the collection to go into circulation? What motivated him to make the collection, what needs did he see it as fulfilling? Was his idea of making it an exceptional piece of creativity (it is, after all, a unique survival), or are we to imagine that collections of this broad type were commonplace? In addition, what underlying grasp of geography can the compiler be reckoned to possess?

Any answers which may be formulated to these taxing questions must by definition depend upon a close reading of the entire collection; this is itself no light task. Invaluable aids for the purpose are Cuntz's index and his sketch map of all the routes in the collection, together with the relevant maps in the *Barrington Atlas*.[7] It is only right to assume, however, that no such aids were available in antiquity; hence any interpretation unduly influenced by the added insight which these modern tools may contribute is to be treated with caution. In particular, it should be stressed that Cuntz's sketch map—for all its intrinsic merits—creates a most misleading distillation of what is to be gained from the work as a whole, insofar as this map by its very nature furnishes the clear synthesis that is emphatically absent from the written text.[8]

No modern reader can fail to be struck by the uniform austerity with which each individual itinerary in the collection is presented. The four indispensable components are a statement of the starting- and end-points and a figure for the total distance between them, followed by a listing of successive

[7] Löhberg (2006) plots the routes of the *Antonine Itinerary* onto base maps from *BAtlas*. For the forgotten map by Pierre Lapie (published 1845) that also plots these routes, see Talbert (2019), 49–60.
[8] This limitation is discussed further below.

intermediate points accompanied by the distance between each (typically stated in miles, although sometimes with a half mile). Knowledge of distance is essential, therefore; in principle, if the distance to or from a place cannot be ascertained, then the place cannot be listed.[9] This is an established pattern for itineraries[10] that dates back to the first century CE at least, when we find it in the *Stathmoi Parthikoi* by Isidore of Charax,[11] for example, on the Vicarello cups,[12] and on the Claudian *Stadiasmus Patarensis.*[13] It would appear that the compiler of the *Antonine Itinerary* had neither the motive, nor perhaps the capacity, to change the pattern or to expand it. In consequence, he supplements the raw data with at most an occasional, stereotyped explanation or forewarning. Thus when he includes two or more alternative routes between a pair of points, they may be variously labeled "recto itinere,"[14] "per conpendium" (shortcut),[15] "alio itinere a . . . ,"[16] "per ripam,"[17] "per maritima loca,"[18] "per mediterranea loca,"[19] or even "per medium."[20] Even so, for travelers seeking to weigh the consequences of choosing, say, the *conpendium*—steeper slopes perhaps, and rougher surfaces—no further guidance is forthcoming.[21] That omission is liable to be felt even more keenly in the few instances where the compiler records as many as three or

[9] This is not to claim, of course, that each distance figure was necessarily accurate, even before copyists' slips were introduced. The combined point-to-point distance figures in each itinerary would, however, need to match the total figure that is always offered in addition. To this extent, distance figures were of greater concern to the compiler of the *Antonine Itinerary* than they were to the maker of the Peutinger map, because the map offers no total figures, and for the most part does not lay out places in relation to the distance between them on the ground. It is just conceivable, therefore, that some of the route stretches without a distance figure on our copy of the map—Vadis Volateris to Aqvas Volaternas, for example (3B2)—reflect, not a copyist's slip, but the lack of any figure even on the original; strictly speaking, completion of the map as we have it was not dependent on the inclusion of these figures.

[10] Note Pliny the Elder's complaint: "Onesicriti et Nearchi navigatio nec nomina habet mansionum nec spatia" (*NH* 6.96). For overview of itineraries, note Fugmann (1998).

[11] Jacoby, *FGrHist* 781; English translation in Shipley (forthcoming), no. 21.

[12] *CIL* XI. 3281–84 with Roldán Hervás (1975), 149–60, and fig. 4.1 above.

[13] See now Şahin and Adak (2004) and figs. 16.4–5 below. The absence of distance figures to accompany the routes listed on the damaged inscription of uncertain date from Junglinster is notable (*CIL* XVII/2.676).

[14] For example, 106.5. For this, and the usages immediately following, Calzolari (1996), 393–94, offers a full listing.

[15] E.g. 82.8–9.

[16] E.g. 419.7; cf. "Aliter a Roma Cosa" (300.1).

[17] E.g. 207.10.

[18] E.g. 90.6.

[19] Uniquely at 231.8–9.

[20] Uniquely at 258.2.

[21] Note by contrast how the *Stadiasmus Patarensis* (note 13 above) lines 30–31 records alternative routes from Oinoanda to Balboura, one through the plain (διὰ τοῦ π[εδ(]ου), the other through mountains (διὰ δὲ τῆς ὀρεινῆς).

four alternative routes between a pair of points;[22] prospective travelers will search in vain for so much as a word to explain why they should even consider the lengthier options. In brief, the most obvious benefit to be derived from assembling alternative routes—the opportunity to elaborate upon their relative lengths and other relevant characteristics—is never exploited.

By the third century, if not long before, there were frequent stopping-points established at least along the empire's main routes; their presence is documented most extensively in the meticulous record of the so-called Bordeaux pilgrim's long journeys in 333.[23] The most modest stopping-points—those that the pilgrim terms *mutatio* as opposed to *mansio*—were not perhaps prime choices for an overnight stay, but his record shows that in a day's journey of twenty to twenty-five miles travelers could expect to encounter one, if not two. In general, by contrast, these more modest stopping-points are notably absent from the *Antonine Itinerary* collection, where the stops listed are more widely spaced, typically fifteen to thirty miles apart, thus an entire day's journey.

How is the absence of the greater detail to be understood? It seems improbable that none of the individual itineraries assembled for the collection ever included more modest stopping-points; and the compiler of any collection intended for use would surely wish to retain such detail in the interests of assisting his audience as fully as possible. The surprising absence of this detail from the *Antonine Itinerary* collection could suggest, then, that it was deliberately excised by the compiler. If so, that is a remarkable piece of intervention, for no discernible purpose, by someone who (as we shall see) was a far from active editor in other respects. He can hardly have been striving to save space by this means, because he blatantly wastes space in other ways. Nor can his purpose have been to achieve uniformity of presentation, when in other obvious respects uniformity is of no concern to him. The question of his purpose in this regard is best set aside for now, however, and reopened later.

Instead, the two features just introduced—wasted space and lack of uniformity—merit closer attention. On any extended reading, both leave a lasting impression. The collection constantly repeats routes in whole or in part, sometimes more than once.[24] In certain instances, to be sure, the

[22] Two examples: Sebastia to Cocuso at 176.3; 178.6; 180.6, in effect repeated at 212.5. Bracara to Asturica at 422.2; 427.4; 429.5; and (placed last by Cuntz) 423.6.

[23] Geyer and Cuntz (1965), 1–26, and essay 7 above.

[24] For striking illustrations of such repetitions, compare 210.5–217.4 with 177.5–210.4. Cuntz's presentation of the itineraries ignores the elements of repetition.

second occurrence of a route comes long after the first, which may provide some justification for the repetition; but more commonly the interval between the two occurrences is short. In certain instances, moreover, on its second occurrence a route is offered in reverse direction from the first time. Differences in the choice of intermediate stopping-points, not to mention in the names themselves, also occur. Such variations can be regarded favorably as contributing a welcome air of authenticity. They foster the impression—no doubt accurate—that the collection preserves the record of journeys actually made.

At the same time these variations can cause confusion and frustration. Without doubt, the south–north route Ponte Aeni to Ad Castra (259.3–6) merits inclusion, but use of the latter name is distinctly unhelpful here, when on its previous appearance—three routes earlier—the place is called Regino (250.1, on an east–west route); only from another source of information can it be established that two different names are being used for the same place.[25] Equally, the traveler planning to go from Tarraco to Caesaraugusta, and concerned at the scanty information about this route offered at 391.1–392.1, is not forewarned that when the route recurs in reverse at 451.2–452.5 it is with twice the number of intermediate points listed. The compiler evidently lacked the knowledge, or confidence, or concern, to standardize either the presentation of each route or the forms of place-names, let alone the case in which names are given.[26] He also declines to link routes by highlighting those stopping-points which are junctions or forks of note, even when a route from there,[27] or through there,[28] is to occur later in the collection.

Without doubt, Romans did not expect written materials to be organized and edited to modern standards, so in all likelihood ancient readers of the *Antonine Itinerary* collection would be less perturbed than their modern counterparts by the differences it fails to reconcile. Even so, the excessive repetition of routes in whole or in part remains difficult to account for. This prominent feature cannot have escaped the compiler's notice as he was writing. It is a puzzle that he did not act to reduce it, if only to save himself much laborious copying in the process. What benefit did he believe was to

[25] Even an experienced mapmaker could be misled in this way: evidently the maker of the Peutinger map did not become aware from the materials he assembled that Prvsias and Cio, for example, are the same place, not two different ones (8B2).

[26] As is typical of itineraries, the form of most names is accusative or ablative (signifying motion to or from) rather than nominative: see further Calzolari (1996), 401–402.

[27] For example, Ulbia at 79.4 and 80.8.

[28] For example, Aquis Regis at 47.3 and 54.2.

be gained by retaining so much redundant material? Eventually, on Cuntz's page 46, he does state that on the Via Lavicana from Rome, after reaching Fregellano, the route continues to Benevento "mansionibus quibus et in Prenestina."[29] Even thereafter, however, he uses this handy formulation only four more times—once on a route from Mediolanum to northern Spain (387.6) and three times on routes within Spain.[30] If he was prepared to do so in these instances, then why not in many others?

The puzzle of repetition leads on to the puzzle of the collection's arrangement, and it in turn raises the issue of the character of the individual itineraries. A certain logic can be found in the arrangement, but its application is far from consistent. Reasonably enough, the start is the southwest of Mauretania Tingitana, through North Africa, to Libya (1.1–78.3); then Sardinia, Corsica, Sicily (78.4–98.1); then Italy, including the Via Appia, and southward generally from Rome (98.2–123.7). An astonishing trek follows: Rome–Ariminum–Mediolanum–Aquileia–Byzantium–Antioch–Alexandria–Hiera Sicaminos, with several routes in Egypt appended (123.8–173.4). Next, Thrace very briefly, followed by central Asia Minor eastward (175.1– 217.4). From there, a shift into the Danube lands which eventually brings a return to Italy, especially the north and down to Rome, including several more named roads from the city (217.5–317.2). Then Greece sketchily, with a single no less sketchy foray into western Asia Minor (317.3–339.5). Then several Alpine crossings from Mediolanum into Gaul and Germany, and routes within those two regions (339.6–387.3). Finally, the Iberian peninsula (387.4–463.2), and Britain (463.3–486.17).

The most disruptive component here is the splitting of the routes in Italy between two widely separated sections; in principle these clearly merit combining into one, with some consequent elimination of repetition. It should be noted, too, that the collection's coverage is nowhere near complete (not that it makes any claim to be). Large areas barely feature or are altogether absent: among them, Palestine and Arabia; southern and western Asia Minor; Cyprus and Crete; the Balkan lands all the way from the Danube south to the Via Egnatia; the west of Gaul from south to north.

It is appropriate to wonder why there should be these particular substantial omissions. At the same time it should be recognized that coverage of those areas which do fall within the collection's scope varies immensely.

[29] 305.5. The Via Prenestina is in fact the immediately preceding route.
[30] 439.11–14; 446.2.

Perhaps the most successful single section happens to be the penultimate, for the Iberian peninsula. It begins with two "backbone" routes, the first from Mediolanum to Legio VII Gemina (387.4–395.4), the second from Arelate to Karthago Spartaria and Castulo (396.1–402.5). Then follows a variety of routes within the peninsula in a generally comprehensible sequence, with widespread coverage (albeit not complete) and relatively minimal repetition.

The practice of opening with a "backbone" route and then branching out from there is to be found in several other sections of the collection too. The first section, for North Africa, is a clear example, although its later part offers much less widespread coverage than there is for the Iberian peninsula.[31] Other limitations emerge in this first section. Circuitous routes are offered without warning. An egregious example is Tacapae to Lepti Magna, which does duly warn that it will proceed along the *limes Tripolitanus* (73.4; i.e. rather than along the coast), but adds nothing about its opening stages which reach Agma after a 182-mile loop via Turris Tamalleni, when the direct route along the coast would be only 25.[32] Routes that in fact radiate from a single center are dispersed rather than grouped together, and not necessarily all offered in the same direction.[33] Some starting-points are obscure to modern experts, and must have been to most ancient readers too: Turres Caesaris,[34] for example, somewhere south of Cirta, or Assuras,[35] south of a main junction at Musti, which would seem a more serviceable choice of starting-point.

Perhaps the least satisfying section of the collection is that for Thrace and central Asia Minor eastward (175.1–217.4). All the limitations just cited recur there. The repetition of routes toward the end of the section is excessive, and there is not even a single "backbone" route to give the section some cohesion.[36] Rather, the opening Thracian part comprises no more than two insignificant, disjointed routes, from Cabile to Hadrianopolis (175.1–6), and from Plotinopolis to Heraclea via (with no warning of the detour) Traianopolis (175.7–176.2). For an informed compiler to make the two into one (by linking Hadrianopolis and Plotinopolis)[37] should not have been difficult, but this

[31] Almost no coverage for the densely settled region south of Carthage, for instance.

[32] Cf. 59.7.

[33] Note, for example, four routes in and out of Theveste starting at 33.2; 46.2; 53.5; 54.8 (the last two arranged consecutively).

[34] 34.7. *BAtlas* Map 34 (*Directory*, p. 527) is unsure of the location.

[35] 47.6. *BAtlas* 33D1.

[36] However, one such route—from Byzantium to Antioch (138.5–147.1)—does form part of the astonishing trek from Rome to Hiera Sicaminos.

[37] *BAtlas* 51H1–G2.

is not done.[38] Immediately following (!), readers are offered three different routes in succession for traveling from Sebastia (in Cappadocia) to Cocuso. This, too, is useful information in its own limited way, but still it can hardly have been a prime concern of most Roman travelers in Asia Minor, especially if they were in need first of guidance—never offered by the collection—into its interior. The next itinerary is a route from Arabissos—no more than fifty-two Roman miles from Cocuso, although its relative proximity only emerges later[39]—to Satala. Then two alternatives in succession for Germanicia to Edessa are offered. However, after two other, unrelated routes have been presented next,[40] there follows a third alternative from Germanicia to Edessa, which the compiler never troubled to marshal with the previous two.

*****

Limitations of this type could be illustrated further, although to little purpose beyond reinforcement of the widespread opinion that the *Antonine Itinerary* collection is patchy in its coverage, loose in its organization, confusingly repetitive, and uninformative where it offers a choice of routes. It is more important instead to pursue questions which arise from the very negativity of the assessment.

From what sources might all these individual itineraries have been gathered, and for what purpose? At least some of the "backbone" itineraries were no doubt "official" Roman reference documents accessible without undue difficulty to staff in the service of the emperor or governors or procurators, as well as to soldiers.[41] Other itineraries may derive from locally authorized "signposts" (the Latin technical term is unattested) that were erected at certain junction cities.[42] Perhaps such signposts were the source of the miscellaneous itineraries that are straightforwardly local,[43] in some instances even "one stop."[44]

[38] This instance of separate itineraries which could usefully have been combined is far from unique in the collection. For example, 90.5 merits adding to the end of the next itinerary at 93.1.

[39] 214.11–13.

[40] Antiochia–Hemesa (187.2–188.3), and Arbalisso–Muzana (188.4–6).

[41] It seems natural to imagine that official record-keeping was one reason for Trajan to order that certain roads be measured ("mensuris viarum actis") and milestones erected (*AE* 1969/70. 589, milestone of 114/115 from the Peloponnese).

[42] Atuatuca Tungrorum, for example (*CIL* XVII/2. 675).

[43] For example, Aquis Regibus to Sufibus in forty-three Roman miles with a single stopping-point (47.3–5); Litirno to Miseno in twelve Roman miles with two stopping-points (123.4–7).

[44] For example, "Item a Luca Pisis m. p. XII" (289.1), "Ab Hispali Italicam m. p. VI" (413.6). For a "one stop" itinerary on a very different scale, compare "A Salacia Ossonoba m. p. XVI" (418.6), where the figure is reckoned to be a copyist's slip for CXVI.

As suggested above, other itineraries bear the mark of an individual who made a particular journey. One plausible indicator is an untypical starting- or finishing-point.[45] Itineraries may also suggest an individual's record when the intermediate points on a route that is repeated differ (a feature mentioned earlier) or when the choice of stopping-points seems idiosyncratic. It is striking, for example, that once the long itinerary Mediolano–Ad Columnam[46] turns inland from the Adriatic coast road,[47] there are at least three notable cities on the way that it omits to mention.[48] Conceivably, this itinerary is the record of a traveler who (for whatever reason) sometimes preferred to stop for the night at road stations rather than in cities, and thus had no reason to list the latter despite their significance in other respects.[49] It is a frequent occurrence in the collection to discover—without warning, as usual—itineraries so roundabout as to be most ill-suited for travelers intent upon reaching one city from another with the least amount of difficulty and delay. Thus these itineraries, too, give the impression of being the record of individuals who—for whatever reason—diverged from a direct route.

Why did individuals record their journeys in this way, and where would the compiler have obtained such material to assemble into a collection? The individuals are perhaps most likely to have been officials or soldiers,[50] men dispatched on duties that involved a tour of cities, or a round of forts or the like, who thereafter submitted their itinerary to some headquarters office in order to document their absence or to claim expenses, or both. Evidently all that mattered for the purpose was a record of their principal stops and of the distance covered between each.[51] There was no call for this record to be notably full or accurate, nor were comments or explanations required.[52]

[45] The two routes in Thrace, for example, already mentioned; Dolica to Seriane (194.7–195.3; *BAtlas* 67E2–68E3); Eumari to Neapolis (195.9–197.4; *BAtlas* 68D5–69B5).

[46] 98.2–106.4.

[47] At Aterno vicus (101.5).

[48] Corfinium (*BAtlas* 42F4), Aquilonia (45B2), Vibo (46D4).

[49] Compare the listing of Salernum as only "in medio Salerno" (with no distance figure), between Nuceria and Ad Tanarum in the itinerary headed "Appia" (109.4).

[50] In several instances, the military presence at a stopping-point on a route is noted briefly, in particular between Viminacio and Novioduno at 217.5–226.1. Compare the header at 266.8–9: "ab Acinquo Crumero que castra constituta sint."

[51] Compare the records to be found in the archive of Theophanes for the journey he made c. 320, especially *P.Ryl.* IV.627–28 and 638, and essay 10 below.

[52] When a provincial boundary is crossed, for example, it is rare for that to be noted specifically. The references to the Malva river as the divider of the two Mauretanias (12.1–2), to "fines Marmariae" and "fines Alexandriae" (70.7; 71.8), and to "Sedisca, fi. Ponti" (217.2) are exceptional. To be sure, the name Fines or a variant is common, but the significance of the boundary goes unexplained: see further Calzolari (1996), 416–17, and more broadly essay 6 above.

At one time or another the compiler no doubt worked in such headquarters offices, and some of the journeys in the collection may have been made by his own acquaintances. He presumably made journeys himself, and in so doing copied down records of routes and distances that came to his attention. Conceivably, it was also part of his responsibilities to devise itineraries for journeys that had to be undertaken,[53] such as those of soldiers referred to by Ambrose in a sermon that impresses upon the faithful Christian the value of observing God's law:[54]

> The soldier who is setting out on a journey does not organize his own travel arrangements, nor does he choose his route by his own judgment, nor look for charming shortcuts, in case he gets separated from his unit. Rather, he receives his itinerary [*itinerarium*] from his commander and adheres to it: he proceeds according to the prescribed stages, marching with his weapons, and he makes the journey by the correct route, so that he may find the support for his travel already prepared. If he goes by another route, he receives no rations, and does not find lodging [*mansio*] prepared for him. In view of the commander's order that all preparations be made for them, those who follow do not diverge to right or left from the prearranged route.

I see our compiler's aims in assembling his collection of itineraries as modest indeed. The sheer quantity and variety of information about routes, stopping-points, and distances that he could gather simply intrigued him. It was of no concern to him that his collection comprised an incomplete jumble of components, poorly organized, almost unedited, remaining full of repetition and inconsistency. The collection was in any case not a finished product, but open-ended, a work still in formation, which he might be able to supplement and improve from time to time in the future.

Such circumstances offer reason to consider whether it was ever the compiler's intention to put his collection into circulation for use by others. Scholars have perhaps been too ready to assume that he planned it, and

[53] Certain routes in the *Antonine Itinerary* are unquestionably impressive for their ability to proceed to a distant destination through a succession of junctions, for example 356.1–363.2 Mediolano–Gesoriaco.

[54] *Expositio psalmi CXVIII* 5.2.1–2: "Miles qui ingreditur iter viandi ordinem non ipse disponit sibi nec pro suo arbitrio viam carpit nec voluptaria captat conpendia, ne recedat a signis, sed itinerarium ab imperatore accipit et custodit illud, praescripto incedit ordine, cum armis suis ambulat rectaque via conficit iter, ut inveniat commeatum parata subsidia. Si alio ambulaverit itinere, annonam non accipit, mansionem paratam non invenit, quia imperator his iubet haec praeparari omnia qui secuntur nec dextra nec sinistra a praescripto itinere declinant."

issued it, as a collection of routes designed to be useful to long-distance travelers on the pattern of other surviving itineraries.[55] As explained above, if that were the intention, then I think he failed badly. To be sure, the collection can be made to serve as a traveler's guide, but it is a far from satisfactory one. Rather, its true character is that of a hobbyist's assemblage. The compiler was enchanted by itineraries of various kinds and their associated data, and it became his hobby to collect them as best he could, arranging them in a rough order, but otherwise preserving each more or less just as he acquired it. Since all this was done with an eye to no more than private satisfaction (and perhaps personal reminiscence in certain instances), the collection's multiple limitations were of no consequence to its compiler.

It is important to recognize that the manner in which many of the journeys are recorded is all too likely to disappoint the modern scholar who turns to the collection for information about geography and related topics. The fact is, however, that the records typically comprise just the minimum required for a timesheet or expenses claim, where thoroughness or accuracy count for little. Thus the record for Taurino to Vindobona (242.1–248.2) stands out for its unusual fullness in including many intermediate stopping-points—names followed by "in medio," without a distance figure—between the main ones.[56] Generally speaking, as already mentioned, no warning is given of a significant detour within an itinerary,[57] nor is a transfer from land to water necessarily noted.[58]

There has been much discussion of whether the compiler took material from a map, or related the itineraries that he assembled to a map. This method of working has most recently been advocated by Mauro Calzolari,[59] but I remain skeptical. Reference to a map suggests a compiler with concerns for coverage, organization, and use by others, which are largely missing here. This is not to deny that the compiler possesses a grasp of geography, sketchy though it may be. He is at least aware of many of the empire's principal regions and provinces, as a glance through Cuntz's *Conspectus Itinerum A*[60] soon demonstrates. Equally, the collection's arrangement, as well as the wording of

[55] See the review of opinions by Calzolari (1996), 376–80.
[56] For sporadic instances elsewhere of "In Medio" as a stopping-point listed in its own right, with distance figure, note for example 188.5; 189.4; 212.6 and 8 (compare 180.6–181.2).
[57] For the long loop to Agma, see note 32 above. Compare, for example, the detours through Hadrianopolim between Traianopoli and Perintho Erac. (322.4–323.5); to Mutina (282.1); to Germe (335.4).
[58] In Egypt, for example (152.4–171.4).
[59] Calzolari (1996), 382–84.
[60] Cuntz (1929), 103–106.

many of its headings,[61] presuppose knowledge of the locations of regions and provinces relative to one another.[62] Likewise its compiler knows of the Alps and its Cottian, Graian, Maritime, and Pennine ranges.[63] The same can be affirmed of four great rivers: Danube, Euphrates, Nile, Po. Even though only one of these is ever named—the Nile, once, in a heading[64]—large parts of the collection presuppose awareness of one or another of the four.[65]

It seems only right to assume that the compiler recognizes, however hazily, the significance of major starting- or ending-points that have found a place in the collection's itineraries, but about which there is no further clue here: Clysmo, for instance, or Beronicen, or Trepezunta.[66] By the same token, a level of geographical awareness has to be associated with his specification of water crossings. An accurate map reveals that some of these crossings are unavoidable (as across the English Channel),[67] while others are practical rather than quite unavoidable (as along the coast of North Africa from Tingi to Portus Divinos);[68] whether the compiler always appreciated that difference is unclear.[69] Likewise elusive is his grasp of the relative importance of stopping-points in areas which only a single itinerary in the collection traverses, so that no point's importance here can even be inferred from its recurrence in other itineraries. Consequently, in a listing such as "... Pompeis, Naissi, Remisiana, Turribus, Meldia, Serdica, Burgaraca ...," any sense of how one point differs from another has to be sought from other sources.[70] The descriptive phrase "caput Germaniarum" added to the starting-point Lugduno

[61] In general, I regard the headings as an integral part of the original work rather than additions by a post-Roman editor.

[62] For example, "de Pannoniis in Gallias" (231.8), "a Durrachio per Macedoniam et Trachiam Bizantium usque" (317.3–4). How such knowledge might be acquired remains an open question raised in essay 7 above.

[63] See index, s.v. Alpes, in Cuntz (1929).

[64] 164.1 (a notable omission from Cuntz's index).

[65] In particular, the references to itineraries "per ripam" are otherwise mystifying: 207.10 (sc. Euphrates); 217.5 (Danube); 241.1 (Danube).

[66] 170.4; 173.4; 216.4 respectively.

[67] "A Gessoriaco de Gallis Ritupis in portu Brittaniarum" (463.4–5). Also Ad Columnam to Sicilia ("Traiectum Siciliae," 98.4–5), Bizantio to Calcedonia (138.5–139.2), Brundisio/Hydrunti to Durachium/Aulonam (317.5; 323.9–10; 329.1–2), Callipoli to Lamsacum (333.2–3; 333.9).

[68] 9.1–13.7. Compare from Pola "traiectus sinus Liburnici Iader usque" (272.1–2).

[69] Comparison of his fourth route from Bracara to Asturicam (423.6–425.5, partly by water) with his third route immediately preceding (429.5–431.3, which proceeds by land where the fourth sails) clearly does confirm the water crossing in the fourth as an alternative. The same is true of the two successive routes from Isca to Calleva, the first of which proceeds entirely by land (484.10–485.7), while the second includes a "Traiectus" across the mouth of the Severn (485.8–486.7).

[70] 134.4–135.5. For whatever reason, a limited number of itineraries do gloss some or all their stopping-points as *castra, civitas, colonia, mansio, municipium, vicus,* and the like, but this is not the norm; see further Calzolari (1996), 398–401.

(368.3–4) stands out as unique. With this phrase, the compiler reminds himself (or any readers) that this point is to be distinguished from Lugdunum Convenarum and Lugdunum caput Galliarum, both of which form part of other itineraries in the collection.[71]

Who in the Roman world might be expected to have made this collection? Not an educated member of the upper classes, surely. To compile such an unpolished assemblage of tedious detail about unmemorable places would hardly engage anyone at that level. A scholar or teacher seems equally unlikely. There is no denying that, despite their learning, these men could produce writings of poor quality and muddle their facts,[72] but at least they were conscious organizers, concerned first to separate the important from the trivial, and then to present the results for others in an accessible form.[73] The compiler of the *Antonine Itinerary*, by contrast, displays no such talents or concerns. The business of assembling the collection may have enhanced his worldview, but it is hard to imagine that this was his principal motivation. He should be considered an individual of middle rank at best, whose career in the service of the Roman state involved administrative responsibilities, perhaps as a *beneficiarius*[74] or a centurion (or both in turn). His collection may be regarded as an unusually substantial survival among texts for which Romans of middle rank or lower were responsible.

The maker of the collection certainly recalls one type of historian imagined by Lucian,[75] "a soldier or workman or peddler following the army," whose written record was a bare collection of events, "pedestrian and plodding." Equally, he recalls those individuals of various types who loved nothing better than to reminisce about their extensive travels both within the Roman empire and beyond,[76] and sometimes even had these publicly commemorated. The veteran Aurelius Gaius was one such.[77] From an inscription in Greek

[71] See index, s.v. Lugdunum, in Cuntz (1929).
[72] For a clear instance, note the *Liber Memorialis* by L. Ampelius (ed. Arnaud-Lindet [2003]).
[73] Those are also the concerns of Vibius Sequester, *De Fluminibus, Fontibus, Lacubus, Nemoribus, Paludibus, Montibus, Gentibus per Litteras*, for example, among other authors in *GLM*.
[74] For the wide range of their roles, see Ott (1995), especially Teil B.
[75] *Hist. Conscr.* 16: Ἄλλος δέ τις αὐτῶν ὑπόμνημα τῶν γεγονότων γυμνὸν συναγαγὼν ἐν γραφῇ κομιδῇ πεζὸν καὶ χαμαιπετές, οἷον καὶ στρατιώτης ἄν τις τὰ καθ' ἡμέραν ὑπογραφόμενος συνέθηκεν ἢ τέκτων ἢ κάπηλός τις συμπερινοστῶν τῇ στρατιᾷ.
[76] As Horsfall (2003), 114 observes, this predilection is recognized both by Plutarch, *Quaest. Conv.* 2.1.2 (630 B–C) and by Macrobius, *Sat.* 7.2.6: "Those who have crossed seas and lands are delighted when asked about a largely unknown land location or sea cove. They respond with pleasure, and describe the sites now in words, now with a stick, considering it a matter of pride to set before the eyes of others what they had seen" ("nec non et qui obierunt maria ac terras gaudent cum de ignoto multis vel terrarum situ vel sinu maris interrogantur, libenterque respondent et describunt modo verbis modo radio loca, gloriosum putantes, quae ipsi viderant aliorum oculis obicere").
[77] See Drew-Bear (1981) for comparable instances.

erected after his retirement home to Cotiaeum in Phrygia (and now damaged), we may deduce that he had served successively among forces led by the Tetrarchs Galerius, Diocletian, and Maximian between 294 and 298. He boasts that he had made "a circuit of the empire"(τη[ν ἡγεμον]ίαν κυκλεύσας), and his listing of the locations to which he had traveled takes the form of a loop by sea and land (*periplous, periegesis*) familiar from classical literature:[78]

> . . . Asia, Caria, [Phrygia?], Lycia(?), Lycaonia, Cilicia, [Coele Syria?], Syria Phoenice, Arabia, Palaestina, [Aegyptus], Alexandria, India, [Osrhoene?], Mesopotamia, Cappadocia, [Armenia?], Galatia, Bithynia, Thracia, Moesia [Inferior?], Carpia, [Dacia, Moesia Superior?], Sarmatia four times, Viminacium [----] Gothia twice, Germania [] Dardania, Dalmatia, Pannonia, [Noricum, Italia?], Gallia, Hispania, Mauretania, [Africa].

By naming provinces or regions for the most part, rather than cities, Aurelius Gaius furnishes a most expansive impression of where he has traveled (fig. 9.1). His choices not only extend to "exotic" India, wherever he means by that name, but they also strongly project Roman imperial pride. Moreover, this pride assumes a traditional, even nostalgic, character in its surely conscious avoidance of any of the new names for provinces that the Tetrarchs' reorganization of the empire's administration had instituted.[79]

Our compiler was of course no innovator in assembling texts or extracts of a specific type into a collection. It is harder to say, however, what *collections* of *itineraries* had been made before he took up the idea. In the late fourth century, Vegetius' idealizing *De Re Militari* does at least urge that a commander (*dux*) ought to be able to consult a set of itineraries for relevant war zones:[80]

[78] This translation and my map follow the interpretation proposed by Kevin Wilkinson (2012). Note that, although Aurelius Gaius includes Sarmatia, Gothia, and Germania, he does not consider them integral to his circuit. Rather, during his service in the Danube lands, he had traveled there more than once.

[79] "Il est plus remarquable de noter qu'en dépit des variations des designations officielles, l'usage populaire reste étonnamment stable. Aucun des noms nouveaux créés pour désigner des provinces nouvelles n'est entré dans l'usage courant," observes Sartre (1983), 31. Compare the choices made by Optatus, quoted in essay 7 n. 19 above.

[80] 3.6: "Primum itineraria omnium regionum, in quibus bellum geritur plenissime debet habere perscripta, ita ut locorum intervalla non solum passuum numero sed etiam viarum qualitate perdiscat, compendia deverticula montes flumina ad fidem descripta consideret; usque adeo ut sollertiores duces itineraria provinciarum in quibus necessitas gerebatur non tantum adnotata sed etiam picta habuisse firmentur, ut non solum consilio mentis verum aspectu oculorum viam profecturus eligeret."

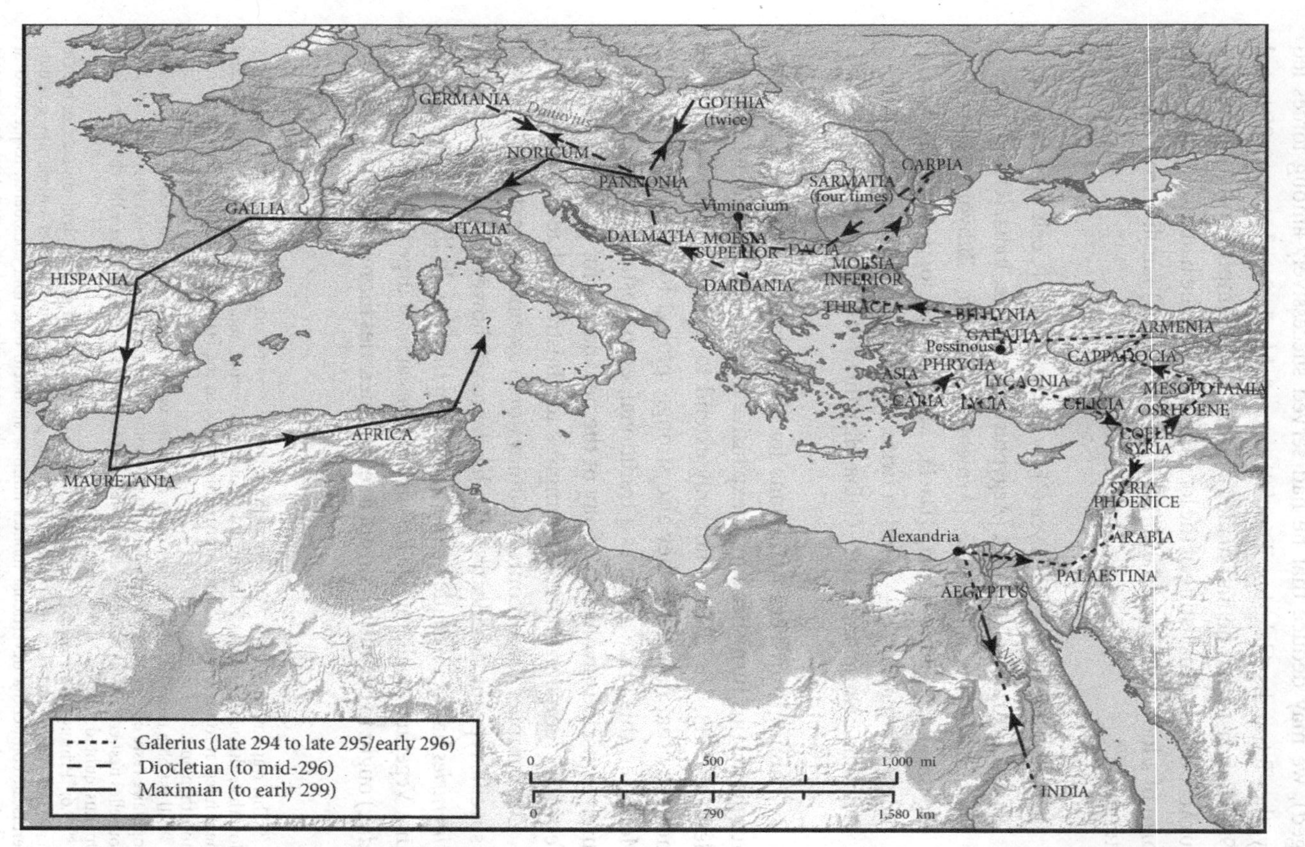

Fig. 9.1 Circuit of the Roman empire by Aurelius Gaius, 285–299 CE, traced on a modern map base. He served successively in forces led by the Tetrarchs Galerius, Diocletian, and Maximian. Map: Ancient World Mapping Center.

In the first place a commander ought to have itineraries of all the regions in which war is waged written out [*perscripta*] very fully so that he may gain a firm grasp of the distances between places (and not only from the number of miles but also the condition of the routes), and may take into account shortcuts, byways, mountains, and rivers accurately recorded. Moreover, for those provinces where crises were occurring, we are assured that the more able commanders had itineraries that were not just noted down [*adnotata*] but also in picture form [*picta*]. Thus when setting out he would choose a route not only by a mental check but visually.

However, there is no knowing to what extent such sets of itineraries really were made, and (if so) how far back in time the practice dated. If prior to the fourth century there already existed a format that was acknowledged as standard for a collection—as in the case of various types of *fasti*, for instance[81]—the compiler of the *Antonine Itinerary* seems unaware of it or ignores it. While plenty of single itineraries survive (direct routes, without detours), and even regional groups as on the Claudian *Stadiasmus Patarensis*,[82] no other Roman collection of them is extant, nor is one even attested, "official" or private.[83] The term *itinerarium* itself seems to be attested in only one (nonofficial) instance before the fourth century.[84] On a memorial at Smyrna,[85] the Greek book-titles *Asias Stadiasmon* and *Europes Stadiasmon* feature among the historical and geographical output of Hermogenes, a prolific doctor and author, but their meaning remains elusive. If they are records of travel (by Hermogenes himself?)[86]—as opposed, say, to learned discussions of the length assigned to the stade by different communities—measurement in stades might suggest sea voyages rather than land journeys. While we should never underestimate the huge amount of written material

[81] See *BNP*, ss. vv. Calendar, Fasti.
[82] See note 13 above.
[83] For long after antiquity, too, itineraries more or less on the Roman model continued to be compiled, but whether they were deliberately collected is another matter. See in general Friedman and Figg (2000), s.v. Itineraries and *Periploi*. Note now the valuable annotated listing, century by-century, of all geographical and topographical texts in Greek, Latin, and Syriac from 1 to 700 CE (also astrological and cosmographical ones), appended to Johnson (2016), 139–56.
[84] On three of the four Vicarello cups (Roldán Hervás [1975], 149–60, and essay 4 fig. 4.1 above): *itinerarium* (3281), *itinerar* (3282), *itinerare* (3283). The earliest attestation of *itinerarius* used adjectivally now seems to be *AE* 2000.1191, a second-century dedication to "diis itine[rariis]" discussed by Kolb (2005).
[85] *IK Smyrna* 23.1 no. 536 (ed. Georg Petzl). The date of the memorial is uncertain—Petzl hazards no more than "1. Jh. n. Chr. (?)"—as is the identification of this Hermogenes with any of the known figures so named.
[86] Jacoby's suggestion in *FGrHist* 579T1.

associated with Roman imperial administration that is now lost,[87] there is a case for claiming that in the present state of our knowledge the compilation of the *Antonine Itinerary* deserves to be regarded as a more original and untypical initiative than we might otherwise assume.

In consequence, greater caution may be called for on the part of those scholars who rely heavily upon the *Antonine Itinerary* to promote the sweeping generalization that Romans' worldview was primarily a "hodological" one. To quote C. R. Whittaker: "space itself was defined by itineraries, since it was through itineraries that Romans actually experienced space, that is, by lines and not by shapes."[88] There are already other grounds for maintaining that Roman worldview was more probably molded by a variety of means, among which itineraries featured, but did not dominate.[89] The prospect that the *Antonine Itinerary* may be an unusual compilation, not a routine one of a common type, only serves to reinforce this broader interpretation.

If the *Antonine Itinerary* is unusual, the creation of the Peutinger map—at around the same period perhaps—may seem all the more remarkable.[90] There is no clue to what individual or team made the map, or where their workplace was. Quite clearly, however, he (or they) was able to exploit a far more extensive set of itinerary data than the compiler of the *Antonine Itinerary* assembles; this set permits more or less comprehensive coverage of the entire Roman empire, together with some routes that even proceed beyond it into the Persian empire and India. How much trouble it was for the cartographer to assemble and organize all that data for his ambitious purpose is yet another open question. We cannot say how usefully the itineraries at his disposal were organized,[91] nor can we be sure what previous maps of the Roman world had been made which showed land routes in this way, although none is attested.[92] We should at least recognize, however, that if the Peutinger map's attention to these routes is innovatory, and if they had to be plotted from a mass of compilations as raw for the most part as those assembled in

[87] The point is well made by Eck (2002), 131–33.
[88] Whittaker (2002), 102.
[89] For discussion, see essay 7 above.
[90] See further essays 13 and 14 below.
[91] For certain, however, these materials in turn had their shortcomings, and did *not* comprise a full, consistent listing of direct routes.
[92] With such maps at his disposal, he might not have duplicated routes in the area beyond the eastern Mediterranean in the puzzling way that he does. Earlier maps, with or without routes marked, would have been one source able to assist him in establishing the relative importance of settlements.

the *Antonine Itinerary*, then the cartographer's labor would have been immense and the result merits all the more admiration.

It is hard to envisage the *Antonine Itinerary*, with its manifest limitations and idiosyncrasies, holding much attraction for any contemporary in antiquity beyond its compiler. Rather, its appeal only develops later, after the crumbling of the Roman empire and the collection's own miraculous survival. Then, from the Middle Ages onward, the collection is esteemed as a uniquely informative evocation of the amazing journeys that had once been possible under Roman rule.[93] As such, it is loosely comparable on an empire-wide scale to the mid-twelfth-century *Mirabilia Urbis Romae*, which (among much else) brought ancient Rome itself back to life for visitors to the city's ruins.[94] It is ironic that posterity, with its thirst for detailed knowledge and its dearth of sources, has accorded the *Antonine Itinerary* a respect and a value far beyond the private interests of its compiler, who barely had any reader in mind beyond himself.[95] This personal, introverted quality of his terse and uneven record is overdue for recognition. The chance survival of the collection preserves for us an oblique and tantalizing glimpse into the routine administrative rhythm of Roman officialdom.

[93] For travel on the empire's highways, see further essays 16 and 17 below.
[94] See Friedman and Figg (2000), s.v. *Marvels of Rome*.
[95] Of course, such generous revaluation of an ancient text is far from exceptional. Compare the claim made by Brennan (1996) that in its surviving form the *Notitia Dignitatum* represents a nostalgic, ideological compilation, not one for practical use.

10

John Matthews' *The Journey of Theophanes: Travel, Business, and Daily Life in the Roman East* (2006)

For sheer quantity of surviving documentation, there can hardly be a journey in classical antiquity to match that of Theophanes and his entourage, who in the early 320s (perhaps 322 or 323) went from Hermopolis Magna in Egypt to Syrian Antioch, and back again after a two-and-a-half-month stay. Not only does John Matthews brilliantly exploit the dry accounts kept on this journey, but his study also serves as an instructive paradigm of striking shifts that have occurred in the study of classical antiquity over the past half century. Embedded within this book lie several more journeys, both literal and metaphorical, beyond the one made by Theophanes himself. In recent years, the scholarly community has complained time and again that colleagues entrusted with unpublished ancient materials are too slow to issue them and thus to share them with others who may be interested. To be sure, the Greek papyri comprising the so-called Archive of Theophanes were themselves slow to reach publication; but ever since they appeared in 1952 the level of interest in them has remained minimal.[1] A total of around 1,500 lines documenting in relentless detail where the travelers stayed each night en route, and above all what they paid for their food and drink daily plus some other expenses, hardly seemed treasure trove fit to attract the ambitious researcher (fig. 10.1).

True, some attention has already been given by a handful of scholars to the terminology used for items of food and drink (with clear signs of Latin infiltrating the Greek East), as well as to the prices recorded in comparison with other sources during a period of monetary inflation. It is Matthews' achievement, however, to realize that this material offers unique potential to illuminate social and cultural history across a far wider spectrum than has hitherto been recognized. For these purposes, his book offers eight chapters,

[1] Note, however, Kirsten (1959), overlooked by Matthews.

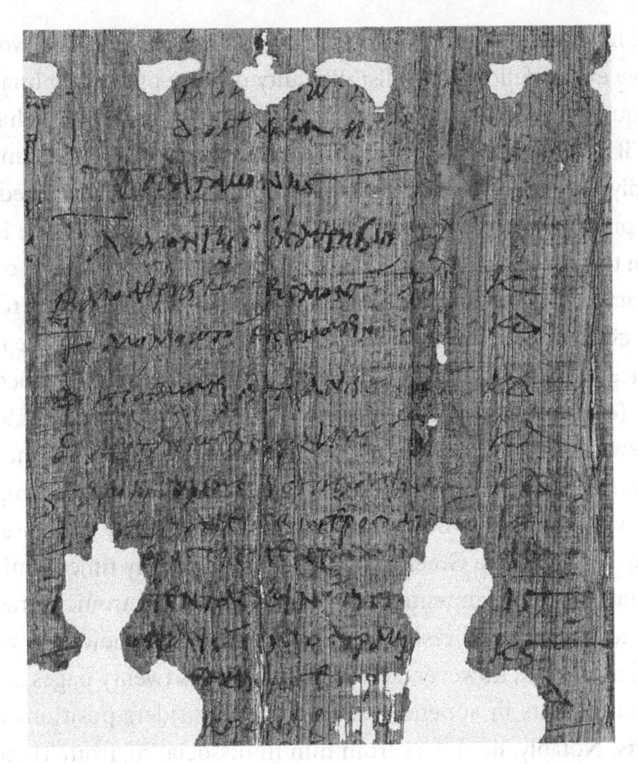

Fig. 10.1 Greek papyrus record for part of Theophanes' outward journey to Antioch. The nine stages here—from Nikiou (in the Nile delta) to Ostrakine—are numbered A to Θ at the left, with distances in Roman miles to the right (K = 20, KΔ = 24). *P.Ryl.* 627 verso, col. ii lines 223–33, translated in Matthews (2006), 56, 59. Photo: © The University of Manchester.

covering Theophanes' home circumstances in Hermopolis; the party's travels to and from Antioch; their stay there; costs and prices; food and diet. Four appendices include textual notes on the papyri, and an indispensable tabulation of expenditure item by item. The main text of the book is lucid and well pitched to engage a very diverse audience, despite the intricacies of the unfamiliar material. The range of informatively captioned illustrations and maps is a further enhancement.

One major reason why a very diverse audience can engage with the papyrus record here is that Matthews has had the courage to follow the still relatively novel fashion of presenting it all in English, piece by piece, at the appropriate stages of the book. How different from the original publication in 1952, where Colin Roberts presents nothing but the Greek, together with a concise introduction and the bare minimum of terse notes on words that might give even a fellow specialist difficulty or were previously unattested.[2] In consequence, even at that time, the potential readership must have been tiny and it has only continued to decline, now almost to vanishing point. Admittedly, in 1952 a full translation would have further increased the expense of producing what was already a costly, delayed publication in financially lean times. Roberts' preface and his compressed presentation of certain badly damaged documents (for example, *P.Ryl.* 639, 643–51) duly recognize the latter circumstances, ones with which he was to become all the more familiar once he made the unusual career shift from Oxford Classics don to Secretary (equivalent to chief executive) of the University Press in 1954.[3]

Matthews, for his part, is surely right to have concluded that he did not need to re-edit the papyri or to reproduce Roberts' Greek text. He might have gone on to point out that today everyone with the capacity to access Google can in any case read the Greek text of *P.Ryl.* gratis at any time thanks to the Duke Databank of Documentary Papyri, one of the electronic miracles that now acts to advance and reshape our scholarship. Matthews has vigilantly reread all the Greek, however, and offers specialists twenty pages of shrewd editorial comments in appendix 2. He also reconsiders positions adopted by Roberts. Notably, he differs from him in dissociating from Theophanes' journey altogether the inventories of clothing and other household articles in *P.Ryl.* 627; indeed, he adds (p. 44), the items listed may not even belong to Theophanes. Matthews' caution here is justified in my view. At the same time he offers good reason for being more convinced than Roberts was that the draft accounts in damaged *P.Ryl.* 639 do relate to Theophanes' stay at Antioch (98–99).

The hazardous journey of the papyri themselves from unknown findspot to publication and eventually translation is itself a long, obscure one that underscores the quirkiness of fate. The bulk of them, it seems, were bought from a dealer in Egypt around 1896 by Arthur Hunt, acting on behalf of Lord Crawford, from whom they were acquired by the John Rylands Library in Manchester in 1901 (the year after it opened), where they still are. At one stage or another, however, certain items somehow became separated from

[2] This was for long a standard style of presentation, it should be remembered, which was maintained even in the first British sourcebooks for students' use in ancient history courses: see Talbert (2006).

[3] For the careers of Roberts and of others, attention could usefully have been drawn to entries in Todd (2004).

the main group, including in particular a fine Latin letter written by one Vitalis commending Theophanes to the governor of an eastern province; this found its way to Strasbourg, where it was published by Harry Bresslau in 1903. A second such letter in the same hand addressed to a different governor—but in poorer condition, and missing Theophanes' name—remained in the main archive, and was the key item that enabled Roberts to link the archive to Theophanes when he began work on its publication in the mid-1930s. Meantime, no later than 1907–1908, further letters connected with Theophanes had been brought to Queen's College, Oxford, by Hunt and his associate Bernard Grenfell, both of them research fellows of the college. Here these papyri—placed between fine-quality pages of the weekly *Oxford University Gazette* in order to "relax" them—evidently lay neglected in a couple of tin boxes for at least half a century until rescued and published by Brinley Rees in 1964 (*P.Herm. Rees*).

Matthews was himself an undergraduate at Queen's, and later fellow, living and working—did he but know it, as he reflects on page 7—in close proximity to where these latter Theophanes papyri had so long been "relaxing." Still less can he have known when he first learned of the Theophanes Archive as a graduate student, from reading an article on imperial bureaucrats in the Roman provinces by Yale professor Ramsay MacMullen, that he would himself later make the journey there to fill the latter's place on retirement. Such a professional move from one country or continent to another, once a relatively rare event in the classical field, today borders on the commonplace and makes a valued contribution to the internationalization of approaches to antiquity.

Matthews' professional life has journeyed not only literally in space, but also in focus. He is known for his studies of western aristocracies and the imperial court in the late fourth and early fifth centuries, of the historian Ammianus Marcellinus, and of the formidable Theodosian Code. All this work addresses what his Ammianus would approve of as "negotiorum celsitudines" (26.1.1), elevated matters worthy of historical inquiry. The record of Theophanes' journey, by contrast, is one of cabbages, not kings. Until recently few historians would dare to turn their gaze thus, and certainly not in the early 1950s. It was in 1951, after all, that A. H. M. Jones famously identified a papyrus scrap as a more or less contemporary official copy of part of a letter of Constantine quoted in Eusebius' *Life* of the emperor, a work long doubted by many to be genuine. With such weighty controversies to pursue, who would choose to concentrate instead on Theophanes' quotidian trivia? It is a mark of how we, and our scholarship, have shifted that today the scholar who does so can be applauded.

We have grown to recognize how much we are what we eat, and how food, drink, and everything associated with their consumption can have extensive social and cultural significance. In the context of discussing how best to translate the meal termed *ariston* in Theophanes' accounts, Matthews recalls encountering for the first time as an Oxford freshman—his "provincial origins" behind him—"dinner" used for the evening meal and "lunch" for the midday one (92 n. 9). To purists, he might have added, the latter term itself was a deplorable abbreviation of "luncheon." More recently, ours has become a world where the diet-conscious all love their salad, the leafy sine qua non of today's American main meal. By contrast, who in Britain in the aftermath of World War II took salad seriously? In condemning the "dull, tasteless and old-fashioned" meals served to himself and other patients at the exclusive London Clinic in the late 1990s, Alan Bennett scathingly characterizes its notion of a salad as "what one would be given in Leeds in 1947, namely a piece of lettuce, a slice of tomato and another of cucumber."[4]

Now Matthews can offer an entire learned appendix on the identification of *kemia* and *kemoraphanos* in Theophanes' accounts (cf. 216), the former (he suggests) "greens from a variety of radish or similar root crop," the latter "the young shoots of some member of the cabbage family" (234). These identifications are advanced with reference to Greengrocers' English and Hungarian remedies, and a similar one earlier (of *tourtion*, a type of bread, 191) with citation of a recipe found in *Bon Appétit* magazine no less. What would Matthews' own father think, who "would sometimes bring home half a sheep's or pig's head . . . claim[ing] it was the best part of the animal"? (91 n. 8; trotters of animals unspecified occur several times in the accounts, a head once, 212). Yet, ironically, it is in 2006 and from Matthews' home town of Leicester that *The Cambridge Dictionary of Classical Civilization* primarily comes. Three of its four principal editors are professors at the University of Leicester, one of whom Lin Foxhall (American by birth and training) contributes an invaluable entry (with full-page table) on vegetables. This entry in this new dictionary, and others under such categories as "Environment", "Food" and "Foodstuffs" (cf. xxv, xxxii–xxxiv), plainly reflect

[4] Bennett (2005), 610.

novel perspectives on antiquity that Matthews and many others are now opening up in most rewarding ways.

Matthews' pathbreaking study offers so much of value that I hesitate to question its balance, and I mean no disrespect by so doing, especially when little more than subjective taste may be at issue. Chapter 2, for example, on Theophanes and his friends at Hermopolis is entirely appropriate in its placement and execution. Even so, it remains disappointing that the background material about the city and its religious life cannot be closely linked to Theophanes or his journey, and that the surviving assemblage of letters hardly affords more than slight acquaintance with the man or his family. What then follows after this slow start—treatments of the journeys out and back, and of Antioch itself—proves far more satisfying by contrast. Here Matthews covers in depth the routes taken by boat and by road (fig. 10.2), the imperial transport service (*cursus publicus*) which Theophanes evidently used, and the great city of Antioch itself, with a marvelous appreciation of its life as depicted in the topographical border of a mid-fifth-century mosaic found in a Roman villa at suburban Daphne. The concluding pair of chapters—on costs and prices, followed by food and diet—are no less masterly treatments of their topics, addressing such matters as the level of inflation in the first twenty years of the fourth century, and seeking to distinguish between items purchased for the more privileged members of the party and those for the lower servants' table. The complete menus for meals eaten at a specific time, day, and place are even reconstructed from soup to nuts—or rather "ab ovo usque ad mala" (176)—with vermouth *absinthion* as aperitif (90). No chef can be identified, however.

There are, to be sure, frustrations which Matthews rightly acknowledges cannot be overcome. How irksome it is, for example, that there is no knowing the size of Theophanes' entourage at any point. Ideally, the accounts might assist a reckoning, but they seldom attach a quantity to the price paid for an item (92, 154). For the outbound journey, Matthews (68, 94–95, 165) must be correct to imagine a group of ten or even more persons conveyed in no less than three heavy vehicles. Perhaps the group accompanying Theophanes on his return in late July was smaller (168); without question, it proceeded surprisingly fast and without rest days (130, 132). Altogether, further comparative data on these aspects would have been welcome. Chapter 16, "Time under Way," in Michael McCormick's important book *Origins of the European Economy: Communications and Commerce, A.D. 300–900* (2001) has much to contribute on speed of travel. Moreover, his observation (449) that travelers

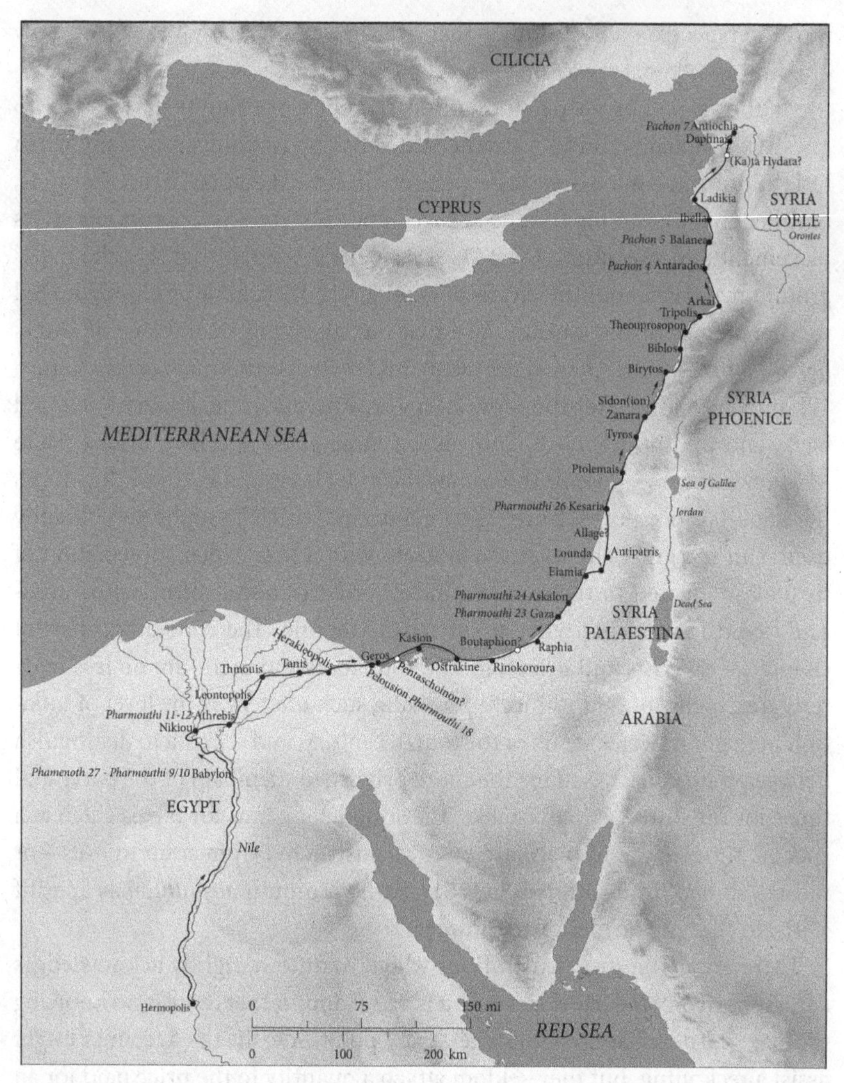

Fig. 10.2 Theophanes' outward journey traced on a modern map base, with dates (in italic) where recorded, according to the Egyptian calendar. Map: After Ancient World Mapping Center, *Theophanes, Journeys between Hermopolis and Antioch* (awmc.unc.edu/wordpress/free-maps/maps-for-texts/).

might prove fearful of diseases connected with warm weather could conceivably help to explain why Theophanes dashed home at such a breakneck pace. Ten may indeed be too low a total for the outbound group. The upper servants alone who accompanied the emperor Tiberius' *dispensator* Musicus

Scurranus from Lugdunum to Rome numbered as many as sixteen (*ILS* 1514), and the *cohors* accompanying the proconsul of Macedonia, P. Antius Orestes, to Samothrace in 165 CE totals around three times that figure (*AE* 1965.205). In both these instances, the principal traveler was of course an official who desired or felt obliged to display his position. But Theophanes' journey, too, called for much handling of the heavy baggage—at every *mansio* from one set of vehicles to another presumably—and for countless other mundane tasks to be performed (cf. 165).[5]

Did Theophanes himself do the shopping, for example? A gentleman certainly might do so. Among the witty one-liners of Polemo that Philostratus considered memorable was his quip on encountering a fellow sophist buying sausages, sprats, and other cheap cuts: "You can't possibly act out the arrogance of a Darius or Xerxes convincingly if this is what you eat" (*Vit. Soph.* 541). There need be little doubt that Theophanes personally purchased the extravagant hat accounted for at Pelusium on the outbound journey (51, 226), for instance, as well as the Silenus-figure wine-jar accounted for there also on the return (125, 136).[6] Even so, it is hard to envisage a man of his status and loftier preoccupations bothering to go to market daily to make the routine purchases of bread, greens, firewood, and the like for the party, as Matthews evidently assumes (9, 90–91, 123, 164).

This said, another most frustrating feature of the entire record is Theophanes' own sheer elusiveness, so that it is virtually impossible to imagine anything about him with confidence. Matthews recognizes this difficulty. We can readily conclude with him that Theophanes and the superior members of the party "enjoyed a typical Mediterranean diet of the time, varied and well furnished in an upper-class fashion" (178), with plenty of meat and good wine but surprisingly little fresh fish (170). We gather that Theophanes had the status of *scholasticus* or practicing lawyer (8, 36). But we simply do not know why he made this five-month trip beyond the fact that

[5] For the last day of the outbound journey, when an advance group from the party covered as many as sixty-four Roman miles, and six "Sarmatians" were specially hired, it seems more probable that these would be outriders to clear the way, or baggage-carriers, rather than an escort of guards as Matthews proposes (9, 50, 56, 61, 67), despite his assurance elsewhere (62) that banditry on this route was not to be expected.

[6] This Silenus is recorded as being "from Tyre," which Matthews takes to signify a delayed accounting for a purchase made there. Perhaps, alternatively, "in Tyrian style" may be meant, or an item imported from Tyre? Such a "typically tacky tourist's purchase" (125) merits a reference to Künzl and Koeppel (2002). Was there a store perhaps, or even an assistant, in Pelusium that took Theophanes' fancy?

its purpose was serious, possibly a financial issue requiring negotiation with the *vicarius* of the diocese of Oriens at his Antioch headquarters on behalf of a city or province (38–39). The contrast with the well-known journeys of Cicero through Cilicia and of Pliny the Younger through Bithynia, for example, could not be starker; in these instances it is the everyday details that remain a blank, but the purpose is amply documented.

There is no reason to imagine that on the road Theophanes displayed the gregarious vulgarity of the emperor Vitellius, who early in the day liked to ask *muliones* and *viatores* with a hearty belch whether they had had their breakfast yet (Suet. *Vitellius* 7). However, to judge by Theophanes' accounts one could be forgiven for suspecting that on the whole he is really rather a dull cove. Even in the throbbing metropolis of Antioch with all the attractions underlined by Matthews (77–79), he appears not to go to the baths especially often (93, 225–26) or to attend entertainments or festivals. He evidently never hangs out at the barber's or buys cosmetics. (How gleeful Pliny was to discover that Hostilius Firminus, natty scoundrel and proconsular legate, had credited a bribe to himself as *unguentarius*, *Ep*. 2.11.23.) Was Theophanes' mission too time-consuming to permit leisure pursuits? Surely not for all of two and a half months, especially when we recall the limited duration of the typical ancient workday. Was he too stretched financially to enjoy more cakes and ale? Again, I doubt that. Matthews' speculation (39) that Theophanes had meticulous accounts kept in hopes of reclaiming expenses from whoever commissioned his travels seems wide of the mark. More probably, as Matthews also considers (39), the trip was euergetism on Theophanes' part, a generous, voluntary public service undertaken by a man who could afford it and conceivably had some personal interest in the outcome.

Last but not least, my sense is that Matthews' inevitably rather unrewarding focus upon the figure of Theophanes diverts him from giving the real heroes of the surviving record their due. They go unsung, and are taken largely for granted along with the rest of the unidentified minions (94). I mean the actual compilers of the accounts.[7] As Matthews does recognize (98, 100), their drafts and fair copies display unfailing professional devotion, almost pedantry in places. Hence the entry "gourds for cooking" at Antioch on Pauni 21 is over-written with the correction "green vegetables" (194,

[7] Matthews sometimes implies that Theophanes would have compiled the accounts himself, mistakenly in my view: contrast 108 and 150.

215)—as if it seriously mattered, especially when the cost of this purchase is piffling. But of course a careful record of how the master's money has been spent had to matter, which is why normally (as already noted) the quantity of a purchase is disregarded, and only the price is given. How closely the master would ever peruse the accounts might be unpredictable—Pliny once admitted to inspecting such documents "with reluctance and cursorily" (*Ep.* 5.14.8 "invitus et cursim")—but still pride and caution required that the work be done scrupulously. To a modern auditor, these accounts (which were, of course, for the master alone) must appear distinctly unsatisfactory insofar as dozens of trifling expenditures minutely recorded are interspersed with substantial sums lacking further detail (cf. 227–31). A few of these latter might even revive hopes that Theophanes did see some fun in Antioch after all. In particular, the coy payment to an unnamed widow made just before he departs there for home invites speculation in this regard (117). The amount must be large, although damage to the papyrus at this interesting point (cf. 4) removes the figure. Matthews conjectures it to be a lump sum for rent. But in that case why not name the owner of the property outright, and would not a substantial down payment have already been made?

In any event, I suggest it to be only likely that further accounts of a comparable nature were kept on Theophanes' journey in addition to those which happen to survive. More broadly, Matthews' exemplary presentation of the latter acts as a vivid reminder to us of a fundamental part of the everyday routine of all well-to-do households throughout the Roman empire. Alongside MacMullen's "epigraphic habit," it is time for the less visible but even more pervasive "bookkeeping habit" to be given due recognition too. Pliny learned young how revealing accounts could be (*Ep.* 7.31.2), and later he used those of Hostilius Firminus (above) and those of Caecilius Classicus, "written in his own hand" (*Ep.* 3.9.13), to devastating effect. Equally, when Pliny undertook the defense of Varenus Rufus, it was the latter's accounts that his Bithynian accusers sought (*Ep.* 7.6.2). For many a bright young slave (like Trimalchio), the job of *dispensator* had unmistakable attraction. Whoever in Theophanes' party fulfilled this function—and, incidentally, continued likewise back home (cf. most, if not all, of the fragmentary *P.Ryl.* 640–51)—served him well. Somewhere in this book they merit at least a section devoted specifically to themselves and their procedures. In this key respect Matthews fails to draw inspiration from contemporary trends in historical research. Accustomed to occupying the scholar's traditional place at High Table (166), he is not gripped by the postmodern urge to penetrate below stairs and to reassess the world anew from there. That seems a pity. For it is, after all, largely a cookhouse-eye view of the unnamed master's journey to Antioch and back that has endured, and that Matthews has brought so splendidly and so unexpectedly back to life. Enjoy.[8]

[8] For travel on the Roman empire's highways, see further essays 7, 9 above, 16, 17 below.

PART II

MAPS FOR WHOM, AND WHY

11

The Unfinished State of the Artemidorus Map

What Is Missing, and Why?

One notable feature of the remarkable "Artemidorus papyrus," first published in 2008,[1] is that the varied contents include an unfinished map. Consideration of what is missing from this map can provide an instructive means to address many fundamental issues about its nature and the purpose it was intended to fill once finished.[2]

First, however, a word is called for about the possibility that the map is not ancient at all, but the work of a nineteenth- or twentieth-century forger, who very consciously drew only what we see, and quite deliberately proceeded no further. I have no means of decisively refuting this hypothesis, nor am I qualified to make an unequivocal claim that the map must be ancient work rather than modern, or vice versa. I merely state that, on the basis of my own competence such as it is, I share the view of the editors in 2008 that the map is genuinely ancient. On the papyrus, a substantial gap is evidently left on purpose for inserting the map between two text passages. The subject of the passage to the right is the Iberian peninsula, so it is natural to think that this region somehow features on the map. In that event a nineteenth- or twentieth-century forger, I suspect, would all too readily—perhaps even unconsciously—have introduced at least some resemblance to the modern standard view of the Mediterranean or part of it, with its north orientation. The influence of this view is plainly evident in the often-reprinted reconstructions of "the world according to Herodotus" (or Dionysius Periegetes, or Strabo, and other ancient authors) from the works of such late nineteenth-century scholars as Carl Müller and Edward Bunbury.[3] I find no such tendency toward it detectable

[1] Gallazzi, Kramer, and Settis (2008).
[2] In what follows some points are repeated from Talbert (2009).
[3] Note the references assembled by Brodersen (1995), 12, and the illustrations in Harley and Woodward (1987), 135, 172, 175.

here, however, as it *is* detectable by contrast in the allegedly fifth-century BCE "Soleto map," a tiny piece of potsherd said to have emerged during an excavation at Soleto in the far south of Italy in 2003.[4]

I would also have expected a forger to furnish even one more clue, no matter how tenuous, so as to reduce the sense of incomprehension which to my eye the map in its present unfinished state creates. To be sure, forgers delight in fabricating puzzles, but at the same time they are also notorious for deliberately planting hints of one kind or another that invite solutions to their teasing. If there be any such caprice lurking in the map, it is striking that it seems to have escaped notice so far. The nineteenth-century figure Constantine Simonides, who has been proposed as fabricator of the entire papyrus, quite plainly loved to tease and fool experts and collectors. To quote the recent characterization of him by Richard Janko: "S[imonide]s' brazen ambitions to outsmart the *Philologen* knew no bounds of time or space. . . . [H]e constantly planted clues within his fabrications, as if taunting scholars to find him out."[5] Some of his clues were indeed far from subtle—for example, use of the English name Charles Stewart, with its initials C.S. the same as his own. Were even a mere one or two place-names (or even parts of such names) marked on the Artemidorus papyrus map, without question the effect would be to intensify the speculation that its unfinished state provokes. For a forger like Simonides, the temptation to add a few letters for the purpose would have been hard to resist. The fact that there are none is certainly no guarantee in itself that the map must be ancient, but it does (I think) at least reduce the likelihood of forgery.

If we now proceed, reckoning the map to be ancient, still there would appear to be a daunting number of components missing for one reason or other. First, it is not even possible to establish the map's overall dimensions. According to the editors,[6] the length is to be reckoned as not less than about 99 cm, although it could have been as much as 113. Where the map would have ended, or had to end, at the left is patently uncertain. Only the situation at the right is evident, and here the map seems to stop a little short of how far it could have extended in that direction. By contrast, the map runs notably, even perilously, close to the upper and lower edges of the papyrus, exploiting more or less its full height of about 32 cm. A frame to enclose the map might

[4] For an image, see Gallazzi and Settis (2006), 95; Wikipedia, s.v. Mappa di Soleto.
[5] Janko (2009), 410.
[6] Gallazzi, Kramer, and Settis (2008), 275.

be expected, but there is no sign of one. Conceivably, this was a component to be added later in a colored ink, although it might seem more natural (and more prudent) to mark out a frame first.

The role that color was to play in the map's presentation is a further puzzle. Granted, color is not a requirement, but its use would be the greatest help to the mapmaker in engaging viewers and in rendering the work more versatile and more effective. Even sparing use of a single color could prove a valuable enhancement, as seen in the Romans' widespread use of red on monumental inscriptions as well as on the Severan Marble Plan of Rome (*Forma Urbis*).[7] It seems reasonable to imagine that some color was to be added to the Artemidorus papyrus map. At the least, one or more colors would complete the various pictorial "vignette" symbols. At present these are just light sketches (a normal preliminary stage in copying), to which further elements were no doubt to be added. Certainly, with the map in its present state, we lack an accurate impression of how the symbols were to look once finished.

Again on the matter of color, I concur with the editors in believing that the parallel lines which they identify as A, B, C, D, and E are meant to be associated as pairs.[8] I further agree that in all likelihood color or shading was to be added through each of these five pairs so as to form broad swathes across the map. However, because that step must remain an inference, the version of the editors' outline which I present here (fig. 11.1) deliberately stops short of infilling these swathes as theirs does.

It is impossible to say whether color was somehow to be added to any of the single lines—which are everywhere uniform in both type and weight—or, say, in the loop between the editors' *via* 8 and *via* 9. If nothing else, such enhancement could certainly dispel the current unsatisfying impression that the single lines all signify the *same* feature—all routes, or rivers, or boundaries, for instance. Color would provide the handiest means for the mapmaker to create distinctions. In this regard, we may note a striking contrast with the practice of the Peutinger map (*Tabula Peutingeriana*), a creation that admittedly survives only in a medieval copy; the original cannot be earlier than the second century CE. Some of the surviving copy's linework is conspicuously colored, but each line-type is also distinct; so even a color-blind user could not confuse coastlines, rivers, and routes on the Peutinger map.[9]

[7] See http://formaurbis.stanford.edu with Trimble (2007) and essay 18 below.
[8] Gallazzi, Kramer, and Settis (2008), 287–89.
[9] See ibid., 301, fig. 3.6a, and Talbert (2010), 97–98, with essays 13–15 below.

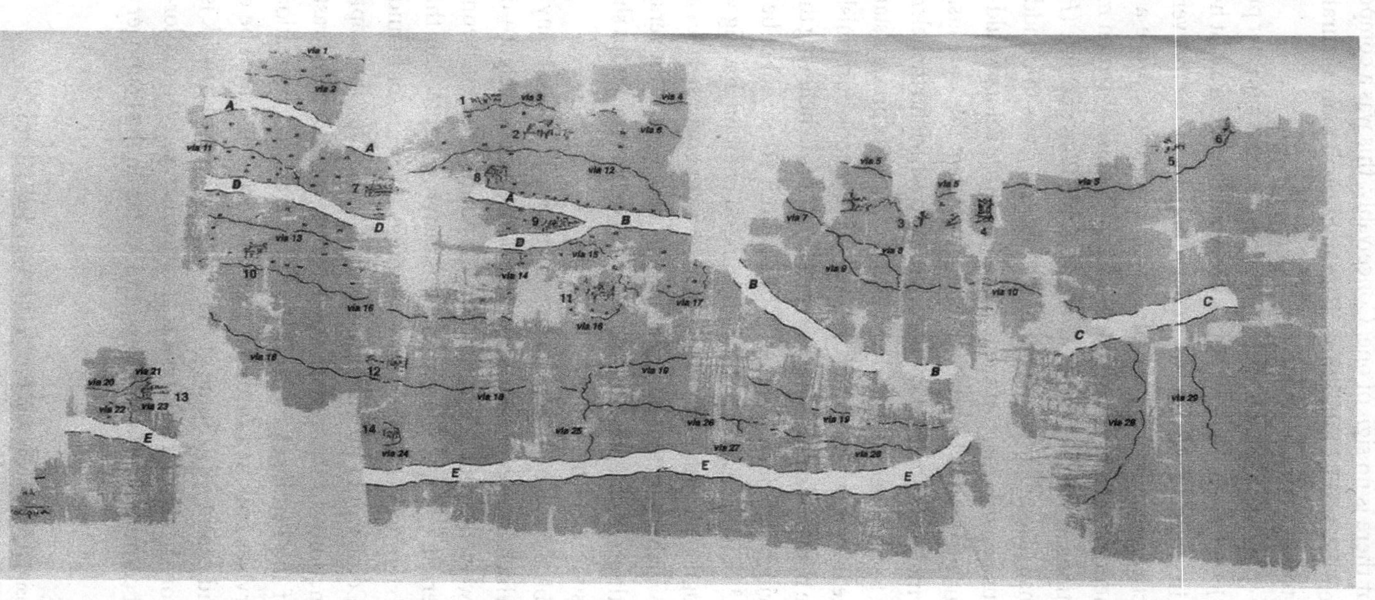

Fig. 11.1 The editors' map redrawn in grayscale, and without filler for the pairs of parallel lines lettered A–E. See Gallazzi, Kramer, and Settis (2008), 296–97. Map: Ancient World Mapping Center.

In the case of the Artemidorus papyrus map it is impossible to say whether there were also to be features—wholly missing at present—that were still due to be added only in a colored ink. Possible such features might include lakes, mountain ranges, routes; on the Peutinger map all the linework for routes, for example, is just in red.

As well as color, another missing component that surely remains to be added is lettering. Strictly speaking, it too is not essential, although the value of a map without it is liable to be diminished, and the significance of certain symbols might in consequence escape viewers. In the present case, if entire names were to be added rather than just a very few letters to represent numbers (whether in Greek or Latin scripts), then the issue arises of how these insertions would be made neatly in some instances, especially for symbols 9 and 13. The principles which the maker of the Peutinger map evidently sought to maintain whenever possible for the placement of lettering, even though they have no applicability to the Artemidorus papyrus map, are nonetheless suggestive. These principles are that the name for a place marked with a symbol should go immediately above the symbol; lettering should be kept clear of water or mountain range or any other feature; and lettering (other than display capitals sometimes) should be laid out as horizontally as possible.[10]

Quite clearly, to name the symbols on the Artemidorus papyrus map according to such principles would be impossible for the most part. Moreover, if its numerous tiny rectangular symbols—the *piccoli quadrati* in the editors' terminology[11]—are also to be named, then by any reckoning the mapmaker faces acute difficulty at the top left because so many such symbols are clustered there. Elsewhere, it would just about be possible to name each of those in the sequence down to the right from symbol 8, if the mapmaker were willing to slant their names almost vertically. For all we know, however, there was never the intention to name *any* of the *piccoli quadrati*, or only some at most. Or maybe some were to be numbered. Or maybe a cluster of them was to share a "group" name. If numbering were to be employed on the map, then either references to these numbers in some associated text are to be expected (although there is no sign of any in the surviving texts on the papyrus) or the provision of some kind of Key.

[10] Talbert (2010), 100–101.
[11] Gallazzi, Kramer, and Settis (2008), 286.

A further puzzle is how each type of vignette or "icon" on the map should be understood. There is no shortage of stylistic matches for such symbols in other Roman artwork, and I agree with the editors[12] that the larger ones are most naturally interpreted as urban settlements: 2, 9, 11, 13, perhaps also 7 and 12. The absence of uniformity in the rendering of the sketches for these symbols is notable. Symbols 8 and 10 could be settlements too, smaller evidently. On the other hand, 4 without a doubt has a rather different appearance; the editors are prepared to see it as a temple or monument, but I would hesitate to be so specific. I feel less certain still about 6, which may not even be preserved complete; the suggestions of stele, monument, or milestone that have been made all seem very speculative. Each visibly different again are 1 and 3, but whether they even represent physical features like mountain or forest (as the editors believe),[13] rather than manmade ones, remains impossible to judge in my view. I find 5 and 14 too scrappy to identify at all; neither appears to be a city.

As to the *piccoli quadrati*, the editors suggest farms or some sort of rural settlement;[14] again, I am unsure. Do all these *quadrati* signify the same type of feature in fact, or would distinctions be introduced somehow with one or more colors? The varied and uneven distribution of the *quadrati* is striking. At the map's top left the makings of a checkerboard pattern can be glimpsed, and there is possibly another top right from symbol 11. Above the broad swathe A–B down right from symbol 8, however, there is only a simple sequence just above the swathe's upper edge, and then no more than a few scattered *quadrati* above symbol 8 and to the left of 9, as well as a mere couple beyond 13 on its hard-to-place fragment. In every instance does each individual *quadrato* count, and is its specific placement very deliberate? Or are the *quadrati* always to be interpreted as just a generalized area-symbol, a scatter of them indicating no more than, say, "dense rural population hereabouts" (or mine shafts, or some other feature)? Clues are lacking. Equally, it is conceivable that some, or all, the *quadrati* are to be linked by a colored linework that would serve to equip them with a meaning which at present eludes the viewer; but how such linkage would be accomplished remains anybody's guess.

How is the linework to be understood? On an ancient map, we may reckon, there is nothing that it could realistically be *expected* to signify other

[12] Ibid., 282–87.
[13] Ibid., 287.
[14] Ibid., 286.

than a boundary of some kind, or route, or watercourse (be it river, canal, or even aqueduct). Accordingly, as stated above, I believe the editors may be correct to interpret five pairs of parallel lines as a large river (A–B–C) with two tributaries (D, E)—or branches, depending on the direction of flow, which remains indeterminable. This said, the uniformly exaggerated size of all these watercourses is a surprise, especially when neither tributary/branch is rendered a "lesser" stream to the main river. This feature touches upon the issue of scale, to be addressed shortly.

To turn first, however, to the single linework, I think it a misstep on the part of the editors to identify it all as roads or routes (*vie*).[15] Routes can be expected to demonstrate greater connectivity and crossover than is visible here, and to link principal settlements more. On the ground, routes also typically run through ancient settlements instead of bypassing them, as the editors seem to envisage in the cases of symbols 13 and 14; the "ring-roads" that they associate with each are hard to credit. It remains entirely possible that some of the single linework does signify routes. But what most, if not all, of it seems more likely to represent is watercourses, not roads. Even so, I am the first to acknowledge that in its current state the single linework has blatant shortcomings as hydrography. How is *via* 12 to be understood, for instance, which appears to flow both from and into the same broad stream (A–B or B–A)? *Via* 25–19, too, even with its right-hand end missing, appears to flow both from and into stream E, as do *vie* 25–26 and 27–26. Moreover, where are the sources of all these rivers great and small? Do we infer that the appropriate mountains, lakes, or springs are still to be added in colored ink? Why does stream C end so abruptly at the right? Where does the sea appear, if at all? "On the missing left" is no doubt the most plausible supposition; but it is equally conceivable that the map's coverage remains limited to an inland area, without ever extending anywhere to the sea.

Two pieces of information vital to our understanding cannot be recovered from the map as we have it: its orientation and its scale. Of course, no ancient observance of scale can be expected to match modern-style precision and consistency. Nonetheless, every map by its very nature has to reflect some sense of scale. At the same time, those features which are selected for highlighting will inevitably appear larger than they should in strict conformity with a map's overall scale. On the Artemidorus papyrus map the point applies most strikingly to the streams A–E as well as to the most

[15] Ibid., 289–91.

substantial symbols. The degree to which they together dwarf the rest of the map's features is surprising and baffling.

This selectivity leads into the larger issues of what the map is designed to achieve or to illustrate, and for what audience. Is it, or is it not, related to the adjacent text by Artemidorus about the Iberian peninsula? If there *is* a connection, does the map somehow take in the entire peninsula, or at least a large part of it? Or is the coverage within it much more localized? Is a major river the core of the intended scope, possibly the Ebro or the Baetis, as the editors conjecture?[16] Could the map's rendering of the landscape (both physical and manmade) be deliberately more idealized than realistic, making it comparable in this respect to the Palestrina Nile mosaic (c. 100 BCE?),[17] say, for a region, or to the Colle Oppio fresco (late first century CE) for a city?[18] Alternatively, if realism is intended, is this a "snapshot" map—representing an area at a specific date, or during a limited period only? Or is the mapmaker less historically conscious, and thus content to set features side by side that were never in fact contemporary? Such merging is plainly visible on the Peutinger map, for instance, and even on Rome's Marble Plan to a limited degree.[19] Is the mapmaker somehow selectively compressing the desired area for coverage, because he feels otherwise unable to accommodate it within the squat format of the standard papyrus roll for a text? Such selective compression—with the associated inevitable distortion—is obviously fundamental to the Peutinger map.

If the copyist of the map in its present state still had plenty to add by way of color and lettering (but for some reason never added either), then, with so many components missing, my regretful conclusion is that it remains impossible to identify what the map was intended to represent. This said, I do believe that the map represents only a limited area at best, perhaps even a very limited one, given how large the principal streams and symbols are drawn. Assuming this perception to be accurate, permit me first to draw attention to an eighteenth-century map which may by chance have some value for envisaging the Artemidorus papyrus map as it might have appeared once finished; and next to propose an alternative, extreme hypothesis about its final appearance.

The eighteenth-century map (fig. 11.2), measuring 161 cm wide by 46, is in the collection of the British Library, London.[20] A *plano* dated to a specific day

[16] Ibid., 294–95.
[17] See Andreae (2003), 78–109.
[18] See La Rocca (2000).
[19] For the former, see Talbert (2010), 134; for the latter, Trimble (2007).
[20] *Additional MS.* 17,661.b.

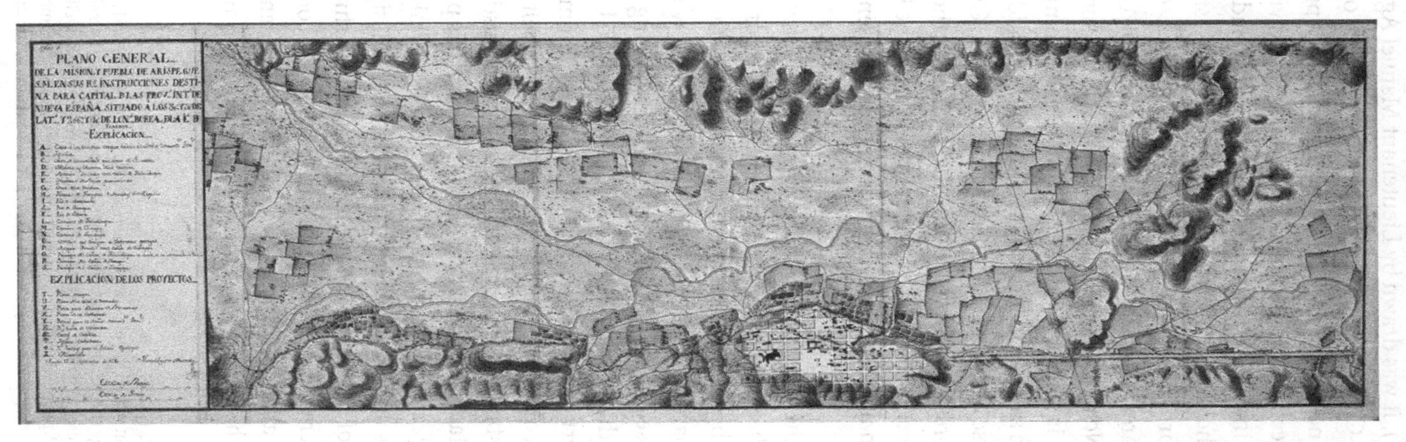

Fig. 11.2 "Plano" for urbanizing the mission and village of Arispe (Mexico), drawn by Lieut. Manuel Agustín Mascaró, September 12, 1780. Approximate scale: 1:3,000. Photo: © The British Library Board, Add. MS. 17,661.b.

(September 12, 1780), it was drawn by Lieutenant Manuel Agustín Mascaró of the Spanish Royal Corps of Engineers,[21] at the large scale of ninety *varas*, or forty *toises*, to an inch (approx. 1:3,000). Oriented East, it presents an idealistic vision (never realized) for urbanizing the mission and village of Arispe on the Sonora river in northern Mexico, a location that had been decreed in 1776 to become the capital of the "Interior Provinces"—stretching from Texas to Baja California—of New Spain.[22] River-courses, canals, ridges, fields, buildings, squares, a park, and roads are all shown, and color is integral to the presentation. No names appear on the map itself, but some features are lettered, and a key is included at the left. This map came to my attention by random chance when a departmental colleague, Cynthia Radding, happened to show it in the course of a presentation about her research. Clearly, it would be inappropriate to press any comparison with the map in the Artemidorus papyrus. Even so, it seems far from absurd to imagine that the latter's appearance, had it ever been finished, could bear some general resemblance to this detailed *plano*.

My extreme hypothesis by contrast calls for reckoning that the map *is* almost finished, and that in consequence only minimal color or lettering or numbers remain to be added.[23] In these circumstances the map is to be regarded as one prepared for just a single individual, who would recognize what he (or she) was seeing with little, if any, need for naming or other guidance. It seems to me that a map of this type is most likely to be representing a landed estate or part of one, either real or imaginary. From written testimony, art, and archaeology we gain a vivid impression (albeit not a detailed one) of the most outstanding Roman *horti* (Lucullani, Sallustiani), of Nero's Golden House and Hadrian's villa at Tivoli, and more.[24] It is possible to perceive corresponding features of their landscapes in the Artemidorus papyrus map. Manmade lakes and watercourses stood out in the design of such estates, as did a variety of residences great and small, temples, shrines, grottoes, monuments, belvederes, fountains, and the like. It is conceivable that some, though hardly all, of the map's *piccoli quadrati* represent statue bases. According to Pliny the Younger's description, in the *horti* of M. Aquilius

[21] See in general Reinhartz (2005).
[22] For the circumstances in which the plan was drawn, see Fireman (1977), 161.
[23] I gratefully acknowledge the development of this hypothesis from remarks by Michael Crawford.
[24] Note, for example, the essays by Littlewood (1987) and Beard (1998), as well as others in these two volumes. The variety of properties that could be termed *hortus*, and the importance of distinguishing between them, are emphasized by Purcell (2001) in his invaluable discussion of the latter volume.

Regulus (*PIR*² A 1005) across the Tiber from the city there were statues along the river bank and a very extensive area covered by immense colonnades.[25] Equally, a network of *piccoli quadrati* on the map could represent an elaborate pergola or hanging garden. The map's single linework in its turn does not necessarily all represent the same one feature (as maintained above). In the case of an estate map, some differentiation was to be made perhaps—by means no longer evident—between minor watercourses, say, and paths or tracks.

In any case, a patron who viewed this landscape could comprehend it with minimal need for coloring or lettering, because the map would be representing what was already under discussion with contractors, architects and artists; this could be either actual work in progress, or a vision, realistic or otherwise. As such, the map is perhaps to be somehow related to the other artistic components of the papyrus rather than to either or both of the text passages, as has been generally assumed. Even so, to be sure, the larger puzzles still remain of how such a curious mix of items comes to be found all together on both sides of a single papyrus, and why the map should be placed where it is between the two texts.

To conclude briefly: only for the sake of argument do I introduce the extreme hypothesis that the map is almost finished and that it sketches a single estate. I hesitate to endorse this hypothesis as yet without further consideration. I do firmly believe that the map forms a large, important component of the papyrus. Moreover, what the map represents, and how it relates to the other components, are issues meriting close attention. Analysis of everything to be seen on the map must be accompanied by alertness to what is evidently missing, and why. At the same time the exceptional prominence of certain features, and possible reasons for it, should not be left out of account. Given the state of the material, satisfying answers to the many questions which arise may inevitably prove elusive. Even so, I am convinced that this cautious methodology—where every effort is made to minimize reliance upon speculation—offers the best prospect for achieving fuller understanding not only of the map itself, but also by extension of the entire Artemidorus papyrus and of the range of ancient cartography.[26]

[25] *Ep.* 4.2.5: "He remains established in his gardens across the Tiber, where he has covered a very extensive area with immense colonnades and the bank with his statues" ("Tenet se trans Tiberim in hortis, in quibus latissimum solum porticibus immensis, ripam statuis suis occupavit").

[26] For the latter aspect, see further essays 12 and 18 below.

12

Claudius' Use of a Map in the Roman Senate

In response to a variety of influences, recent years have witnessed one especially welcome development in the ancient field: an unprecedented and wide-ranging preoccupation with geographical awareness. Pietro Janni, who introduced the concept of "hodological space" to classical scholars, is rightly recognized among the leaders of this productive and still ongoing advance, so that the opportunity to contribute to a volume in honor of his eightieth birthday is deeply appreciated. Janni's 1984 monograph *La Mappa e il Periplo. Cartografia e Spazio Odologico* offered a formative—not to say provocative—stimulus, the force of which continues to be felt.[1] Among much else, this work encouraged scholars to seek out previously unnoticed testimony to engagement with maps, such as the instance from the *Anthologia Palatina* identified by James Scott: a tapestry world-map of indeterminate size presented in circumstances unknown to a Roman emperor no later than Nero.[2] I put forward here a further, intriguing instance for consideration, one that may appear more surprising than Scott's because the reference to this map has somehow eluded attention, despite its occurrence within a very well-known and much studied text. If my identification be accepted, this instance may prove notably instructive, insofar as we can classify the map as likely to have been produced for a specific need. Moreover, the precise circumstances of its commissioning and use can be recovered.

The relevant text is the emperor Claudius' so-called Lugdunum Table speech, delivered to the Roman senate in 48 CE and preserved verbatim, albeit incomplete.[3] The clue to the overlooked reference comprises the two

[1] See especially essays 1 and 7 above.

[2] *AP* IX.778 (Philip of Thessalonica), discussed by Scott (2002), 5–22. An actual map, although unfinished, has of course emerged in the Artemidorus papyrus, attracting immense interest: see essay 11 above.

[3] *ILS* 212 = *FIRA*² I, 43. I quote from the text in Smallwood (1967), no. 369. For bibliography, see the overview of the speech and its reworking by Tacitus (*Ann.* 11.24) in Riess (2003). Although the actual speech lies outside its scope, note also *FRHist*, no. 75 introduction and bibliography.

words "digito demonstrem" in column 2, line 27. By this stage in the speech, if not earlier, the young men from Gallia Comata whose request to be considered for admission to the senatorial order Claudius is endorsing have been brought forward, and he clearly registers their presence by resuming after a self-referential aside: "Tot ecce insignes iuvenes, quot intueor . . ."[4] *Ecce* and *intueor* are the key words for the purpose. He continues:

> "Quod si haec ita esse consentitis, quid ultra desideratis, quam ut vobis digito demonstrem solum ipsum ultra fines provinciae Narbonensis iam vobis senatores mittere, quando ex Luguduno habere nos nostri ordinis viros non paenitet?"
>
> "But if you agree that this is so, what more do you want than that I should point out to you *with a finger* that the land beyond the borders of the province of Narbonensis already sends you senators, since we do not regret the fact that we take members of our order from Lugdunum?"[5]

Here the words *digito demonstrem* in turn deserve the due force and meaning that have been denied them. Philippe Fabia and Kenneth Wellesley[6] both maintained that the two words signify a gesture by Claudius toward those senators from Lugdunum who are present. While this interpretation cannot be ruled out, it hardly gives special purpose to the gesture. Rather, we would do better to believe that the references to *solum* and *fines* (as well as to *terminos* in the following sentence)[7] are to be taken not just figuratively,[8] but also more literally. I suggest that Claudius is pointing with his finger at a map which actually shows these features.

Such use of a map gives added significance to the conclusion of the aside earlier:

> "Tempus est iam, Ti. Caesar Germanice, detegere te patribus conscriptis quo tendat oratio tua; iam enim ad extremos fines Galliae Narbonensis venisti."

[4] Col. 2, line 23, "The full complement of outstanding young men that—behold (*ecce*)—I am looking at (*intueor*) . . ."

[5] Translation by Braund (1985), 201, which makes nothing of *digito* (hence my addition in italics).

[6] Fabia (1929), 125; Wellesley (1954), 23.

[7] Col. 2, lines 30–31, "Timide quidem, p.c., egressus adsuetos familiaresque vobis provinciarum terminos sum . . ." In my view, Claudius means the terms *fines* and *terminos* to be understood synonymously.

[8] As mentioned in essay 7 n. 15 above.

> "Tiberius Caesar Germanicus, now is the time for you to disclose to the conscript fathers the direction that your speech is taking, because by now you have reached the furthermost borders of Narbonese Gaul."[9]

By the same token, in the sentence opening the previous paragraph, specific purpose is given to the otherwise seemingly redundant *ecce* if it, too, signifies a gesture toward a map where Vienna (modern Vienne) is marked:[10]

> "Ornatissima ecce colonia valentissimaque Viennensium quam longo iam tempore senatores huic curiae confert?"
>
> "Behold the most splendid and most robust colony of Vienna: for what length of time now has it been supplying this House with senators?"

Still earlier, Claudius' reference to his irrepressible pride in having begun the conquest of Britain ("iactationem gloriae prolati imperi ultra Oceanum")[11] would offer an outstanding opportunity to gesture toward a map, and the next sentence ("Sed illoc potius revertar") may indicate an intention to return to this theme.[12] Were Claudius to return to it in this speech (rather than on another occasion altogether), we could certainly expect him to make the observation that, with Roman rule now established in Britain, Gallia Comata can no longer be dismissed as a mere frontier region on the empire's edge. A map would be ideal reinforcement of the point.

To indicate a person or object with a finger was normal Roman practice.[13] For public speakers and advocates to exploit visual aids—either objects already in place (such as statues in the vicinity) or items prepared specially for the occasion—was likewise normal.[14] Quintilian mentions with disapproval advocates' practice of displaying in court pictures that re-create a crime being committed:

[9] Col. 2, lines 20–22. Although proof is impossible, I regard this aside as self-referential: see Talbert (1984), 265–66.

[10] Col. 2, lines 9–10.

[11] Col. 1, lines 37–40.

[12] *OLD* s.v. *illoc* seems to understand the sentence in this way. Fabia (1929), 82 and Levick (2000), 179 both differ. The latter translates: "No, I shall return to the point instead" (i.e. to the theme of changes in magistracies and their authority treated earlier in the paragraph). The incomplete state of the speech as it survives, and of this paragraph in particular, makes it impossible to establish Claudius' intended meaning beyond all doubt.

[13] This instance, however, is ignored by Sittl (1890) and more recently by Aldrete (1999) and Corbeill (2004). See further Bablitz (2007), 188–90.

[14] See in general Aldrete (1999), 3–43; Bablitz (2007), 193.

> "But I would not therefore endorse the practice—which I read about and have also observed occasionally—where a picture of the incident has been painted on a panel or banner to impress its frightfulness upon the judge. How inarticulate an advocate must be who thinks that this mute picture will deliver a more forceful plea than his own speech!"[15]

For a member of the senate to illustrate his speech with a map is not otherwise attested. Even so, such an unusual step on the part of the scholar-emperor Claudius—who had himself portrayed at Patara as "the emperor of the known world" (ὁ τῆς οἰκουμένης αὐτοκράτωρ)[16]—does not seem odd at all. It is conceivable that he merely referred to a map displayed permanently in whatever chamber the senate held this session. But no such map is attested in any of its known meeting-places during the Principate,[17] and it seems more likely that Claudius commissioned one for his specific needs on this occasion.

To be visible to a body of several hundred members, Claudius' map must have been large. In the Curia Julia at least, it could have been accommodated satisfactorily if placed, say, up on the wall behind the president's platform.[18] In scope the map perhaps spanned as far as southern England (*ultra Oceanum*) in one direction, to northern Italy or even Rome itself in another, marking at least some settlements (most obviously Vienna and Lugdunum?) and some provincial boundaries, with Gallia Narbonensis clearly differentiated from the rest of Gaul. There is no clue to the map's orientation. In general, however, Claudius must have reckoned that senators were familiar enough with maps—displayed in public or otherwise[19]—for this one to serve as an effective, meaningful reinforcement of his support for the Gauls' request. This map, and the use to which Claudius put it, thus add instructively to our appreciation of the place of maps in Roman culture.[20]

15 *Inst.* 6.1.32 Winterbottom: "Sed non ideo probaverim, quod factum et lego et ipse aliquando vidi, depictam <in> tabula sipariove imaginem rei cuius atrocitate iudex erat commovendus: quae enim est actoris infantia qui mutam illam effigiem magis quam orationem pro se putet locuturam?"

16 Şahin and Adak (2007), 36–37 (left face, lines 3–4).

17 Talbert (1984), 113–30. In addition, there is no testimony to the senate ever meeting in the Porticus Vipsania, where Agrippa's famous world-map was displayed. For this structure, see *LTUR* s.v.; Boatwright (2015). For discussion of the map's character, see Arnaud (2007–2008).

18 A member might then have been seated as far as about twenty meters from the map: see Talbert (1984), 121, with figs. 5 and 7.

19 Publicly displayed images of this type are usefully surveyed by Holliday (2002), especially 104–107.

20 See further Salway (2012) and essay 18 below. Remarks by Mary Boatwright served to prompt the framing of this contribution; my thanks to her, as well as to Jerzy Linderski for generously commenting on a draft.

13

Cartography and Taste in Peutinger's Roman Map

When colleagues ask what is now engaging my attention after the completion of the *Barrington Atlas of the Greek and Roman World* (2000), and I mention Peutinger's Roman map,[1] it is clear that the response puzzles them. They regard the map as a thoroughly studied document from which little more is likely to be learned. Their impression is understandable, but in fact misplaced. Rather, the Peutinger map belongs in that deceptive category of ancient texts or monuments so familiar to everyone in our field that a satisfactory edition or presentation is generally assumed to exist, when really there is none. To be sure, in this instance there has been one outstanding attempt, a monumental volume containing much of lasting value, but old now and inevitably long overdue for replacement. This massive *Itineraria Romana* by Konrad Miller (1844–1933) was published in 1916 of all tragic years, in Stuttgart, and reflects work done intermittently as far back as the 1880s, including a revision of the standard engraving of the map issued by Franz Christoph von Scheyb in 1753. Even after the publication of full-size photographs of the entire map in 1888 (in grayscale), Miller continued to rely upon, and to reproduce, his own revised engraving. In the longer term, reliance upon color photographs would naturally have been still more desirable for study, as soon as their publication became practical. It was not until 1976, however, that color photographs of the Peutinger map appeared.[2]

Miller was neither a trained classical scholar, nor even a professional academic. Instead, it was newfound enthusiasm for Roman archaeology that attracted him to the map.[3] Thus, in line with the choice of title, his *Itineraria*

[1] I fully support Oswald Dilke's call to abandon the traditional name "Peutinger Table" or *Tabula Peutingeriana*, "to avoid any misconception that the original image was somehow carved on a table or was like a statistical table": see Dilke (1987a), 238 n. 25.

[2] See further now Talbert (2010), 30–36, 62–72, and essay 15 below.

[3] Note the two memoirs by his grand-niece, Gertrud Husslein, both valuable not least for reflecting his astonishing energy, and the breadth and depth of his talents and interests: see Gaube (1986), IX–XIII, and Husslein (1995). In the first memoir, she slips in stating (XI) that Miller's imaginative

Romana aims primarily to test the accuracy of the routes and distances shown on the map against surviving material or documentary testimony for conditions on the ground. For ease of presentation, he divides up the map in accordance with the boundaries of the Roman empire's fourth-century administrative dioceses, and then follows routes ("Strecken") within each diocese, place by place. It must be appreciated, however, that often the courses determined for individual routes are purely of his own devising, and that neither the names nor boundaries of dioceses appear on the map. In the same prescriptive vein, Miller readily rewrites lettering and redraws linework to eliminate what he perceives to be error or confusion.[4]

Since 1916, scholars with an interest in a specific route or region have consistently paid due attention to the Peutinger map,[5] but there has been no comprehensive reappraisal. The only attempt I am aware of is one by Otto Cuntz (1858–1932), but it was evidently abandoned after his death.[6] During the past half century in particular, the difficulties of replicating Miller's achievement have become quite overwhelming, and they only increase annually with the continuing high level of activity in archaeology and epigraphy. Today, without a reliable and reasonably up-to-date synthesis of routes to

reconstruction of the Peutinger map's missing left-hand parchment originally appeared as early as 1888 rather than a decade later.

[4] This impatient, opinionated approach was perhaps characteristic. Compare the observation by Westrem (2001), xxvi n. 35: "Miller, by his own admission, never saw the Hereford map. His published facsimile [1896], which is 3/7 the size of the map itself, includes altered text that corresponds to his readings (incorrect, and apparently derived from what he thought a reading ought to be); in some cases his readings and the facsimile clearly disagree." Miller did at least examine the Peutinger map himself.

[5] In *L'Année Philologique* under "Auteurs et textes," search such entries as "Tabula Peutingeriana" (or vice versa) and "Itineraria et geographica." Note also, for example, the extensive bibliography appended to Calzolari (2000), and now Diederich (2018).

[6] In the brief Praefatio to his *Itineraria Romana* vol. 1 (1929), Cuntz records that the original intention was for the Roman itineraries to be edited jointly by Wilhelm Kubitschek (1858–1936) and himself. But once Kubitschek had become otherwise preoccupied, Cuntz determined to press ahead alone, and following this first volume of the pair he next meant to issue the second, containing "tabulae Peutingerianae itinera, Geographum Ravennatem, itineraria minora." However, when a second volume did appear, in 1940, the sole editor, Joseph Schnetz (1873–1952), explained at the beginning of the Praefatio that Cuntz had died in December 1932, leaving "nihil nisi quae ad editionem Tabulae Peutingerianae praeparaverat." Schnetz, because of his prior work on the Ravenna Geographer, preferred to limit the second volume to this text and that of Guido, *Geographica*, with a third volume to follow; it never has. The second volume even lacked an index until this was added by Marianne Zumschlinge, when both volumes were reissued with some more recent bibliography (in vol. 1, by Gerhard Wirth) in 1990. From his own allusion in vol. 1 (p. VII), the impression must be that Cuntz's approach to the Peutinger map would have matched Miller's, with close attention to names and distance figures, and variants in the recording of both. As his severe reviews in (1917) and (1917–1918) show, Kubitschek was not slow to find much fault with Miller's work; even so, it has proved indispensable and has yet to be replaced.

serve as foundation, even a team of scholars would be hard-pressed to cover the entire map in the way that Miller did single-handed. Fortunately, since 2000 the *Barrington Atlas* and its accompanying *Map-by-Map Directory* have provided such a foundation. Among much else, between them they do represent, and document, current knowledge of principal routes, so that to make a comparison with the Peutinger map in this regard at long last becomes practicable.

Meantime, other developments of importance have occurred. The traditional, narrow preoccupation with whether an ancient historical or geographical text, or map, is "accurate" has been overtaken by a broader concern to approach it on its own terms, and to appreciate what it reflects of the author's purpose and taste, regardless of its "accuracy" from a modern perspective.[7] More specifically, during the past fifteen years a debate (still ongoing) has arisen about ancient map-consciousness, and in particular about the extent to which Romans made maps and used them. As a remarkable example of Roman mapmaking, the Peutinger map has no match, and could thus fairly be expected to play a key role in the discussion. In the event, however, that has not been the case. Not only are the map's sheer size and rich detail forbidding but also there is simply no study which has sought to evaluate it as a piece of creative cartography. Moreover, the mere fact that it is a map has inclined revisionist participants in the debate (myself among them) to focus attention elsewhere, as part of their effort to rethink the established, modernizing assumption that among Romans maps were made and used much as they are today.

It is natural enough, therefore, for the fullest contribution to the debate—Kai Brodersen's invaluable 1995 monograph *Terra Cognita*—to devote barely two pages (186–87) to the Peutinger map, on the grounds that it offers no more than a diagrammatic representation of information already available in purely written lists, above all in the so-called *Antonine Itinerary*.[8] Both in his monograph, and elsewhere more recently, Brodersen characterizes the Peutinger map as a "route diagram" in a style matching that of the modern London Tube map, or a European communications outline map.[9] My own current thinking, by contrast, is that this view seriously undervalues the Peutinger map from a cartographic perspective. I would suggest that whoever

[7] Note, for example, Wood (1992), Yee (2001), and the pointers offered by Burnett (2000), 5–7; further, essay 18 below.

[8] See essay 9 above.

[9] Brodersen (1995), 64; (2001), fig. 2.4 and p. 19.

created it was no newcomer to mapmaking, but must have had extensive prior knowledge and experience of such work; no less can the intended users or recipients have been unfamiliar with maps.

There should be no further delay, however, in turning to the Peutinger map itself—held by the National Library in Vienna, Austria—a copy made on parchment around 1200 of a Roman original of indeterminate date, although the fourth century seems most likely.[10] From here on, simply for convenience, I refer to the mapmaker as a male in the singular, while readily acknowledging that "he" could just as well have been a team,[11] and not necessarily an all-male one. I also leave open the unanswerable question of whether it was the mapmaker himself who determined the map's basic design features, or whether these were prescribed by a prospective user who commissioned the work. Either way, we need to appreciate that three bold, interlocking decisions were taken at the outset, which together pose formidable challenges for the mapmaker. They are, first, the map's height; second, its coverage; and third, the placement of Rome. I comment on each in turn.

First, the height is not to exceed about 30 or 31 cm, with an extra two or three to allow for margins as a frame top and bottom (fig. 13.1). Such a modest choice creates a notably compact and portable object, as well as permitting the map to be laid out on a series of mass-produced, standard-size papyrus sheets, which just need to be pasted together left and right to form a roll of whatever length is wanted (assuming that the map either was a papyrus roll originally, or was designed to resemble one). This was the regular format for a text, and it would appear that papyrus sheets were typically never made much taller.[12]

For a map with the coverage expected of this one, however, the standard text format seems far from ideal, and why it was made the choice is matter for speculation. Taller maps as such were by no means unknown,[13] and some must have been drawn on papyrus. In all likelihood, the world map for which

[10] For summary discussion of the manuscript and its history since its rediscovery early in the sixteenth century, see the excellent *Kommentar* volume issued by Ekkehard Weber to accompany his color reproduction (1976).

[11] Note that the verses attached to a (lost) world map revised and copied for Theodosius II in 435 specifically refer to lettering as the contribution of one worker, landscape that of another (". . . dum scribit pingit et alter"): *GLM*, pp. 19–20, and essay 15 n. 12 below.

[12] The height of a roll 27.3 cm tall is described as being "at the upper end of the range" by Parsons (2001), 153. Traianos Gagos informs me that he is not aware of any papyrus sheet taller than approximately 40 cm.

[13] For example, the largest (B) of the three stone cadasters at Orange must have measured at least 7 x 5.5m tall, and Rome's Marble Plan (*Forma Urbis*) approximately 18 x 13m: see Dilke (1987b), 223, 226, and essay 18 below.

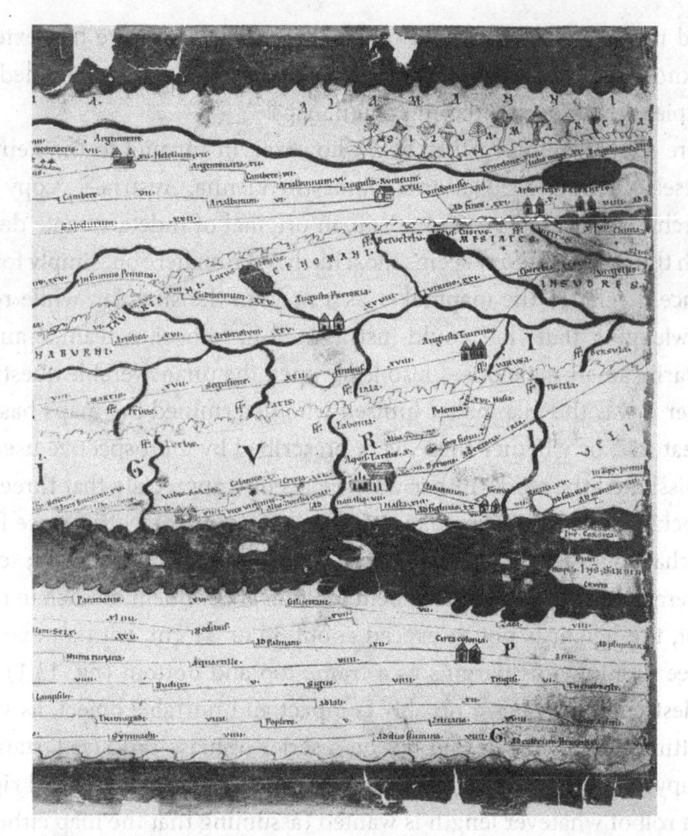

Fig. 13.1 Right-hand end of Peutinger map parchment 2. Photo: *Peut.* Map B.

Ptolemy offers specifications in his *Geography* would need to be at least one meter high by two broad.[14] By definition, the production of such a tall map on papyrus would call for standard-size sheets to be pasted together horizontally as well as vertically, and the stylus would then have to negotiate more sheet-joins. Even so, we might expect any anxiety over such minor obstacles to be outweighed by the greater scope gained for layout; but in the case of the Peutinger map that was evidently not a decisive consideration.

If the map's maker somehow or other had its height imposed upon him, he may well have felt yet further challenged when he learned (second) the required coverage. Eastward, we know that this extends to the Indian subcontinent and Sri Lanka. Westward, the limit of coverage remains indeterminable because that end is lost (fig. 13.2). At a minimum, however, there must be a strong

[14] Calculation by Berggren and Jones (2000), 47. Strabo (2.5.10) gives instructions for drawing a map that is to be about two meters or more in length.

Fig. 13.2 Left-hand end of Peutinger map parchment 1. Photo: *Peut.* Map B.

likelihood that England and Wales were included (only southeast England survives), and that coverage of North Africa and mainland Europe continued to the Atlantic Ocean (including the Iberian peninsula, which is lost almost in its entirety). The map as it survives runs to eleven pieces of parchment, each about sixty cm wide, or actually 674 in total; they made up a single seamless roll until 1863, when the pieces were separated for better preservation.[15] Miller was bold enough to attempt a reconstruction of the lost western end, but he proved strangely cautious in reckoning that only a single further piece of parchment would suffice; this would give an overall length of about 735 cm.[16]

[15] The fact that the map was copied thus in this instance still leaves open the question of how it was produced originally. At least there can be no question that it was designed to be read as a single, seamless whole.

[16] Miller (1898), Taf. 5. A clear illustration of his undue caution is to be found in his bunching of the capitals AQV here, to link up with ITANIA spread across 1B1–1B5. It is uncharacteristic of the mapmaker to lay out a region name of this type with notable variations in the spacing between each of its

I certainly echo the doubts voiced about whether the mapmaker would have confined himself to using only one parchment to represent all the land area likely to be missing,[17] especially if the coverage extended beyond the Roman empire here in the West, as we see that it did in the East. At the western end, Scotland, Ireland, Madeira, and the Canaries could all have been included. Moreover, conceivably there was also some combination of captions, lists and a dedication placed adjacent to the map, but still forming part of the design. We may recall that Ptolemy, for his main map, supplies a caption that amounts to three full pages of a recent English translation.[18] It is natural to think that the Peutinger map was produced for one or more patrons, or at their request, so that a dedication would by no means be a surprise. Equally, lists or a table of total distances between main settlements could be placed adjacent to the map. Such total figures are a typical, and obviously useful, component of written itineraries (including the *Stadiasmus Maris Magni*),[19] but are absent from this map. Altogether I think there is ample cause to reckon that two, or even three, parchments have been lost at the western end rather than just one; their restoration would make the overall length 795 cm or even 855 (at approximately sixty cm per parchment; fig. 13.3).

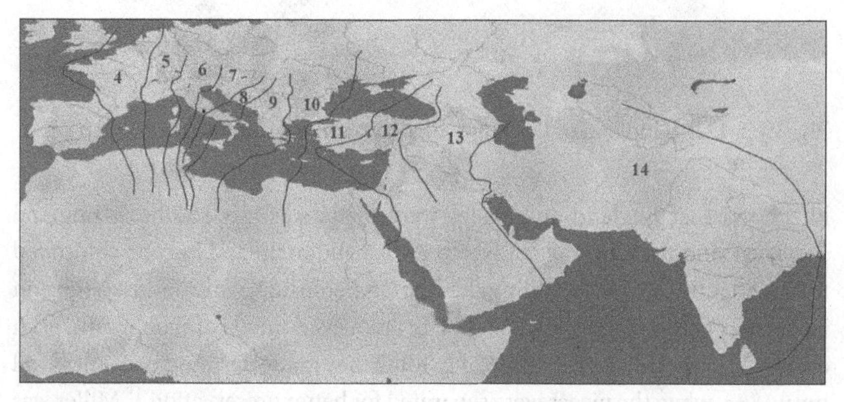

Fig. 13.3 Spread of the Peutinger map's surviving parchments (numbered 4–14, with three assumed missing at left-hand end) sketched on a modern map base. Map: Christos Nüssli, Euratlas.

letters. We can fairly assume that the spacing of the lost letters AQV would correspond more closely to the layout of the remainder of the name on parchment 1 than to Miller's reconstruction.

[17] See, for example, Weber (1976), 13.
[18] *Geog.* 7.5.1–16 = Berggren and Jones (2000), 108–11.
[19] See, for example, the points made by Salway (2001), 32–34.

To both south and north, the map's coverage for the most part matches the furthest recognized land routes. Consequently, in Africa it stops short of the desert, except in Egypt; within mainland Europe, the Rhine and Danube rivers serve as northern limits west of Dacia. In short, the map spans the Roman *orbis terrarum*. The mapmaker, however, may have been alternately horrified or intrigued at the prospect of having to compress it all—from Britain to as far south as India—within a frame no taller than about thirty cm. In whatever way he normally visualized or represented this span, it was surely not in such a flattened form; the same doubt might be expressed about the vision of whoever was to acquire the map. The "normal" vision of both (whatever form it took exactly) is likely to have been a less flattened one, which was now to be adapted to fit a frame of distinctly restrictive height. As to the length that the mapmaker might determine for the map, this would by definition also have to take into account the nature and density of physical and cultural data selected for inclusion; but otherwise he may have been relieved at the prospect of gaining some scope to fix the map's length himself.

Or, maybe, he was offered less scope in this regard than he might have hoped for, because (third) he may well have been told where Rome had to be placed, and (if so) his task would only become still more complicated. Miller's reconstruction (supplementing what survives by no more than a single parchment at the western end) sets Rome at the center of the Mediterranean. However, the most ambitious alternative hypothesis (supplementing what survives by as many as three parchments) is even more attractive in that it sets Rome precisely at the center of the entire map. Either way, it seems appropriate to conclude that the placement of Rome is a very deliberate one,[20] and a key consideration in settling the layout of the map as a whole (fig. 13.4).[21]

Next, the mapmaker's choices of physical and cultural data for inclusion should be briefly noted. The red linework for routes stands out clearly, with the names of over 2,700 settlements or stations along the way, and a figure for the length of each stage. In principle, written sources alone, like the *Antonine*

[20] It no doubt contributes in turn to the prominence accorded to Italy, which benefits from exceptionally generous coverage in relation to its land area (about three and a half parchment-widths, from the far right of parchment 2 well into parchment 6).

[21] We may observe—with Weber (1976), 13—that only a slight further increase in length from 855 to 888 cm would bring the map to thirty Roman feet precisely. There is no knowing what significance, if any, the mapmaker might have attached to the achievement of this round figure. At the least, it is not hard to imagine him welcoming a little extra space at each end, especially to the right, where his design surely incorporated vertical margins (now missing) to complete the frame running along the top and bottom of the map.

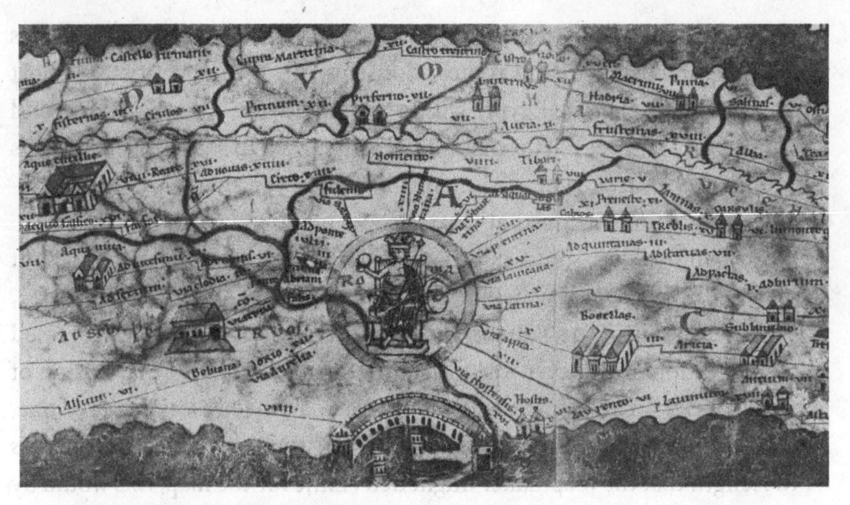

Fig. 13.4 Rome at the presumed center of the Peutinger map.
Photo: *Peut.* Map B.

Itinerary and "signposts" at major intersections, could supply all this data.[22] Even so, that does not make it appropriate to regard the Peutinger map as merely a diagrammatic representation of routes and their associated place-names and distances. This is simply Miller's choice of perspective that, under his influence, has been adhered to universally ever since, to the point where (as noted above) Brodersen likens the map to modern schematic ones such as the London Tube diagram. But that type of map derives much of its success and popularity from the deliberate exclusion of as many "irrelevant" or "extraneous" features as possible.[23] The Peutinger map, by contrast, is nowhere near so focused. Rather, the routes are set within a physical representation of the main landmasses, marked out by their shorelines and manipulated to fit the most elongated frame imaginable. This manipulation by the mapmaker merits recognition as an astonishing feat of creative design. From whatever source materials the physical landscape was derived, naturally they needed to be cartographic rather than textual.

A truly inventive stroke was to reduce bodies of open water to a minimum, retaining enough of them to separate the different landmasses from one another, but otherwise freeing up space badly needed for cramming

[22] In this connection note Salway (2001), 54–60.
[23] Note that the only associated feature retained on the London Tube diagram is the river Thames.

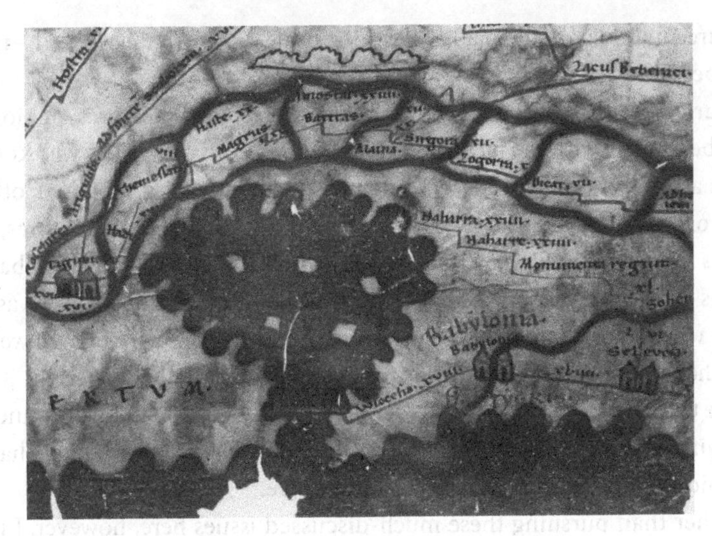

Fig. 13.5 "Persian Gulf" on Peutinger map parchment 10. Photo: *Peut.* Map B.

in more land. Even though a distressing amount of the black lettering inscribed over the green water tint has become illegible, it is quite clear that there was a deliberate concern to name bodies of open water, as well as to mark and name many islands, principally within the Mediterranean, although not exclusively. Elsewhere, for example in 10C4, it is striking to see five islands each carefully marked and named within a "Persian Gulf" to the left of Babylonia (fig. 13.5). Elevation is not shown with much sophistication. Only the principal mountain ranges and areas of high ground are marked, and even they appear in just a very generalized form. That said, the "mountain range" symbol does occur very often. Concern to show rivers is more conspicuous, though also uneven across the map. The last type of data to note is cultural, a great variety of names for peoples as well as for regions or provinces. In the case of the latter two especially, widely spaced lettering often requires the eye to follow with care over a considerable distance when identifying a name.[24]

All the physical and cultural data shown, it should be stressed, reflects considered choices on the part of the mapmaker, made in conjunction with the basic design decisions already noted. It surely has to be by choice, for instance, that many more peoples are marked to the north and east of the Roman world than to its south in Africa. In any event, the range and quality

24 PROVINCIA AFRICA extends furthest, from 2C5 to 6C3.

of source materials at the mapmaker's disposal—wherever he worked—must have been impressive.

I turn next to consider (again, briefly) the mapmaker's presentation of data, beginning with the pictorial symbols. The Peutinger map is well known for its attractive range of these, some plainly intended as individual, others more or less standardized. Altogether they mark 557 places or features, not always according to criteria that are self-evident today. The group of baths/spas is readily identifiable, for example, although questions can be asked about why it was chosen for inclusion (using such large symbols), as well as how the selection was made, and from what sources. More generally, it is a puzzle that certain supposedly insignificant places are given the prominence of a symbol at all,[25] when some notable cities have none,[26] and others have a very modest one in relation to their importance.[27]

Rather than pursuing these much-discussed issues here, however, I turn to a quite different function of the pictorial symbols which seems to have been overlooked: the role of many (though probably not all) as "anchors" in establishing the framework for the network of routes (fig. 13.6). Consider the likely stages by which a map such as this one is created.[28] After first settling its intended scope, dimensions, and contents, the mapmaker's second step (also large) is surely to lay out the shorelines and at least the principal rivers and mountain ranges. This step demands some attention to scale, however variable and imperfect. The same applies too to the third step, which (I suggest) is to mark in relation to the physical base those settlements that form the main junctions for the route network that will follow.[29] Also at this third stage, I imagine, the mapmaker may choose to mark with symbols further principal

[25] For example, Ad Mercvrivm (3C5), Addianam (8C5).

[26] For example, Casarodvno (1B3), Tergeste (3A5), (H)Aila (8C5, just to the left of Addianam). In these instances, as in so many others, there is of course no knowing the degree to which slips or omissions by copyists have served to transmit the original mapmaker's work inadequately; see further essay 15 below.

[27] For example, Mediolanvm (3A2), Chartagine (4C5).

[28] See further essay 14 below. Readers of Weber (1976), 12—quoted in turn by Brodersen (1995), 187—will realize that my reconstruction of the stages by which the map was made is the reverse of theirs, insofar as they mistakenly regard the routes as its underpinning and the physical features as, in a word, "decoration." Compare how a copyist began in the case of the unfinished Artemidorus map, discussed in essay 11 above.

[29] For the most part, a settlement or station where the map user faces a choice of onward routes is marked by a symbol. That is not necessarily the case, however, where two routes merely merge (but do not intersect), in particular when one seems to be only a short link, as between the Sicilibba and Avitta stretches in 4C4, for example. My formation "main junction" excludes these instances. For an intersection-point not to be marked by a symbol is most unusual (but note Dertona in 2B5, for example; in all likelihood Casarodvno in 1B3 was originally marked by a symbol).

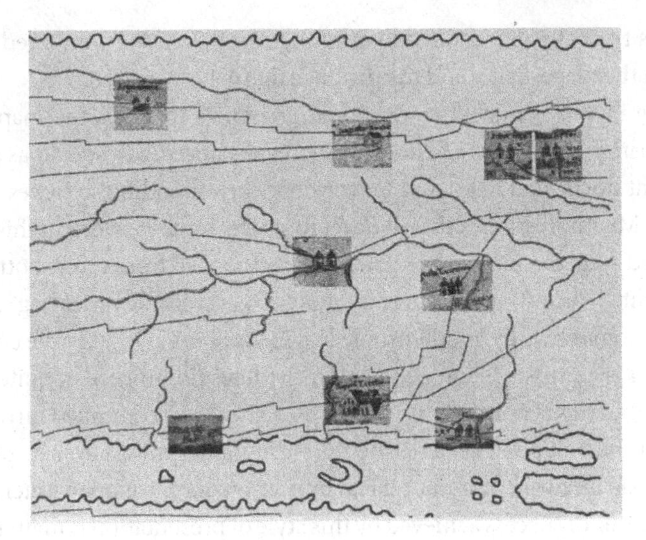

Fig. 13.6 Linework and symbols on part of Peutinger map parchment 2 (as shown in fig. 13.1). Figure: Ancient World Mapping Center.

settlements that happen not to be junctions, but are somehow considered of sufficient importance to place in relation to the main junctions.[30]

It is vital to recognize that the only settlements or stations which the mapmaker is prepared to locate specifically are those marked by a symbol (with the name normally written above it, if possible). Otherwise each place just has its name on "its" stretch of route, followed by the figure giving the distance from there to the next place. Quite deliberately, however, none of these names without a symbol is marked to pinpoint the location of the place. By the same token, the length of the corresponding stretch of route as shown on the map bears no relation to the mileage given for it; the briefest stretch on the map can represent a great distance,[31] a really elongated stretch a very modest one.[32] So, having selected and then marked his "symbol" settlements (two challenging steps), it only remains for the mapmaker to determine which intermediate places he wants to include, and the distance between each. Scale does not apply, therefore, for everything that lies along the route from one "symbol" settlement at a main junction to the next. Instead, the length of a

[30] Along the upper part of parchment 4, for example, note Pola, Vindobona (A1), Adprotoriv(m), Sabarie (A2), Brigantio (A3), Aqvinco, Sardona (A4), Siscia (A5).

[31] The Chidvm (9B1) and Bitvrs (10B5) stretches, for example.

[32] Several instances in Africa, for example, between 4C2 and 5C5; note also the four miles between Tavrvno and Singidvno (5A5–6A1).

stretch is typically determined by nothing more than the combined total of letters in the place-name and the distance figure for the stretch.[33]

Finally in this connection we should recognize that it is the mapmaker's aim to write place-names and accompanying distance figures so far as possible in straight horizontal lines, and to draw the corresponding stretches of route likewise. My impression is that in developing the route network in this way he proceeded—again, so far as possible—across the map from top to bottom and left to right.[34] Each stretch of route is defined by a "chicane" or "zigzag" at either end,[35] or sometimes by another route leaving or joining it.[36] Gentle curvature or sloping is admissible. On occasion, too, routes do proceed at quite a steep diagonal, but this is normally a last resort, and not to be continued further than necessary. If at all possible, lettering for place-names must never be placed to overlay road linework, nor must it run into river courses or open water.

The degree of success achieved by this style of presentation is matter for debate. To be sure, if it was the mapmaker's intention to distil meaningfully into a single, compact map an accumulated mass of purely written itinerary lists and other such documents—with their inescapable overlaps, repetitions, and lack of connections or sense of direction—then indeed this ambitious goal is achieved, and an entirely new perspective opened up. Even so, the compactness, for all its undoubted cunning, comes at a price. The placement required for some of the principal "symbol" settlements creates a very distorted representation. Pergamum and Alexandria, for example, lie almost one above the other on parchment 8 (fig. 13.7), admittedly with a band of open water in between; by contrast, Hippo Regius and Carthage lie nearly two parchment-widths apart (3C2–4C5).

Even users able to discount such distortions as these could come to regret the additional loss, in effect, of two compass points. Only for general impression is there a north-south dimension to the map. Otherwise the west-east one alone is meaningful, although (again inevitably) what appears to be a west-east route on the map may well not prove to be that on the ground. In North Africa on parchments 1–5, for example, it would be wrong to infer

[33] Hence on a route where the mapmaker remains determined to mark more intermediate places than his layout can accommodate (rather than selecting only as many as can comfortably fit), an untidy jam results. See, for example, the route between Arelato and Valentia (1B5–2B1).

[34] This is Ptolemy's recommendation, *Geog.* 2.1.5–10 = Berggren and Jones (2000), 94–95.

[35] Both terms, as well as the German "Haken" ("hook") used by Miller, can be no more than approximations of the mapmaker's practice. My thanks to Martin Cropp for suggesting "chicane," a sharp bend designed as a challenge in Formula One motor racing.

[36] See, for example, the Fvlgvrita (5C5) and Ypepa-Anogome (8B5) stretches.

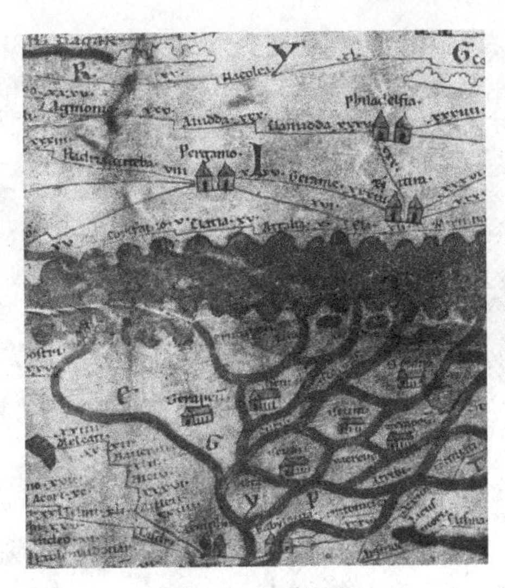

Fig. 13.7 Relative placement of Pergamum (upper center) and Alexandria (lower left; unnamed symbol where the Nile's leftmost branch reaches the sea on Peutinger map parchment 8. Photo: *Peut.* Map B.

that the five or six routes to be seen forging horizontally in parallel across the map do in fact all proceed thus on the ground; far from it. The same caution applies to the three routes proceeding to the right from Trapezvnte (9A2), to cite only one further instance (fig. 13.8).

So was the Peutinger map intended at all for practical, contemporary use as a route map? Its presentation definitely gives that impression, and without question there are certain parts which could be used in this way. Even so, altogether it has too many shortcomings to be practical. Travelers who tried to rely on it for making journeys would quickly be liable to encounter puzzles and frustrations. Among them is the fact that, in our surviving copy at least, there are many instances where no distance figure is attached to a name, as well as many clearly demarcated stretches of route lacking both name and distance figure. It is normally impossible for users to determine whether such a "blank" stretch should be ignored as mere embroidery or slip of the pen, or whether travelers would in fact face an indeterminable further distance.[37] Moreover, in a dismaying number of instances where there is a figure, travelers would hardly feel reassured to read that the distance to the next place is forty, sixty, even one hundred and more miles—further, in other words, than most

[37] For illustration of these dilemmas, see the route Aqvileia–Virvno (3A5–4A2).

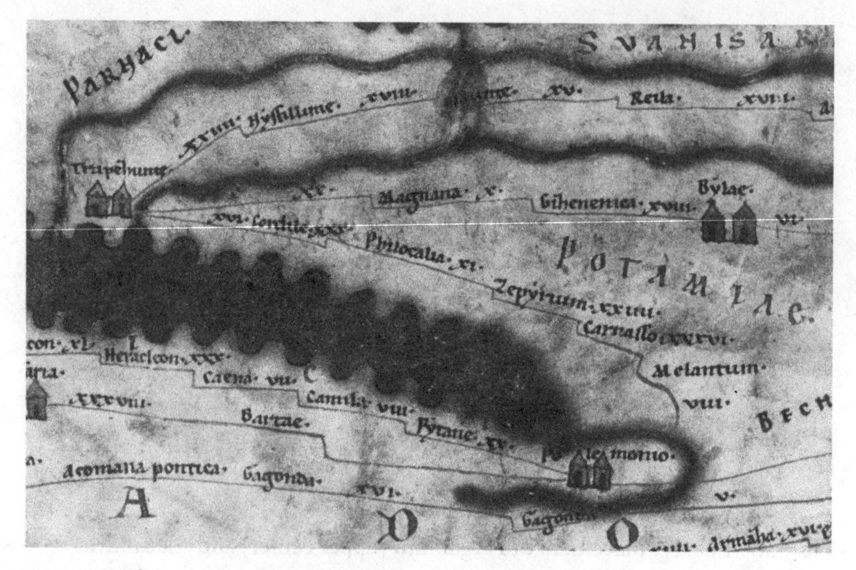

Fig. 13.8 Three routes from Trapezvnte (top left) on Peutinger map parchment 9. Photo: *Peut.* Map B.

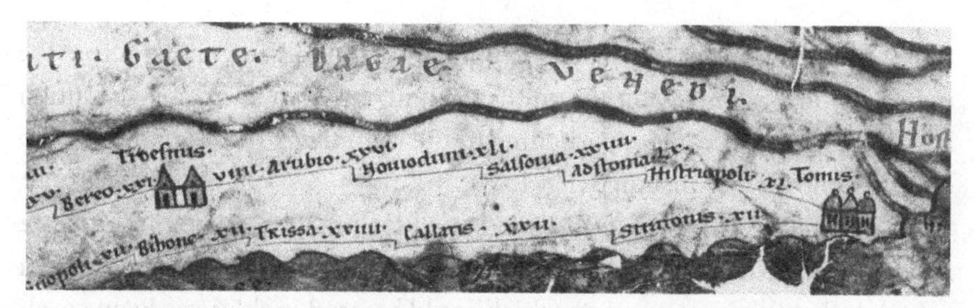

Fig. 13.9 Peutinger map's route Troesmis–Tomis. Photo: *Peut.* Map B.

travelers would be able to cover in a day or even two (fig. 13.9). It is true that several of these long intervals occur in remote regions and so may reflect local conditions faithfully enough, but in other instances some shortcoming on the part of the mapmaker could be suspected.[38] For these reasons among others I see the users that the mapmaker has in mind as more detached—well-educated aristocrats of the Western empire who, whenever they do travel, leave the practical arrangements to their staff.[39]

[38] The *Bordeaux Itinerary* is more reassuring by contrast; it seldom advances further than 16 miles or so without recording a road station, often much less, and 24 (once) is its longest interval. In the *Antonine Itinerary* intervals of over 40 miles are rare, and most are below 30. See now for both Itineraries, Talbert (2010), 206–86.

[39] Consider the *nobilitas* of Rome itself, so mercilessly ridiculed by Ammianus Marcellinus not least for their attitude toward travel (28.4.18).

The mapmaker's main purpose is to boost such westerners' pride in the range and greatness of Rome's sway historically; hence Milan, Carthage, and Alexandria, for example, are all deliberately denied prominence. Only users well grounded in history and geography will properly appreciate the map, whereas others will be confused and misled. At the same time, to a striking degree, its presentation reflects fourth-century intellectual taste in education, art, and literature.[40] Higher learning at this period is unashamedly exclusive. Any users of the map who are not already aware that Hippo Regius and Carthage in reality lie closer to one another, and Pergamum and Alexandria much further apart, or that the three routes emerging from Trapezus do not in fact all proceed east, simply reveal their lamentable ignorance. No less pitiful are those who imagine that Herclanivm (5B4) and Pompeis (5B5) still exist as marked, or that the Roman road networks shown in Dacia and the Agri Decumates still function.

The better informed, by contrast, may revel in the sheer accumulation of names and details that is so fashionable a feature of fourth-century literature (both prose and poetry), with the added pleasure that many of the names are exotic and "difficult." It is a further delight to realize that the place-names and distance figures derive from sources of quite a different type, and are ingeniously reused here in a fresh, more attractive setting (as was done in "cento" verse compositions). One can exclaim over the map's format (which is really a text format), so exquisitely compact, with the space so fully used thanks to the deft elimination of most open water and subtle molding of landmasses. Certain contrasts seem almost to tease. In some sense there is a discernible north-south dimension, for instance, but altogether it is mostly lacking. Equally, in some respects attention is paid to scale, but in others—as often in works of art—adherence to it is manifestly abandoned. A viewer with the same sense of fun as the writer of the *Historia Augusta* would no doubt be intrigued.

There can be little question that the map in its original form fully reflected the fourth century's love of bright colors.[41] It is tempting to go further and liken its pictorial symbols to the jewels or precious stones (*segmenta*), with which the finest masterpieces of the time were studded for heightened

[40] My preliminary treatment of this theme here draws special inspiration from Roberts (1989). In the same connection, note also MacMullen (1990); Maguire (1999).

[41] The exceptionally rare occurrence of blue may appear surprising. A blue tint is only to be seen filling the center of some of the large bath/spa symbols (the instances in 6A5 and 9B2 seem clearest, followed perhaps by 6B4). Whether this coloring was original, however, or merely added by a copyist later as an extra flourish, is indeterminable. It could be argued that the mapmaker would hardly have included blue in his palette, only to employ it so little. Alternatively, he may have deliberately kept such flashes of blue to a minimum. Thereby they are rendered all the more conspicuous, and their incorporation into the bath/spa symbols (in turn made large deliberately) would give added pleasure to patrons with a special fondness for such resorts. All water features are otherwise rendered in green.

brilliance—most obviously mosaics or monuments, but also (figuratively) poems, and (literally) dress. In this latter regard, the portraits of the two consuls which form part of the Codex-Calendar of 354 are particularly striking (fig. 13.10a, b). Constantius II alone, as emperor, has jewels on his *toga picta*, but observe the cameo vignettes on that of his colleague Gallus, and the bands or stripes (*clavi*) on both. The Peutinger map in turn can be viewed as a great, long colorful robe or frieze celebrating Rome and Roman power, with the pictorial symbols as eye-catching *segmenta* and the route linework as *clavi*. Originally, perhaps, the map could just as well have been presented, not on papyrus, but as a wall-painting or tapestry made to resemble a papyrus roll.

Two lesser, related fashions in fourth-century art which the map reflects are, first, the fondness for depicting groups of figures who superficially all look alike, but on closer inspection turn out to have different faces or posture or gesture. This superficial sameness that is not necessarily borne out on closer inspection is a feature of the map's pictorial symbols in particular. The second fashion is the predilection for row formation, the organization of groups in ranks, which matches the deliberately "horizontal" layout of so much of the route network.

More broadly, it is characteristic of a fourth-century work of art (poems, mosaics, buildings both inside and out, etc.) to try and make such a dazzling initial impression that readers or viewers actually become disoriented at first. For contemporary taste this is a merit, not a fault. Once recovered, viewers may turn their attention wherever they wish, and appreciate the work's different elements at their own pace. There ought not to be any dominant element which imposes the adoption of a particular perspective. The Peutinger map fully reflects thinking of this type. It must have been overwhelmingly long, as well as dazzling at first sight. Modern observers have all been conditioned by Miller to focus on the route network as a dominant element, with virtual disregard for everything else. But the fact is that so many more components are present, too, on both land and sea. Even the routes themselves are not presented in a prescriptive way. From whatever point on the network users set out, in due course they reach a junction, where typically two or three onward choices are presented. In each such instance users are left completely free to determine their preference. Nothing is done to classify routes by importance, or directness, or any other criterion. Moreover, the map offers no figures for the *total* distances between "symbol" settlements, nor does it mark any boundary-lines (not even between dioceses or empires,

Fig. 13.10a, b Constantius II and Gallus as consuls in the *Calendar of 354*. Figure: Salzman (1990), figs. 13–14. Photo: Art Resource, NY (452397).

let alone provinces). Rather, it consciously appears seamless, uniform, almost unending.

Without question, the Peutinger map will always be extraordinarily important to anyone with an interest in the route network of the Roman empire and far beyond. But others, too, should take courage to examine it, and to realize how layered a masterpiece it is, capable of being "read" in a variety of ways and on various levels. As a mirror of its age, it is rewarding not only from a cartographic perspective but also artistically and intellectually. It deserves to be restored to the cultural mainstream, rather than to remain an isolated curiosity which is typically considered from no more than a single, modern perspective. The "edition" that I am at present engaged upon is specifically designed to promote greater openness of approach.

My two last words return to the map's cartography, which is so fundamental a component, so woefully neglected to date. First, I now see no means of evading the inference that an ambitious work of such immensity, quality, and sophistication could only emerge from an established Roman cartographic tradition; it seems inconceivable for it to have been somehow created almost ex nihilo. Of course, at the same time it may be a map that also advances Roman cartography rather than simply demonstrating levels of knowledge and accomplishment already achieved. In any event, the current preference for doubting whether Roman cartography ever attained much development seems overdue for some less skeptical reconsideration. In the same connection, second, students of Roman cartography ought perhaps likewise to be readier to see substance in findings by medievalist colleagues who identify elements in maps of their period as Roman in origin. Such traces of Roman cartography, however imperfectly transmitted, now seem more credible, and they merit systematic search and appraisal.[42]

[42] See further now Gautier Dalché (2008).

14

Peutinger's Map

The Physical Landscape Framework

The Peutinger Roman map can make a special contribution to our understanding of how space was perceived in antiquity. In fact its value and importance in this regard are too extensive for satisfactory coverage in a single presentation. Accordingly, I confine the focus here to the overall scope of the map and to its physical landscape framework. Readers should be cautioned at once that both this focus and my approach are at variance with the current, established view of the map. It sees a diagrammatic representation of route information found in itinerary lists as the map's basis, with physical features only being added thereafter as "decoration."[1] My perception of the map, by contrast, is a holistic one, which regards it as a bold work crafted by a seasoned cartographer to form a single, cohesive creation.

The new, and inevitably controversial, feature of this perception is the long overdue attention and respect that it brings to the map's design. It is remarkable that until relatively recently there was the same inattention to this fundamental aspect of the Bayeux Tapestry too—an English masterpiece of the late eleventh century in all likelihood, and comparable to the map in several notable respects, although much more intensively studied. Ironically, it was an art historian of antiquity who first attempted this approach to the Tapestry in a 1991 article.[2] The Peutinger map—created around 300 CE in my view,[3] but surviving only in a copy made around 1200—still awaits such attention.

In principle, many of the settlements on the route network that are marked by symbols could be considered the basis of the map, insofar as the symbols for these settlements are placed accurately enough in relation to the physical landscape. By contrast the routes that link them, with a string of intermediate stopping-points, are merely adapted to occupy the space available (fig. 14.1).

[1] See, for example, Weber (1976), 12; Wagner (1984); Brodersen (1995), 187; Whittaker (2002), 83, 87, 94–95; and essay 13 above.

[2] Brilliant (1991).

[3] Note now, however, the arguments for a Carolingian origin advanced by Albu (2005; and further 2014), with dissent by Salway (2005).

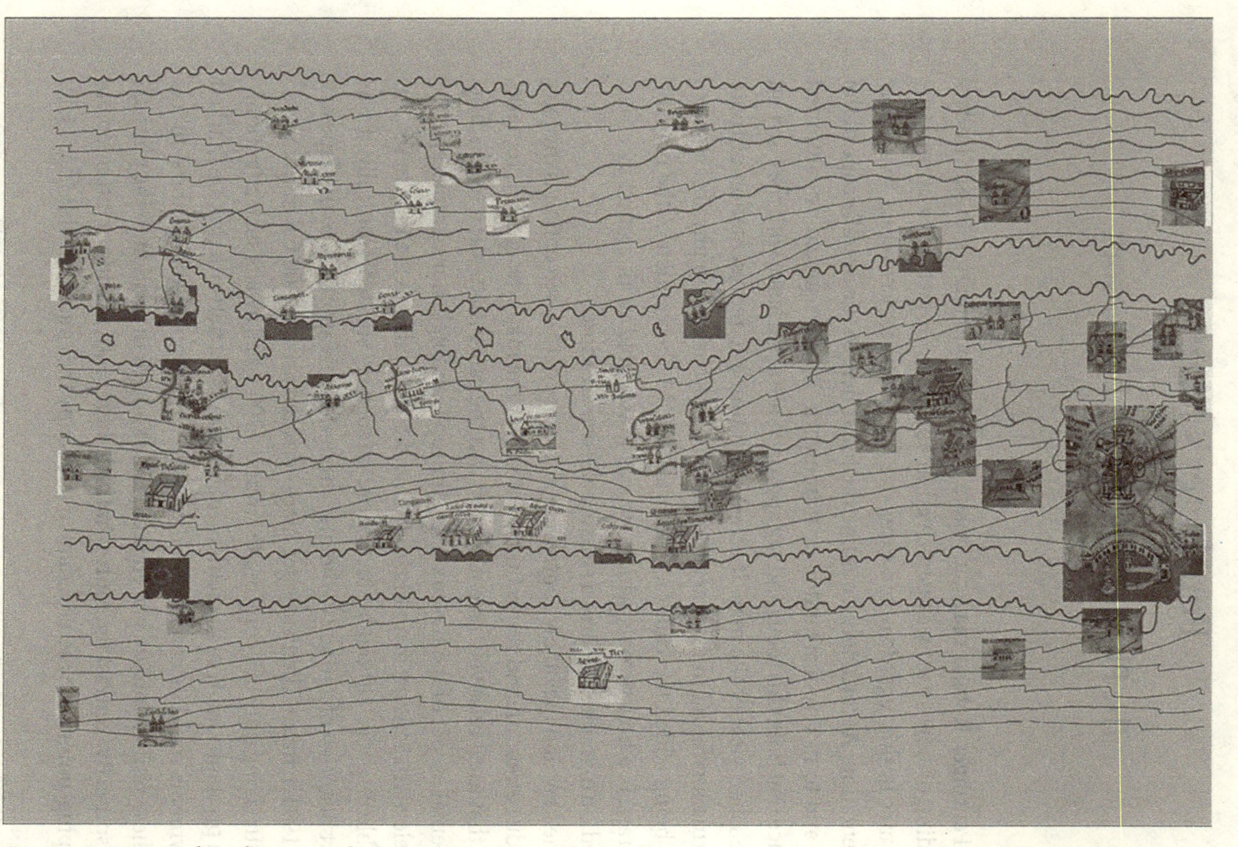

Fig. 14.1 Rivers, shorelines, symbols, and route-stretches on Peutinger map parchment 4 (city of Rome at far right). Figure: Ancient World Mapping Center.

For certain, therefore, the route network as a whole cannot be the basis of the map. Moreover, it seems unrealistic to imagine that the symbols on which this network depends were laid out *prior* to any physical landscape, especially in view of the map's quite unconventional shape. No, the mapmaker's first steps must have been to determine his frame, and then to outline the physical landscape within it.[4] Naturally, some of the decisions taken with regard to these steps may have been imposed by the intended recipient of the map, or by other fixed circumstances relating to its intended placement and use. However, today such distinctions can only be matters of conjecture, which I leave aside here.

The map's shape is extreme. It is everywhere about 33 cm in height for its full surviving width of about 670 cm. (The surviving copy preserves it on a series of eleven parchments more or less equal in size.) There is no knowing to what further length the map originally extended at its lost left-hand end, nor whether the copy's right-hand end was also that of the original. In the former connection, there are sound reasons to reject Konrad Miller's claim (still generally accepted) that only a single parchment has gone missing from the copy at the left-hand end—making an approximate total width of 730 cm for the map, if a parchment width be reckoned as 60 cm. For all its ingenuity, Miller's reconstruction of that lost parchment published in 1898 has distinct limitations. In particular, it unnecessarily compresses the presentation of Britain and the Iberian peninsula, and it sets aside several of the mapmaker's cartographic principles.[5]

In my view by contrast, a fundamental concern on the part of the mapmaker was to site the city of Rome at the center of the map, where it would form the focus from both horizontal and vertical perspectives. On this basis, for the width left of Rome to correspond to what survives to its right (approximately 430 cm), the existing 240 cm to the left would need to be supplemented by a further 190 cm or so, in effect three parchment segments (fig. 14.2).

In what is now missing at the left, the mapmaker surely included more of Great Britain, if not all of it, and completed his coverage of the Mediterranean and its surrounding lands.[6] Equally, proof is lacking that the right-hand end of the map as it survives marked the end of the original design. A vertical

[4] Note, incidentally, that the copyist of the unfinished map designed (we suppose) to accompany text by Artemidorus began by defining the physical landscape: see essay 11 above.

[5] Further discussion in Talbert (2007) and now (2010), 189–92.

[6] See further essay 13 above.

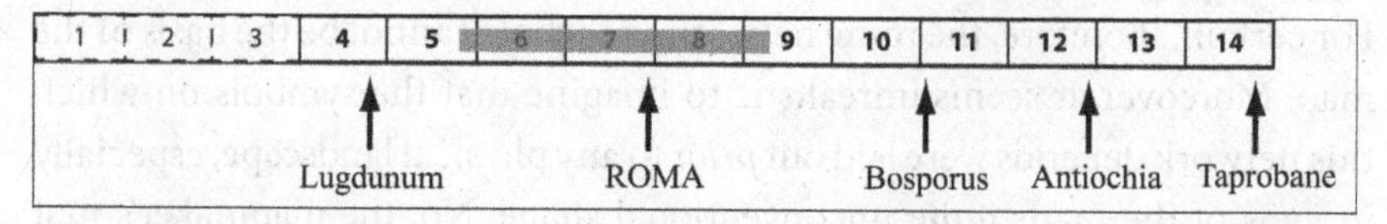

Fig. 14.2 Spread of the Peutinger map's surviving parchments (numbered 4–14, with three assumed missing at left-hand end). Shading denotes the approximate extent of Italy. Figure: Ancient World Mapping Center.

margin line to link the lines running along the top and bottom could be expected, but the copy lacks one. Roman awareness that there were lands beyond India is not in doubt, so a lost continuation of the map that renders them in some way cannot be entirely ruled out. Text of one type or another might have been placed at either end or both.

Where the width of the map is concerned, it is striking that from midway through parchment 10 to the end of parchment 11 the mapmaker for the first time repeatedly abandons his principle of presenting land routes "horizontally," and instead tilts several or even renders them near-vertical.[7] A likely cause of this compression, which saves space, may be his awareness that the end of his frame is fast approaching, so that he feels constrained to confine the lands beyond the Euphrates river to no more than this unduly small fraction of the map. As it happens, these lands for the most part lie beyond the Roman empire, and consequently may be of lesser interest to Romans.

Broadly speaking, the western limit of the map must have been the Atlantic Ocean, and in our copy the eastern limit is India and Insvla Taprobane. A band of open water (nowhere named) surrounds the entire map as it survives. The lost continuation of Britain was no doubt angled so as to permit this open water to encircle it, and even to extend onward—into the Atlantic—around the missing left-hand end. The open water's outer limit is defined by a margin top and bottom. Shoreline marks its inner limit, except along much of the bottom. Here, from the start of parchment 1 to the start of parchment 9, an unbroken mountain range fulfills this purpose instead, until the open water leads into the Indian Ocean.

The incompleteness of the surviving copy contributes to the difficulty of establishing with certainty the scope intended for the map by its designer. To define the scope as the *orbis terrarum* or *oikoumene* ("inhabited world")

[7] See further below, with fig. 14.7.

might seem too loose a description for coverage that extends no further than India and Insvla Taprobane.[8] After all, as already mentioned, there were known to be lands and peoples further beyond in that direction, as well as both north and south of the map's top and bottom margins; Arabia and East Africa in particular are deftly excluded from the map. However, if the original map did not in fact extend further eastward than the surviving copy, its scope could appropriately be described as *orbis Romanus* or *imperium Romanum*, provided that the force of either noun is taken to be vague rather than specific. The sense is thus "the (part of the) world claimed by Rome," "Rome's dominion," or "Rome's sway," rather than "territory under direct Roman control" as can be meant in statements made, say, by jurists.[9]

The mapmaker's intention in representing open water all around the landmasses on the map calls for comment. He may have hoped that thereby he would render the map more recognizable to viewers who adhered to the traditional belief that the *oikoumene* was encircled by Ocean, one reflected by the geographer Strabo:[10]

> We say that our *oikoumene* is situated with sea flowing around it and resembles an island—as indeed demonstrated by both perception and reason, it has been said.

At the same time, it is hard to credit that the mapmaker himself, or his more educated viewers, shared this belief. Rather, from their perspective the encirclement offered a convenient means to limit the coverage of landmasses at the top, bottom, and no doubt right. It also served as one means, among several, of conveying to viewers the cohesion and unity claimed for the *orbis Romanus*.

Far from being regarded as merely a "decorative" element of the map, the principal elements of physical landscape need to be seen as its fundamental underpinning. These elements comprise at a minimum shorelines (thus including islands), principal rivers, and principal mountain ranges.

[8] This is not to forget that these terms themselves lack precise definition: see, for example, Mattern (1999), 44–47; Harris (2005), 15–16.

[9] Note Ulpian in *Dig.* 1.5.17, "In orbe Romano qui sunt ex constitutione imperatoris Antonini cives Romani effecti sunt"; and Celsus, ibid., 43.8.3, "Litora, in quae populus Romanus imperium habet, populi Romani esse arbitror."

[10] 2.5.5: ἱδρῦσθαι φαμεν τὴν καθ' ἡμᾶς οἰκουμένην περίκλυστον θαλάττῃ καὶ ἐοικυῖαν νήσῳ: εἴρηται γὰρ ὅτι καὶ τῇ αἰσθήσει καὶ τῷ λόγῳ δείκνυται τοῦτο; cf. 1.1.8. For discussion, Romm (1992), 12–17, 33–36, 41–44.

All the map's other features are marked in reference to them, just as Strabo recognized:[11]

> It is above all the sea that delineates the land and shapes it, by forming gulfs and open water and straits, and likewise isthmuses, peninsulas, and promontories, with assistance from both rivers and mountains. These are the features that have molded our conception of landmasses and peoples as well as the favorable locations of cities and the various other elements that fill our geographic map. Among these elements, too, is the plethora of islands both dispersed across open water and all along the coastline.

Generally speaking, it would seem probable that the map's base reflects the unrivaled geographic and cartographic learning that was developed in third-century BCE Alexandria by Eratosthenes and his successors.[12] More specifically, however, the one or more sources of the base cannot be identified. In the likely event that the map's extreme shape does reflect an original creation rather than merely the reuse of a forerunner, it would be natural enough that the designer should seek to adapt the base elements of an existing map—one no doubt less extreme in shape—with sufficiently ambitious coverage.

No such Roman maps survive, but their existence is confirmed by the unique outline description in Eumenius' speech addressed to a provincial governor in Gaul during the 290s. Eumenius begs permission to rebuild the rhetorical school at Augustodunum (modern Autun) at his own expense, and he outlines what a beneficial institution it would be again after the previous half century of disruption and crisis. One feature evidently already in place is an *orbis depictus* intended for display; this will be a pleasure to examine, since everything within it has now at last been recovered for Rome.[13] Eumenius' description of its features is inevitably brief and rhetorical, but we gather that the coverage extends at least from Britain to Egypt and Persia. It is also plain enough that the designer had attached importance to rendering

[11] Strabo 2.5.17: Πλεῖστον δ' ἡ θάλαττα γεωγραφεῖ καὶ σχηματίζει τὴν γῆν, κόλπους ἀπεργαζομένη καὶ πελάγη καὶ πορθμούς, ὁμοίως δὲ ἰσθμοὺς καὶ χερρονήσους καὶ ἄκρας. προσλαμβάνουσι δὲ ταύτῃ καὶ οἱ ποταμοὶ καὶ τὰ ὄρη. διὰ γὰρ τῶν τοιούτων ἤπειροί τε καὶ ἔθνη καὶ πόλεων θέσεις εὐφυεῖς ἐνενοήθησαν καὶ τἆλλα ποικίλματα, ὅσων μεστός ἐστιν ὁ χωρογραφικὸς πίναξ· ἐν δὲ τούτοις καὶ τὸ τῶν νήσων πλῆθός ἐστι, κατεσπαρμένον ἔν τε τοῖς πελάγεσι καὶ κατὰ τὴν παραλίαν πᾶσαν.

[12] For an overview, Geus (2003).

[13] *Paneg.* 9(4).21.3 Mynors (1964), "Nunc enim, nunc demum iuvat orbem spectare depictum, cum in illo nihil videmus alienum." For fuller quotation of the passage, see essay 18 n. 18 below.

the principal elements of physical landscape, which include an encircling Ocean:[14]

> . . . there are pictured in that spot . . . the sites of all locations with their names, their extent, and the distance between them, the sources and mouths of all the rivers, the curves of all the coastline's indentations, and the Ocean, both where its circuit girds the earth and where its pressure breaks into it.

Beyond affirming that the designers of the Peutinger map and of the lost Autun map shared a preoccupation with physical landscape, it is impossible to say how alike their renderings of it were. We do not even know whether the Autun map was oriented north (i.e. with north at the top) like the Peutinger map. The choice of orientation was a fundamental step of importance because there was no recognized standard. It is notable that Ptolemy[15] recommends north, but there is ample testimony to the adoption of other choices: the "Dura shield" (fig. 14.3) is oriented south, and Rome's Marble Plan southeast, while the three Orange cadasters vary in their orientation.[16] In practice, three special considerations made a north or a south orientation the only realistic choices for the Peutinger map. The three are the extreme contrast between the map's width and its height; placement of the city of Rome at the map's center; and a concern to demonstrate the comprehensive network of land routes across the Roman empire and even beyond. While the map's shape thus required the virtual elimination of a north-south dimension, the consequent exaggeration of its west-east aspect should not necessarily be thought unfamiliar or unwelcome to its prospective audience, whose worldview was centered upon the Mediterranean.[17]

It is striking that attention to these three special considerations underpinning the map did not lead to the neglect of physical and cultural information of other types. At the same time, the mapmaker concluded that only by removing much open water could he ensure adequate coverage of landmasses for his representation of routes there. As a result, the Bay of Biscay, Mediterranean, Adriatic, Black Sea, Red Sea, Indian Ocean, and

[14] *Paneg.* 9(4)20.2 Mynors (1964), "omnium cum nominibus suis locorum situs spatia intervalla descripta sunt, quidquid ubique fluminum oritur et conditur, quacumque se litorum sinus flectunt, qua vel ambitu cingit orbem vel impetu inrumpit Oceanus."

[15] *Geog.* 2.1.5.

[16] See in brief Brodersen (1995), 145–48, 235, 223–24, respectively, with figs. 2.5b above and 18.1 below.

[17] See further on this perspective Bowersock (2005).

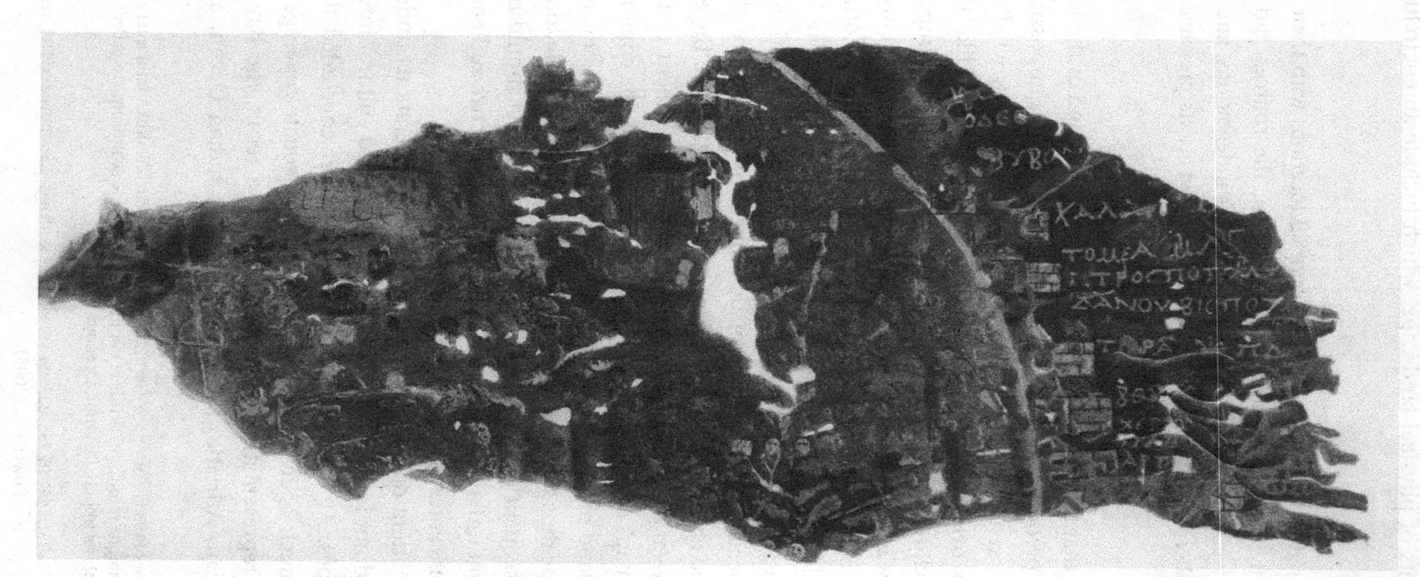

Fig. 14.3 "Dura Shield" map, oriented south. It is no more than a possibility that this parchment fragment (45 × 18 cm) found at Dura-Europos (destroyed around 255 CE) comes from a shield-decoration. Part of the map's purpose may have been to promote the Black Sea (rendered in dark blue) and its region. The west coast is shown to the right here. Successive cities (starting from Odessos at the top) and rivers are named in Greek, with symbols and numerals for the distances between them in Roman miles. Photo: Cumont (1925), planche I.

Caspian Sea are all reduced to relatively narrow channels; the Aegean Sea is compressed, and the Hellespont and Propontis are effectively eliminated. By contrast, the Bosporus between Europe and Asia is made broader than necessary, and thereby emphasized.[18]

Open water is removed, but not necessarily the names for parts of it, nor islands within it. The retention of such names is especially extensive within the Adriatic and Black Seas; the same may be said of islands within the Adriatic and Aegean Seas. The mapmaker clearly attached importance to both types of feature, and he can only have derived them from one or more maps that represented open water in fuller detail. However, most islands are shown no more than small, with a stylized, token shape; an attempt to convey their actual shape, as is made with the islands off the Gulf of Salona,[19] is exceptional and thus puzzling. Sicily, Crete, and Cyprus are the only islands shown large and with a route network in place. In each of these instances, the narrow channel of the Mediterranean is widened to permit insertion of the island; the same is done for the Peloponnese.

It is a surprise that the mapmaker did not present Sardinia likewise. It engages him sufficiently to mark and name four offshore islands to its left and a further two to its right,[20] and it could readily have been extended in length to match Sicily (figs. 14.4 and 14.5).[21] Moreover, the upper shoreline of the Mediterranean could have been raised hereabouts without strain to give Sardinia similar depth to that of Sicily; compare how below Sicily the lower Mediterranean shoreline is depressed to admit the island of Djerba (unnamed).[22] As it is, the mapmaker drastically shrinks Sardinia and foregoes the opportunity to show its extensive route network, which is well attested from the *Antonine Itinerary* and from the survival of over 150 milestones.[23] It is true that presentation of Sardinia's network might be thought to pose more difficulty than those of Sicily, Crete, or Cyprus, insofar as it runs mainly north–south whereas theirs all run west–east; but Sardinia as it appears on the map is already "turned" with its east coast uppermost. Possibly the

[18] The Golden Horn (on which Sycas lies) is in turn exaggerated (8A1).

[19] 5A3; cf. *BAtlas* 20D6.

[20] 2B5–3C1.

[21] Sardinia is shown as approximately 8 cm long by 1.5 deep. Sicily is 21 cm long by 4 at its greatest depth, allowing sufficient space to run three parallel routes from left to right. The map prescribed by Ptolemy (*Geog.* 8.9.1–4) afforded Sardinia and Sicily the same coverage; compare Müller (1901), 14 (incorporating Corsica here too, which Ptolemy in fact assigns to his preceding map); Stückelberger and Grasshoff (2006) 802–803.

[22] 5C5–6C1.

[23] *BAtlas* 48 and *Directory*, p. 736.

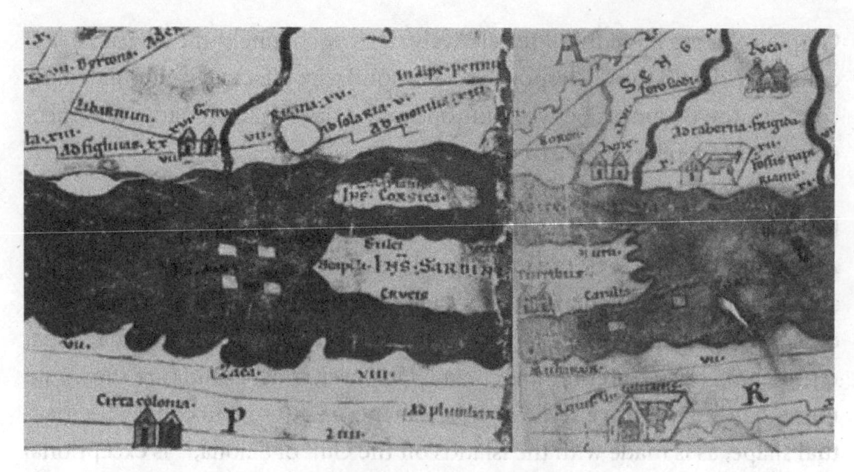

Fig. 14.4 Sardinia (center) on Peutinger map parchments 2–3.
Photo: *Peut.* Map B.

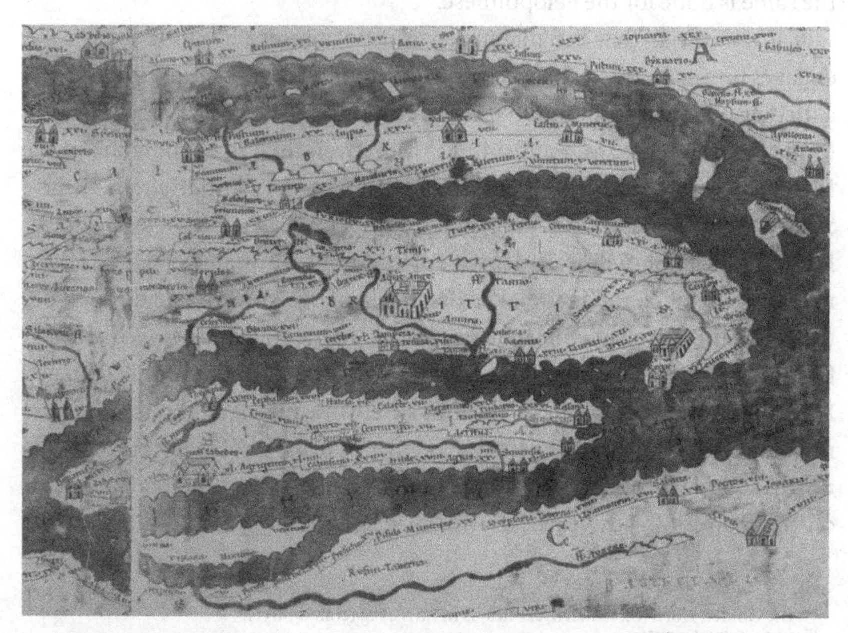

Fig. 14.5 Sicily (lower left and center) on Peutinger map parchments 5–6.
Photo: *Peut.* Map B.

mapmaker was restrained by the obvious need to consider how Corsica should be presented in relation to Sardinia. Even so, he possessed ample ingenuity to devise a placement for Corsica other than its present one between Sardinia and the upper shoreline of the Mediterranean. If he plotted the physical landscape base from left to right,[24] perhaps he lacked the confidence at this relatively early stage to allow for Sardinia to be shown larger, hard though this seems to credit. He restricts himself to marking seven isolated place-names on the island, one of them (Tvrribus) with a symbol. The only other islands on which he marks isolated place-names are Corsica (one) and Djerba (four). The islands of Rhodes (9B1) and of Taprobane (11C4–5) are each shown sufficiently large to accommodate one or more names, but none is marked.

The principal landmasses should be considered next. In order to accommodate these within the map frame, the mapmaker was clearly obliged to manipulate them to a greater or lesser degree.[25] In so doing, he had also to meet the need to provide an equal spread of territory to the left and to the right of Rome as the map's central point. A related concern should have been to ensure that, so far as possible, each region was rendered neither too small to fit all the requisite names, features, and routes within it satisfactorily, nor much larger than was needed for this purpose. There is no knowing just how full or sound a grasp the mapmaker already possessed of routes in particular at the initial stage when he laid out the physical landscape base. It is conceivable that he only gained comprehensive awareness of them as he proceeded, especially if his attempt to show them is a novel initiative.[26] In consequence, at an advanced stage he could do little to adjust for mismatching coverage.

Not only was Italy the Roman heartland, but the unusually dense network of routes here merited marking. In addition, for the mapmaker's purpose it was ideal to give pride of place to Italy—occupying about one-third of the map, and in its very center by my reckoning. Its peninsula in particular, defined by the Adriatic and Mediterranean Seas, was uniquely well suited to elongated presentation on a map of this extreme shape. More generally, Italy

[24] I believe this to have been his procedure; it is also Ptolemy's recommendation (*Geog.* 2.1.4).

[25] This exercise offered him and the viewers of his map a pleasure that would have appealed to those earlier Roman thinkers and poets who contemplated primeval conflicts between land and sea, and the consequences of rejoining separated landmasses. Note the observations by Murphy (2004), 45–48. The map could equally make ambitions of bridging the Adriatic seem feasible (Plin. *NH* 3.101)!

[26] Brief discussion by Talbert (2005), 633–34; and further essay 16 below.

was a vital tool in the struggle to match the lesser extent of territory that could be shown to the left of Rome with the greater extent to its right. In consequence, Italy is laid out to extend over two meters and more, from the right-hand end of parchment 2 almost as far as halfway into parchment 6. This unrivaled central prominence is surely appealing both to the mapmaker and to his intended audience. Toward its end the peninsula splits into two prongs narrower than its main body, and so the mapmaker can readily take the opportunity to fit an elongated island of Sicily here between the lower shoreline of the lower prong and the North African shoreline.[27]

Britain, largely lost in the surviving copy, would need to have been turned on its side to fit the map's shape, with its east coast evidently to the top like that of Sardinia, and thus the north (lost) to the left. Gaul is similarly manipulated, with its north coast turned left to face southern Britain, and then, after a long narrow inlet for the Sinvs Aqvitanicvs, its west coast facing southwest Britain above. The quantity of routes and other cultural data across Gaul is dense overall, but mostly well enough matched to the space assigned for the region.

With considerable ingenuity the mapmaker compresses the regions facing the empire's Rhine and Danube frontiers to occupy the top of the map until the mouths of the Danube are reached near the right-hand end of parchment 7. This vast area gains improved definition from the right-hand end of parchment 3 onward with the appearance of the Adriatic Sea, which continues far into parchment 6, where the Italian peninsula ends. Routes in Dalmatia run immediately above the upper Adriatic shoreline.

Once the Adriatic merges with the Mediterranean (on parchment 6), the mapmaker again has a deeper unbroken landmass at his disposal. He exploits it to introduce a gentle downward curve of the Danube so that the river now flows less close to the map's upper shoreline. Into the space opened up by this means above the river he then fits Dacia and the routes there. Below the river, however, the coverage he wishes to provide of Moesia and Macedonia is notably squeezed and elongated, so that Piraeus (Pyreo) lines up with Tessalonic(a)e and Philippis, and the island of Cytera appears just offshore below Tessalonic(a)e.[28] The part of this region that the mapmaker shrinks most drastically is central Greece, a sacrifice no doubt prompted more by the sheer impossibility of doing it justice (given the map's shape and other

[27] As already mentioned, the North African shoreline is lowered in turn for a brief distance (5C5–6C1) to accommodate the island of Djerba (unnamed).

[28] All in 7B1–3. However, to place Cytera much closer to the Peloponnese would seem to pose no difficulty; so the choice not to do this remains a puzzle.

constraints), than by any special difficulty in securing adequate information about routes there. Meantime the routes laid out within the Peloponnese come close to straining the limited space that it is permitted to occupy.

Thrace is fitted below the Black Sea (which extends from midway across parchment 7) and continues over to the Bosporus (8A1); in consequence this region receives quite ample coverage. Since it requires no great depth, however, below Thrace the mapmaker takes the opportunity to widen the Mediterranean again in order to provide generous space for Crete.[29] The Black Sea—its lower shore broken by a wide Bosporus between Europe and Asia—continues to the middle of parchment 9 and, like the Adriatic earlier, acts as a valuable divider. No route continues along its upper shoreline for a long way beyond Tomis (7A4), placed just to the left of the Danube mouths; a route only resumes toward the far end of this shoreline once Trapezvnte (9A2) is reached.

The mapmaker's choice for presentation of the Black Sea as a narrow "horizontal" channel in effect constrains him from tilting the landmass of Asia Minor in the way that he manipulates Britain, Gaul, Sardinia, and Italy. Instead, he duly sets the north of Asia Minor along the lower Black Sea shoreline (fig. 14.6). That decision seems a natural one, fully justified in itself. No special difficulty is created by presenting Sicily, Crete, Cyprus, and, in particular, North Africa in this way too. In the case of the Black Sea to the right of the Bosporus, however, severe consequences do arise for accommodating Asia Minor thus to the map's extreme shape.[30] The region of the Bosporus itself, as well as of the Propontis and Hellespont notionally, is indeed tilted to face left. But Asia Minor's western[31] and southern coasts[32] are then rendered as one, to form the Mediterranean's upper shoreline. The latter of these coasts is conspicuously compressed, being assigned only about half the former's length. It surely follows that this unsatisfying alignment of Asia Minor contributes to the patent limitations which the presentation of cultural data in its interior exhibits.

The far end of the Mediterranean coincides with the placement of Antiochia (9B4). Immediately prior, the open water channel is widened to allow the inclusion of Cyprus in the same way that Sicily and Crete were

[29] 7B5–8C2.

[30] To the left of the Bosporus by contrast, there are no such consequences to ringing the Black Sea with settlements that actually lie on its west coast (*BAtlas* 22F4 with continuation on 52D2).

[31] From, say, Lamasco (8B2) to Chidvm (9B1 = Cnidos).

[32] From, say, Chidvm to Aregea (9B4 = Aegeae).

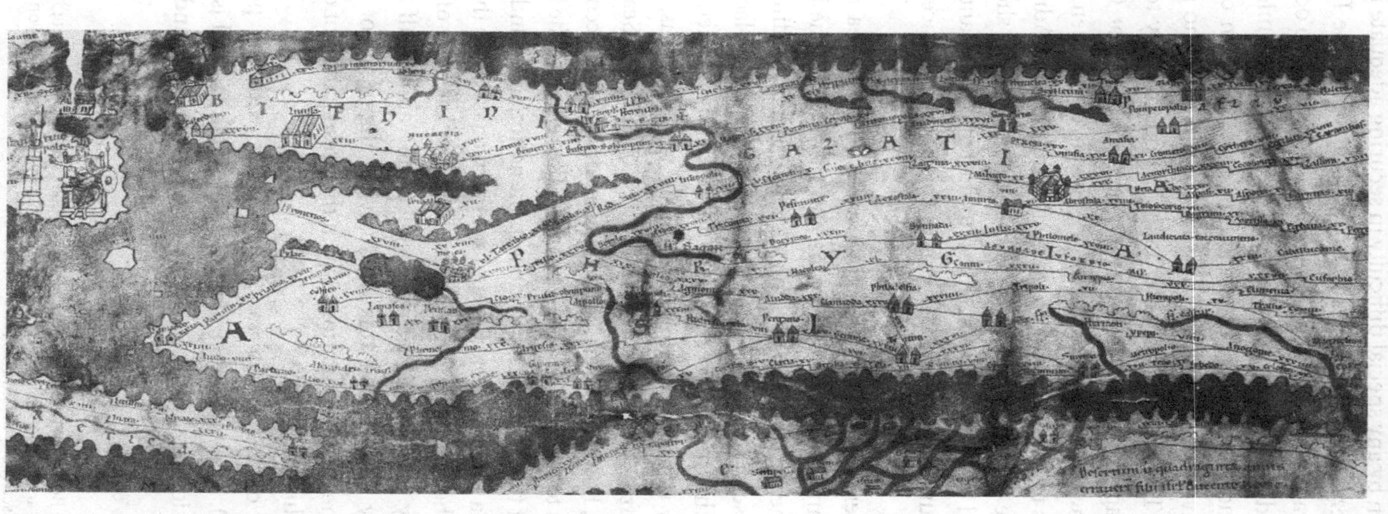

Fig. 14.6 Bosporus (top left) and Asia Minor to the end of Peutinger map parchment 8. Photo: *Peut.* Map B.

handled. Alexandria[33] and the Nile delta as far as Pelvsio (8C4) are intended to feature prominently like Antiochia, although the mapmaker takes the ingenious liberty of placing them, not at the end of the Mediterranean, but earlier. In consequence, the entire stretch of coast from the Nile delta (8C3) to Selevcia and Antiochia (9B4) is shifted to become the final section of the lower Mediterranean shoreline on the map.

North Africa from its lost left-hand end onward as far as Alexandria forms a substantial self-contained segment of the map. Within it, the placements of Alexandria (8C3) and Carthage (4C5, directly below Rome) are surely to be considered the pivotal choices, which in turn determine the layout of the route network. Despite the loss of the network's corresponding "anchor city" at the left-hand end of the map, it is clear that at several stages the width at the mapmaker's disposal to mark the successive stretches on some of his routes in North Africa far exceeds what the relevant names and distance figures require. In consequence, stretches of exaggerated length occur both before Carthage on parchments 2 and 4 and after it on parchment 5.[34] At the same time the map's squatness makes it impossible to reflect the depth of territory traversed in various directions by the routes shown; the impression that they all proceed smoothly eastward is inevitably misleading. Such a false impression is even more patent in Egypt, where the map's shape offers no means to demonstrate the depth of the region, nor the south–north course of the Nile and the long route associated with it, as well as the southeast direction of the route from the river to Pernicide portvm.[35] The river and the route that follows it are both laid out from the left,[36] because the space here[37] was not needed for any other purpose. In addition, this placement permits those who "read" the map from left to right to engage with the river at the earliest opportunity, and to follow it from its source.

Beyond the Mediterranean, the mapmaker exhibits a less confident grasp of the landscape, and he is now seriously hampered both by earlier decisions and by the realization that there is an insufficient space for everything he would like to include here. He may also be calculating that his audience will be least concerned about the representation of this region, and thus willing

[33] 8C3, unnamed.

[34] Elsewhere, too, such stretches can be found in close proximity (along parchment 4 above the Adriatic's upper shoreline, for example, and above Antiochia at the right-hand end of parchment 9), but not to the same extent as in North Africa.

[35] 8C5 = Berenice.

[36] Originating in 8C1.

[37] Unlike to the right in 8C5 and beyond.

to tolerate its especially severe compression in the interests of balancing the coverage of territory to the left and right of the city of Rome at the center. After all, only by reducing yet further the coverage of lands east of the city (or even east of the Bosporus) within the Roman empire, can the design devote more space to lands beyond in that direction. By now the mapmaker has allowed the landscape to lose its cohesion, and it is hard to see how this could be regained. At the top, the uppermost of three routes from Trapezvnte (9A2) proceeds a considerable distance across the map to Sebastoplis (10A2). This is a route that should follow the Black Sea shore;[38] but instead it is placed entirely to the right of it. Below this route, and again to the right of the Black Sea, are laid out routes in eastern Asia Minor extended as far as Samosata (10B3) on the Euphrates. Meantime, further below are marked routes eastward from Damaspo[39] and Antiochia (9B4) to Palmyra (9C5), Zevgma (10C3), and Samosata, a layout severely warped by the decision to situate the eastern shore of the Mediterranean to the left of Antiochia. Damaspo is placed well to the left of Antiochia in consequence.

The fact that the different components of the larger region immediately beyond the Mediterranean fail to coalesce is only the prelude to graver shortcomings. Thereafter, beyond the Euphrates, Mesopotamia is hopelessly compressed, with its northern part (where Roman interests lay) forming the main focus of attention, and Babylonia[40] and Selevcia (10C4) both being situated close to the shore of the Indian Ocean. Regions further east are represented in token fashion at best; both Ecbatana[41] and Persepolis[42] are situated along the shore of the Indian Ocean. The Caspian Sea is shown,[43] but its situation is too remote to assist the mapmaker in presenting related cultural features in the way that he exploits the Adriatic and Black Seas. The Indian subcontinent, reduced to a modest size indeed, is notionally tilted like other landmasses already noted, so that its western shore (the only one documented here in the regular way) runs along the bottom of parchment 11. Ironically, however, at the same time as the mapmaker foreshortens the landscape so much, his presentation of routes beyond the Euphrates suggests that he is struggling against a lack of sufficient space for some of them. To an

[38] See *BAtlas* 87E4–G2; the route could be regarded as the counterpart, so to speak, to the western one from Tomis (7A4).
[39] 9C3 = Damascus.
[40] 10C4 = Babylon.
[41] 11C1 Ecbatanis Partiorvm = Ecbatana.
[42] 11C2 Persepoliscon Mersivm persarvm = Persepolis.
[43] 10A5–11A2 Mare Hyrcanivm (leading into Ocean).

unprecedented extent he now resorts to sloping these vertically rather than maintaining the regular horizontal layout; conceivably, he had not expected to have so much relevant data at his disposal here as turned out to be the case (fig. 14.7).

In the case of both mountain ranges and rivers it can prove difficult, if not impossible, to distinguish between those which are fundamental components of the physical landscape base and those more likely to have been added later after the placement of other features. The distinction is in any event artificial, and in itself of limited significance. More important is an appreciation that certain mountain ranges and rivers do act as defining landscape features along with shorelines. No route crosses the Apennines, for example, between Iria (3B1) and Adcalem (4B2). In addition to the Apennines, the fundamental mountain ranges are the Alps[44] and the Taurus,[45] and in all probability the Pyrenees too, although the map as it survives only preserves their eastern end (to 1B2).

The fundamental rivers without doubt include the longest, such as the Rhine, Rhône, Danube, Po, Tiber, Nile, Euphrates, Tigris, and Ganges. There is every likelihood that the mapmaker marked their courses on the map before any cultural features. How early some other rivers may have been marked, despite their prominence, is not so self-evident: among these are, for example, the Riger (1A5), Garvnna (1B5), [Savvs],[46] and Orontes[47] and the two unnamed ones flowing into the Black Sea either side of Trapezvnte (9A2). The case of the Patabvs—far from unique—serves to illustrate the quandary. It seems likely, but remains impossible to establish, that the mapmaker drew the course of the Patabvs in relation to that of the Renvs, *before* he laid out the pair of routes from Lvgdvno to Noviomagi[48] and the single continuation on to Colo Traiana and Veteribvs (1A5). In drawing the Patabvs, therefore, he may have been purposely demarcating space for these routes.

Altogether, despite our ignorance of the entire context in which the map was originally created, its surviving copy does still permit some appreciation of how the mapmaker addressed the multiple challenges of furnishing the coverage desired—centered upon the city of Rome in particular, and Italy

[44] These are initially a single range (from 2B2), which soon splits; the Apennines continue for far longer (to 6B2) than do the Alps (to 3A2, although with separate offshoots thereafter).

[45] 9B1–11B5.

[46] 3A5, named Fl. only.

[47] Unnamed but for the crossing at 9C5.

[48] 1A1–1A4.

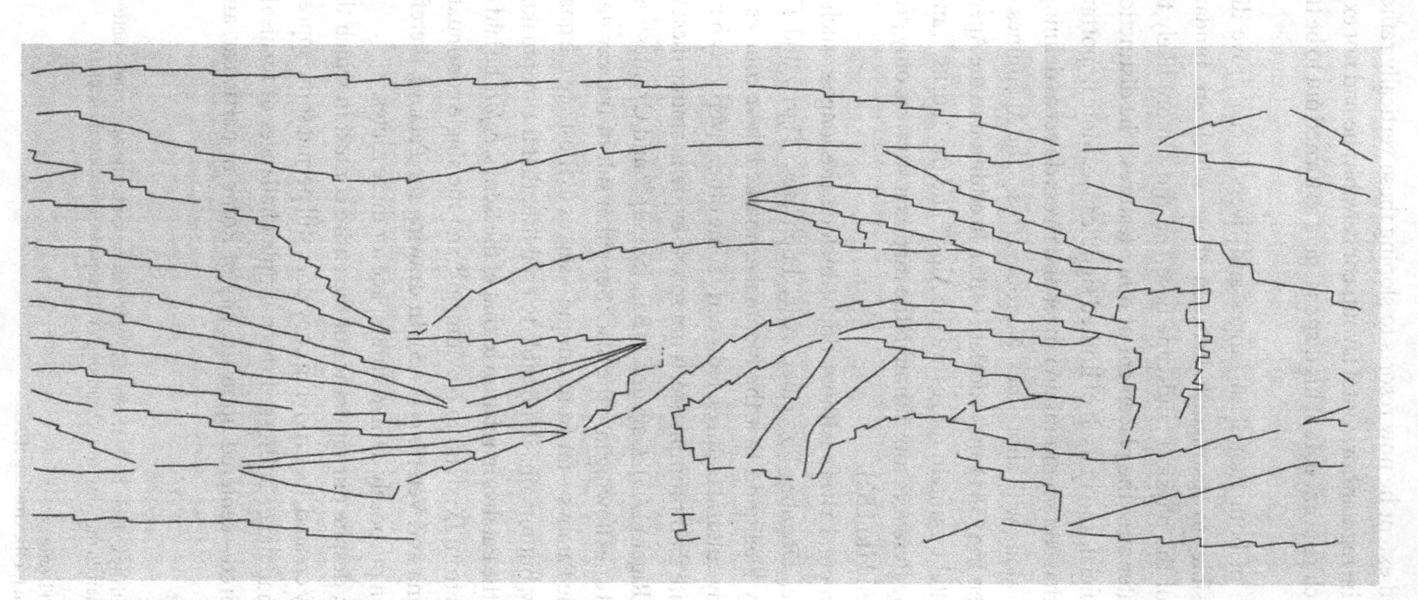

Fig. 14.7 Routes on Peutinger map parchment 10. Figure: Ancient World Mapping Center.

in general—within an extraordinarily elongated frame. He has concluded that certain sacrifices must be made and shortcomings tolerated, such as a lack of depth in North Africa and increasing compression of the lands east of the Bosporus. Some coasts and settlements are situated absurdly close to one another,[49] therefore, while others appear unnaturally far apart.[50] These are consequences of the map's shape, which informed viewers need to understand and allow for; some of them, of course, may even do so with relish.[51] Without doubt, the mapmaker must have exploited earlier maps which showed and named the principal elements of physical landscape with a fair degree of accuracy and without the artificial distortion which his own frame introduces. Whatever the ways may have been in which the map actually advanced cartography, in my view they are to be sought in its representation of cultural geography, not physical. Even so, there is no other Roman map which we can subject to the kind of cartographic analysis attempted here. From the physical aspect alone, it is surely appropriate to conclude that the mapmaker brings remarkable creativity and well-established principles to the achievement of his ambitious goal, and that he is working to please a sophisticated audience within the context of an active cartographic tradition. In consequence, here is good reason for a rethinking of the current widely held opinion that there was little place for mapmaking in Roman culture.[52]

[49] Italy's west coast close to Africa's north coast, for example; Verona (3A3) to Ivavo (3A4), Actanicopoli (6B3) to Leptimagna col. (6C4), Pergamo (8B3) to Alexandria (8C3).

[50] For example, Vesontine (2A1) to Avgusta Rvracvm (2A4), Verona (3A3) to Hostilia (3B4) and in turn to Ravenna (4B1).

[51] Additional limitations arise from the difficulty of accommodating the map's shape to certain principles for the presentation of cultural data. In particular, routes that run parallel to a river on the ground, or settlements situated on a river, cannot necessarily be shown thus on the map: note the instance of the Rhône (1B5–2A1).

[52] See further essay 18 below.

15

Copyists' Engagement with the Peutinger Map

The copying of the Peutinger map is a fundamental issue that should engage everyone who studies the map, even though (I suspect) most of its aspects cannot be definitively resolved. This contribution limits itself to offering some reflections on the copying, and therefore on how far we can, or cannot, rely upon the one surviving copy as the basis for our ideas about the original map.[1] The intention here is not to cast doubt on any particular interpretation of the map, but to urge that we each take the trouble to consider the issue of copying. Needless to add, many have wrestled with it already. Others who have not may hesitate to address it because of the potential unwelcome implications, and because the apparent absence of valid objective criteria causes such frustration. Almost everything that concerns the copying of the map turns out to demand subjective judgment, and there seems to be little hope of establishing a secure foundation on which to build a convincing interpretation.

Apprehension is justified, therefore. Even so, on no account should the issue of copying be evaded. Silke Diederich, in her excellent, painstaking overview of scholarship on the map, duly includes it in her list of most notable controversial issues in her introduction (my italics):[2]

> date of origin and stages of development, design, purpose, correctness and functionality, *mistakes in copying and medieval modifications*, relations to other maps and to written geographical sources.

Further reading of the overview reveals, however, that the only one of these controversies without a specific section devoted to it is "mistakes in copying

[1] For these issues, compare essay 11 above.
[2] Diederich (2018).

and medieval modifications."[3] This contribution is a preliminary step toward addressing the deficiency.

In the first instance, I dare to make an assumption beyond proof, namely that the survival we have is a copy of something, not itself an original creation; it had an exemplar, in other words. This copy—probably produced around 1200[4]—is incomplete at the left-hand end, and teems with more omissions, slips, and muddles than an original creation would seem likely to have.[5] It follows that this flawed state reflects a history of copying which is embedded in the copy. But if we then seek to deconstruct that history by stages or elements, objective criteria seem lacking. In particular, we have no pointer to how many copyings one after another separate our copy from the original. It is only right to acknowledge (I think) that, whenever a fresh copy was made of an existing copy rather than of the original creation itself again, more flaws were then likely to become embedded. Among the most damaging would be those that, when a further copy is made later, the new copyist would almost certainly be unable to correct, and may well not even be aware of: so, for example, a distance figure written XVI on an exemplar, when it was XVII on the original map.[6] In addition, quite apart from flaws that may stem from the shortcomings of copyists, there is the likelihood that they took the initiative to introduce additions and adjustments of their own; to identify these is again inevitably a subjective quest.

Still worse—thinking back to the lost original—is the prospect that it, too, already included mistakes at least of detail, even if its creator was no doubt satisfied with it overall. Thus it is quite possible that some flaws routinely attributed to later copyists may have been present from the very start—for example, the absence of a distance figure from a route-stretch. Plenty of these figures are missing,[7] and while the majority probably do reflect successive copyists' oversights, some could even have been missing from the original. We should be on our guard against idealizing the original map as perfect in every respect.

To be sure, it suits us to envisage the original map as a flawless achievement, which was then repeatedly reduced in quality by one copying after

[3] However, there is a "Vignettes" section.

[4] This broad consensus about its date is accepted by Steinmann in Talbert (2010), 83.

[5] For example, the number of rivers that lack a name is particularly surprising, many perhaps due to copyists' oversight; by contrast, far fewer islands lack names: see ibid., 104, 106.

[6] Ibid., 127.

[7] Stretches overlaid in orange on Talbert (2010), webpage Map A; the number of them in North Africa is notable.

another. But maybe any such drop in quality was not precipitous, especially if the original project to create the map was headed by someone with the skills and motivation of a graphic artist. Alternatively, a scholar with a focus on texts took the lead, but on balance I doubt that. There can be no denying that the involvement of scholars was essential. It had to be they who assembled and organized maps to work from in the first instance, together with all the detailed geographic data (physical and cultural), information about routes and distances, and so forth. Even so, the principal components of the map's overall design—the extreme shape, the remolding of landmasses, the featuring of land routes—are to me an artist's vision, and one intended primarily to make an expansive visual impact, not to achieve a practical or intellectual purpose. The map was never meant to serve as a suitable tool for planning long-distance journeys or military campaigns, let alone for teaching serious geography.[8] Its creator is someone who dares to abuse geography and scale drastically. This creator has also done an immense amount of selecting, and not just of details: Italy is featured more prominently than any other landmass, mainland Greece is shrunk, Egypt swiveled, vast stretches of open water removed, some principal land routes omitted, the widespread presence of the military ignored, and so forth.

This matter of the attitude and motivation lying behind the creation of the map in the first instance is important. So too is that of the surface on which it was drawn, one quite possibly different from the parchment of our copy.[9] Equally important is the question of just who is likely to have copied the map later, and for what purposes. The task was one for a graphic artist, even if it was undertaken at the instigation of a scholar who wished, say, to display the map.[10] It was artists who typically copied maps in the medieval period (and later).[11] Anyone who lacked prior experience in making or copying complex drawings of one kind or other would clearly be challenged to copy the map's physical landscape and linework. Conceivably, at least in some instances when the map was copied, an artist and a scholar collaborated (and maybe more than one of each). According to the verse dedication, such a partnership was formed for making a map presented to the emperor Theodosius II in 435.

[8] Compare, for example, the treatment of how geography was studied and taught in the early medieval West by Lozovsky (2000), 102–55; see further Deluz (2013).

[9] We can only speculate: canvas panels or a wall are obvious alternatives, even tapestry perhaps. See Talbert (2010), 144.

[10] Holes left by nails show that our copy was mounted for display at some stage between 1200 and 1500: ibid., 74–75.

[11] See Woodward (1987), 324.

Moreover, this pair proudly claim in general terms to have taken a proactive approach to their task:

> Supplices hoc famuli, dum scribit pingit et alter,
> Mensibus exiguis, veterum monimenta secuti,
> In melius reparamus opus culpamque priorum
> Tollimus ac totum breviter comprendimus orbem.
>
> Within a few months we humble servants—one of us coloring, while the other writes—restore[d] this work for the better, following memorials from the past, removing previous imperfection, and covering the entire globe in miniature.[12]

Our copy of the Peutinger map could represent collaborative work of the same type. Martin Steinmann, who has made the closest appraisal of the map from the perspective of paleography, remained unsure; sound arguments may be advanced both in favor of collaborative effort and against.[13] In any event, it is my opinion that, just as with the original creation of the map, an artist took the lead when the map was copied, not a scholar.

Whoever it was in the case of our surviving copy specifically, one manifest shortcoming here is carelessness in checking to ensure that linework has been drawn for route-stretches where the names and distance figures are in place—carelessness notably visible in fig. 15.1, for example.[14] To be sure, it may not be our copyist who was responsible for these omissions in the first instance. This linework may already have been missing on the exemplar, and our copyist then declined to take even the minimal initiative of restoring it—as presumably the pair working to please Theodosius II would have done, if we believe their claim quoted above.[15] The same uncertainty over who was responsible for a flaw, and when, surrounds the failure to correct some elementary mistakes or omissions in spelling, even when capitalization of the word makes it conspicuous: note, for instance, PATAVIA (1A1 for BATAVIA), MEDIA PROVI (4A2 for RAETIA PROVINCIA), CAPANIA (5B1 for CAMPANIA). "Campania" and "provincia" in particular are hardly

[12] *GLM* 20 lines 8–11 (but reading *priorum* in line 10).
[13] Talbert (2010), 82.
[14] See further ibid., 129–30. These stretches are overlaid in bright red on the associated webpage Map A.
[15] Some missing route linework has been added in faint pencil or the equivalent before and after the Fl. Rhamma stretch (10B5 with webpage Database s.v.). If this was done before the production of our copy was completed, the added linework was not then inked over in red: Talbert (2010), 85.

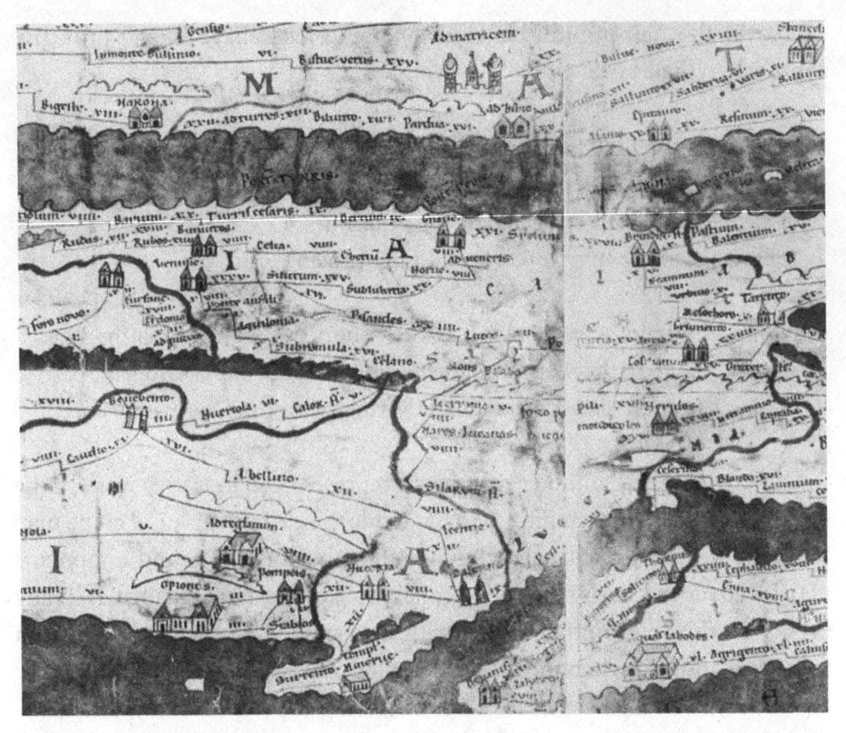

Fig. 15.1 Detail from Peutinger map parchments 5–6. Route linework is missing here most visibly between Narona and Ad zizio (top left); Gnatie and Brindisi (below Ad zizio); to left of Nervlos (below Brindisi); from Ceserina (below Nervlos) down left to Salerno. Photo: *Peut.* Map B.

an unfamiliar name or term, respectively, to anyone with a smattering of knowledge about the Romans and their empire.[16]

While our copyist or predecessors were for whatever reason indifferent to these matters of detail, they did take initiatives sooner or later to alter the earlier (or original) presentation of at least two, and probably three, major cities: Byzantium (or Constantinople) and Antioch, and surely also Tarsus (fig. 15.2a b, c). Our copy lacks a replacement symbol for Tarsus, but its symbols for the other two I imagine to be larger than they were originally (Antioch's has become even larger than Rome's), and there has hardly been

[16] It is true that efforts to correct some other mistakes or omissions in spelling have been made more or less visibly, but again their dating is obscure. Several have usually been thought to postdate the rediscovery of the map by Konrad Celtis: ibid., 84–85, 126.

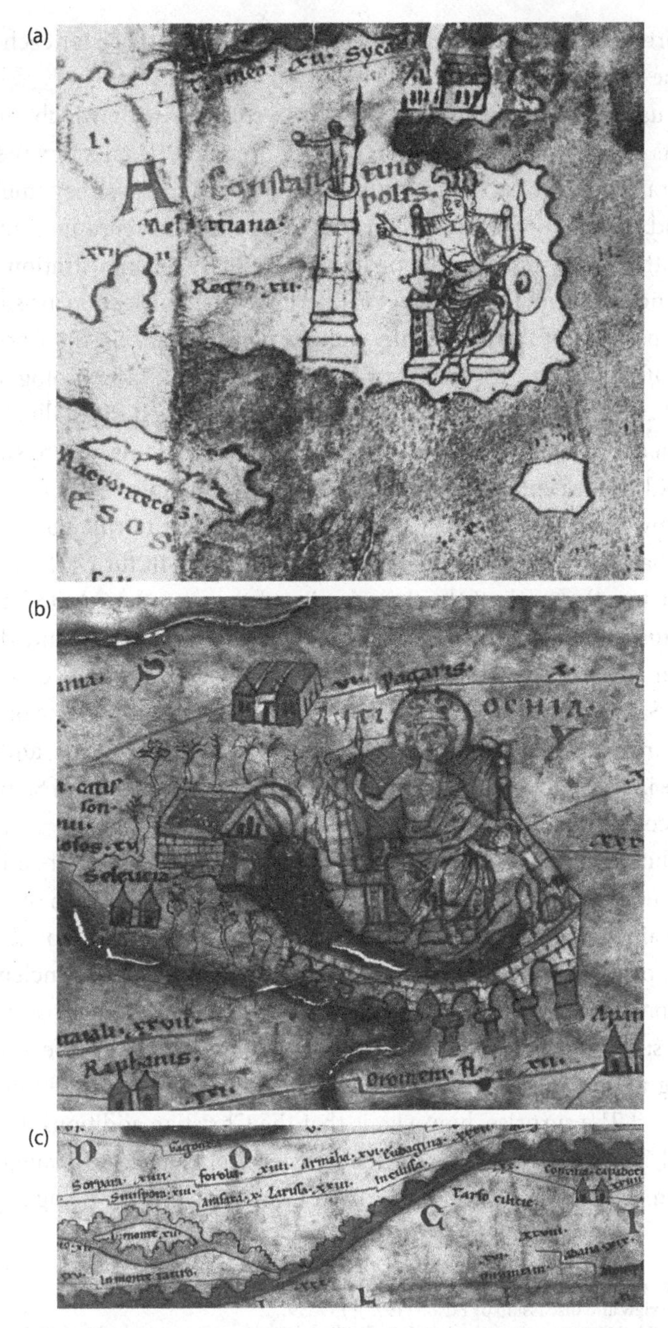

Fig. 15.2a, b, c The Peutinger map's symbols for Constantinople (a) and (Syrian) Antioch (b) on parchments 8 and 9, both evidently reflecting some redesign in which they are not reconnected satisfactorily to the route network. Likewise, an earlier symbol for Tarso cilicie (c, named upper right) on parchment 9 seems to have been dropped and space created for a larger replacement; but this still awaits insertion. Photos: *Peut.* Map B.

an effort to reintegrate either of the new symbols (or space for each) neatly with the route network.[17]

Let us not underestimate the lack of concern with which medieval copyists of maps made changes reflecting their artistic talent or their contemporary interests as well as those of whatever audience they might have in mind. It was by no means considered essential to reproduce an exemplar without any adjustments or improvements. A demonstration of how independently copyists could act is offered by the fourteen maps surviving in manuscripts of the late eighth-century commentary on the Apocalypse (*Book of Revelation*) by Beatus of Liébana in Spain.[18] The prologue to his book 2 confirms that he meant there to be a map (*pictura*) to illustrate the mission of the Apostles. However, the earliest copy of this map to survive is in a mid-tenth-century manuscript. So it is impossible to say how closely this copy reproduces the lost original map, let alone to rule out the possibility that this latter was itself a derivative creation in turn.[19] It is striking nonetheless that each of the three copies of the map thought to be closest to Beatus' own has a different design, and that each marks a quite different number of place-names: ninety-six on the mid-tenth-century copy (made for the San Miguel de Escalada monastery near Léon); 120 on the one in the manuscript dated 1086, now in the cathedral of Burgo de Osma; and 270 on the one in the manuscript produced around 1050 at the abbey of Saint-Sever in Gascony (fig. 15.3).[20]

Of those 270 names in the last case, however, about fifty are relatively recent ones of places marked within a specially large presentation of Gascony (with Saint-Sever's own symbol almost as large as Rome's!). To add recent names, or to give the contemporary name for a place whose ancient name is no longer in use, are typical enough initiatives on the part of medieval copyists; so, too, is it to reflect the impact of Christianity. I suspect that everything identifiably Christian on the Peutinger map is an addition to the original.[21] The surprise, however, is that the Christian additions do remain limited, and that almost no ancient names are changed to contemporary ones. In fact there is just a single case of an updated name being appended

[17] Casarodvno (1B3) has perhaps also lost a symbol it once had: ibid., 124–25.

[18] Overview and discussion by Edson (1997), 149–59.

[19] See ibid., 157, for the hypothesis of John Williams that even the map which originally illustrated Beatus' commentary was not his own, but was derived in whole or part from one that illustrated an earlier (fourth-century) commentary by Tyconius.

[20] Edson (1997) illustrates all three maps: pl. XI, figs. 8.3, 8.4.

[21] Talbert (2010), index s.v. Christianity; cf. Weber (2006).

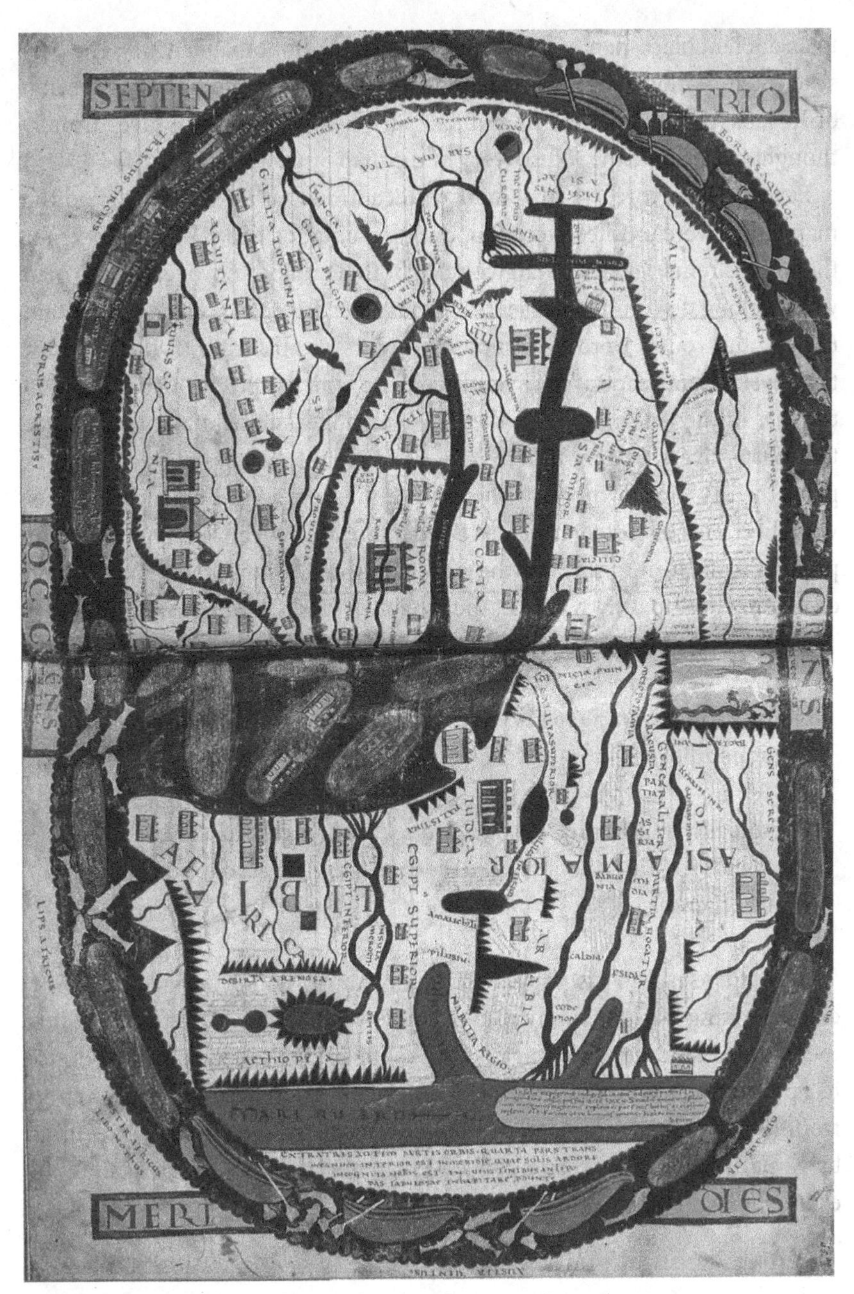

Fig. 15.3 Map (57 × 37 cm) created around 1050 at the Abbey of Saint-Sever in Gascony to illustrate the Apostles' mission in the commentary on the Apocalypse (*Book of Revelation*) by Beatus of Liébana. The large structure upper left here (surmounted by a cross) symbolizes Saint-Sever. The even larger one further down and closer to the center represents Rome. Bibliothèque Nationale de France, MS. lat. 8878, between fols. 45–46. Photo: Art Resource, NY (9441905).

to an ancient place-name: "Gesogiaco qvod nvnc Bononia" (1A2).[22] The persistent temptation to add such updates is nicely illustrated by the addition of Regenspurg below Regino and Salzpurg below Ivavo, initiatives usually thought to postdate the rediscovery of the map by Celtis.[23]

Perhaps, therefore, we may infer with caution that those who copied the map either were uninterested in updating names, or had special respect for the map as remarkable Roman work, or both. Perhaps, too, because the map was so elongated, and because its various lines for coasts, rivers, mountain ranges, and routes were so complex, copyists shrank from trying to recast the physical landscape as they ventured to do with the Beatus map. Rather, in the case of the Peutinger map, they grasped how essential it is to reproduce the coasts, rivers, and mountain ranges as precisely as possible, as well as to place the principal symbols with care.[24] Otherwise, if this extended framework becomes distorted, much of the detail within it can then no longer be placed satisfactorily. Such was the risk that Ptolemy himself warned of when discussing how to draw a map of the inhabited world (*oikoumene*):[25]

> Τό τε γὰρ ἀεὶ μεταφέρειν ἀπὸ τῶν προτέρων παραδειγμάτων ἐπὶ τὰ ὕστερα διὰ τῆς κατὰ μικρὸν παραλλαγῆς εἰς ἀξιόλογον εἴωθεν ἐξάγειν ἀνομοιότητα τὰς μεταβολάς.
>
> Continually transferring [a map] from earlier exemplars to subsequent ones tends to bring about grave distortions in the transcriptions through gradual changes.

It would be easy to cast aspersions on my impression of copyists' restraint, characterizing it as uncannily convenient, given our eagerness to accept that the physical landscape on our copy of the map does adequately reproduce its appearance on the lost original. Even so, my inferences can gain support from the attempts made to copy the map since 1508, when it was acquired by Konrad Peutinger—that is, once we can check each attempt against the exemplar.

[22] Recognition that helyacapitolina had previously been named Jerusalem is of course comparable (9C1). Also, in the case of a people, the formulation CHAMAVI QVIELPRANCI may be intended to record a later name (1A1–1A3).

[23] 3A4, with Talbert (2010), 84–85.

[24] See further essay 14 above.

[25] *Geog.* 1.18.2.

In the first instance it is self-evident why Peutinger chose not to commission a full copy of the map from either of the two (unidentified) artists who undertook to deliver a sample attempt. It was these two samples that Marcus Welser published in despair in 1591, when he thought the map itself could no longer be found (fig. 15.4a, b).[26] In fairness to both artists, we do not know how emphatically Peutinger had stated that he expected a precise copy from them. Certainly, neither artist seems to have aimed for that. Both have taken some trouble over coastlines and rivers, although not very methodically. Neither, however, has grasped the need to represent the routes as a series of stretches defined by sharp downward turns. Both have treated the symbols as open invitations to show their own artistic abilities, the first of the two (fig. 15.4a) notably enlarging them. Because of lack of concern to read the names on the map accurately, both make many spelling mistakes, and the first artist also converts the Latin numerals of the distance figures into Arabic ones. From these two samples alone, what might our assessment of the map be if it had remained lost, as Welser feared?

Peutinger did show sufficient confidence in a third artist, Michael Hummelberg, to commission a copy of the entire map from him. It was evidently made on eighteen sheets, with Hummelberg claiming not to have introduced any corrections or improvements; but he died in 1527 before delivering his copy to be engraved, and the sheets then somehow vanished.[27] So the first surviving copy of the entire map that we can check is the one drawn for Welser in Augsburg by Johannes Moller, as soon as the map was found again in 1597 (fig. 15.5).[28] This was engraved on eight copper-plates, and published at the end of the following year. One considerable limitation of this attempt is that it reduces the map to about half size. Even so, it is immediately apparent that linework has been copied with greater precision than on either of the samples commissioned by Peutinger.

Moreover, Welser claims that he rechecked the engraving against the map itself, and that no corrections or improvements were introduced, let alone mistakes made. This is exaggeration or delusion, however. Welser evidently felt no need to comment on the artistic license of settlement symbols being rendered as sixteenth-century-style turrets, but he does explain that it was impractical to reproduce lettering in the same forms as on the map.[29] His

[26] Talbert (2010), 14–17, with webpage pls. 3a, 3b.
[27] Ibid., 13.
[28] Ibid. 19–23, 173–74, with webpage pl. 4.
[29] These are in any case largely forms to be expected around 1200, not ancient ones: ibid., 81, 100.

(a)

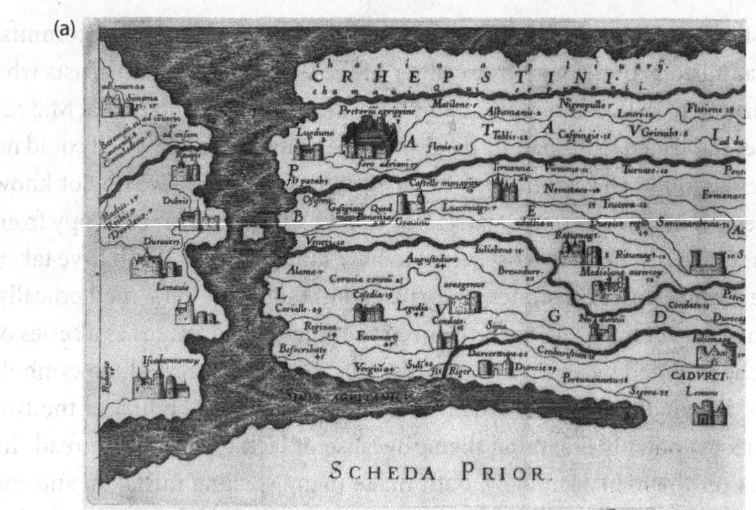

(b)

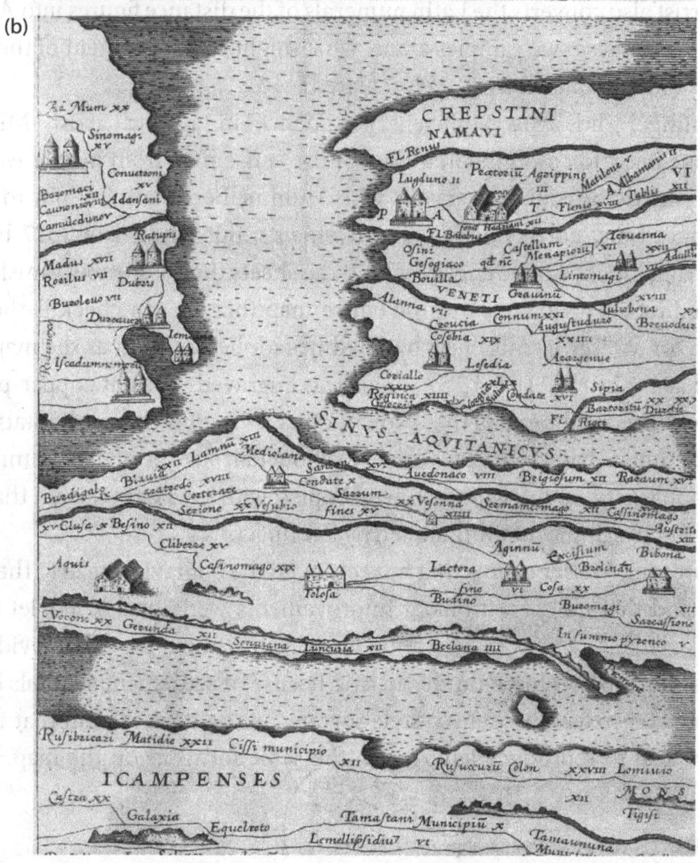

Fig. 15.4a, b Details from sample copies of the Peutinger map's left-hand end made early in the sixteenth century by two unnamed artists for Konrad Peutinger, and eventually published by Marcus Welser in 1591. Maps: Talbert (2010), webpage Plates 3a, b.

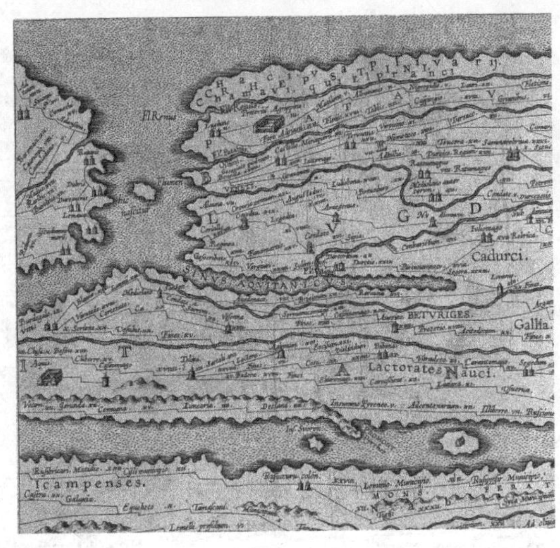

Fig. 15.5 Detail from the left-hand end of the half-size copy of the Peutinger map drawn by Johannes Moller for Marcus Welser, and published by Welser and Johannes Moerentorf (Moretus) in 1598. Map: Talbert (2010), webpage Plate 4.

(to us) most alarming step, however, was to introduce widespread "improvement" of his own by restoring all lettering that he found either illegible or missing, but without indicating just where he had made these improvements. Hence Konrad Miller in the late nineteenth and early twentieth centuries thought it safe to infer that Welser had been able to see at least some lettering which had faded since around 1600 and become invisible;[30] this claim only creates further controversy even when Welser's exemplar can be inspected.[31]

After Welser's half-size attempt, it turned out that only once more would the map be copied afresh directly. This attempt at full size, instigated by Franz Christoph von Scheyb, did not occur until the 1750s, after the map had come to the Imperial Library in Vienna (fig. 15.6).[32] We know that the artist he engaged for the task, Salomon Kleiner, was able to trace the linework by the intrusive method of placing a series of transparent oiled sheets over the map.[33] Von Scheyb, in his grand volume presenting the engravings made from these twelve sheets, boasted at length that he had repeatedly checked proofs for accuracy in every respect. Without doubt, Kleiner's linework is more precise

[30] Miller (1916), LXXVI.

[31] Welser's judgment must be called into question, however, when he is also found altering letters still plainly visible—further, no doubt well-intentioned, "improvement": Talbert (2010), 22.

[32] Von Scheyb (1753), reissued by Ronca (2009); Talbert (2010), 30–35, with webpage pls. 7.1–7.12.

[33] Ibid. 31. At this date, for better preservation the map—with a cloth backing glued to its reverse—was kept on a wooden cylinder which viewers could turn; its eleven parchments were not separated until 1863: ibid., 75–76.

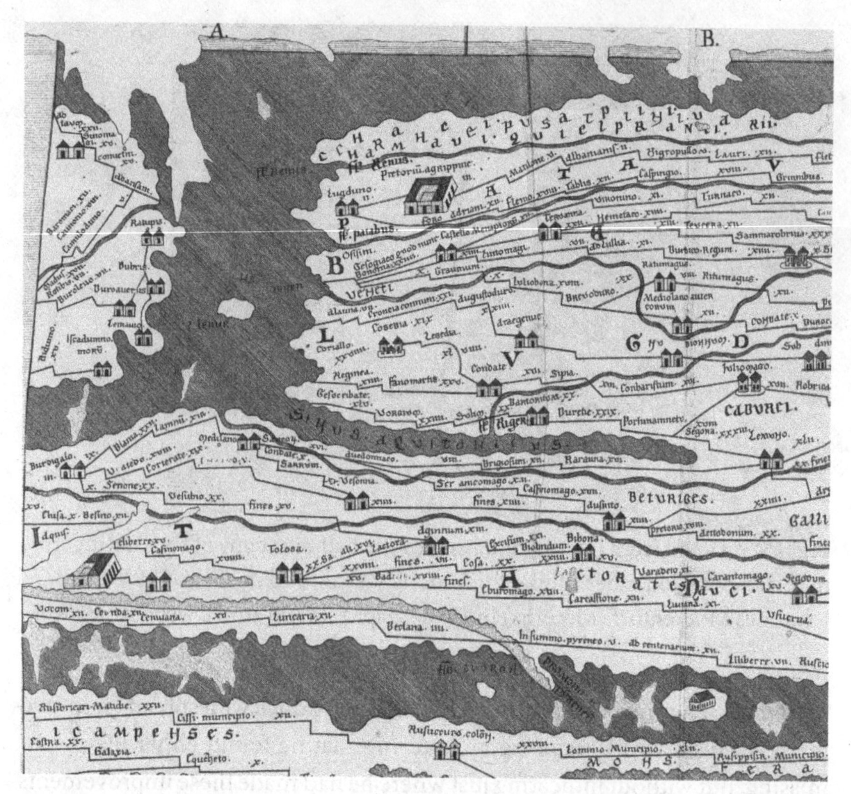

Fig. 15.6 Detail from the left-hand end of the full-size copy of the Peutinger map drawn by Salomon Kleiner for Franz Christoph von Scheyb, and published by von Scheyb in 1753. Map: Talbert (2010), webpage Plate 7.1.

than Moller's was (at half-size) in the 1590s, and it usefully confirms to us how sound a rendering of it could be made. To us, von Scheyb again gains credit because, unlike Welser, he felt no concern to improve the map at all. Where the copy falls short, however, is in von Scheyb's failure to live up to his own boasts. Kleiner evidently did not dare, or did not bother, to consult von Scheyb when he was unsure how to read lettering. Plenty of Kleiner's guesses or assumptions are wrong in fact, and many were not then caught by von Scheyb, whose checking was inadequate; he also overlooked omissions and other slips by Kleiner. So altogether the quality of this copying is mixed. It demonstrates just how difficult the map is to copy accurately in all respects, and raises the possibility—indeed likelihood—that there had been comparable copyings of mixed quality earlier in the map's long history, ones that are sure to have led to loss of data which could not be recovered.

As von Scheyb had intended, his full-size copy stimulated interest in the map, but this in turn triggered a gradual exposure of that copy's shortcomings, once others checked it with greater care.[34] An influential such effort was that of a priest and school principal in Ljubljana, Valentin Vodnik, who assigned two of his pupils to draw a full-size facsimile of von Scheyb's copy in 1809–1810.[35] This facsimile Vodnik then brought to Vienna in 1815 for comparison with the map itself. The seventy-seven corrections he proposed were taken into account for the revision of von Scheyb's copy that the Bavarian Academy in Munich commissioned from one of its expert Fellows, Conrad Mannert, in 1819. Yet Mannert, because of age and other reasons never specified, adamantly refused to go to Vienna to check the copy against the map itself. So instead the Library deputed this task to a very junior, inexperienced member of staff, Friedrich von Bartsch, whose work—done without guidance or supervision—proved far from adequate. But it was simply accepted without question, and the revision proceeded to publication in 1824.[36]

Not until the 1860s was Mannert's revision of von Scheyb's copy closely compared with the map itself by the French ancient historian Ernest Desjardins. He claimed that over 300 further corrections were needed; these he incorporated in the colored lithograph of the map (at full-size) that he published in instalments between 1869 and 1872 to accompany his commentary (never completed).[37] The lithograph's basis, however, remains von Scheyb's copy. Again, when Konrad Miller published his own lithograph (at two-thirds size) in 1888—in turn correcting slips by Desjardins that he found—it, too, was based on von Scheyb's copy.[38] Nonetheless, even after such repeated striving for accuracy, slips remain still. Notable among them is the route linework (without a distance figure) conspicuously linking Avgvsta Tavrinor(vm) and Eporedia (2B5). Although Welser showed this stretch, von Scheyb and Vodnik both overlooked it. Desjardins restored it, but Miller omitted it once more, failing to spot the correction (fig. 15.7a–f).[39]

[34] Note Joseph Heyrenbach's severely critical essay written in 1768: ibid. 37, 181–88.

[35] He took this extreme step because all his efforts to purchase von Scheyb's rare volume had failed, and he could only obtain one on short-term loan. By a miracle, the facsimile survives and is held by the National Museum of Slovenia, Ljubljana. Strangely, Vodnik allowed each pupil to maintain a distinctive style, without requiring consistency: ibid., 44, 175–80, with webpage pl. 8.

[36] Ibid., 36–40, 44–45.

[37] Desjardins (1869–1874). His poignant chance encounter in 1867 with von Bartsch (now on the verge of retirement) almost seems the invention of a dramatist: Talbert (2010), 57–60, with webpage pl. 13.

[38] See ibid., 62–63, with webpage pl. 15; Miller (1916), XXVI n. 1, for slips corrected.

[39] For further slips by Miller, see Calzolari (2003).

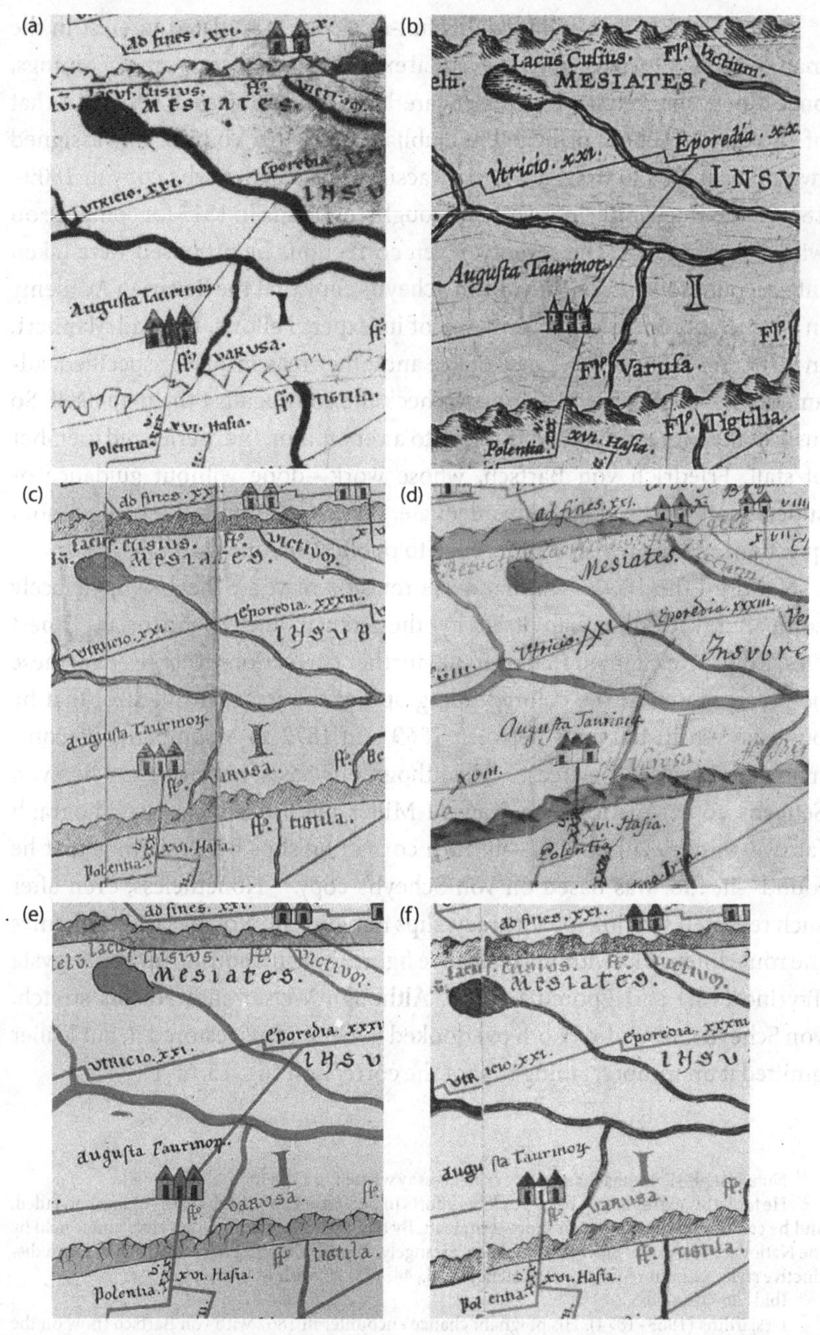

Fig. 15.7a–f Avgvsta Tavrinor(vm)–Eporedia area: (a) at the right-hand end of Peutinger map parchment 2; and as copied successively by (b) Moller/Welser (1598); (c) Kleiner/von Scheyb (1753); (d) Vodnik (1809–1810); (e) Desjardins (1869); (f) Miller (1888). Maps: Talbert (2010), webpage.

There should be no need to press further the point that it is a stern challenge to copy the map without making mistakes or omissions, even when the utmost care is taken and every effort made not to introduce modifications or improvements. The likelihood must be that the map was seldom, if ever, copied in these ideal circumstances before the late nineteenth century—ironically just when photographs (in grayscale only) could first be used instead, although Desjardins and Miller both resisted reliance on them.[40] Hence my call for skepticism about just how accurately our copy made around 1200 reproduces the original creation made centuries earlier. The need I perceive to hesitate over letter-forms, symbols, and anything identifiably Christian has already been mentioned. Similar hesitancy is warranted over the colors used for symbols and mountain ranges on the copy, as well as over the serration of coastlines, lakes, and mountain ranges.[41] Finally, I suspect that copyists added at least some "special notices" of the type *Hic cenocephali nascvntvr* (8C5), and even entire special features such as the SILVA MARCIANA and SILVA VOSAGVS, with their extensive embellishments (both on parchment 2).[42]

In all such instances, however, subjective judgment must be exercised, and inferences drawn that have to stem from individual scholars' visions of the character and purpose originally intended for the map. In this latter connection, moreover, there seems reason to imagine that whoever made the original map might anyway have been indifferent to the flaws in many details that we detect in our copy and fret over. This said, in the absence to date of any comprehensive focus on the map's flaws, a methodical cataloging and judicious analysis of these could make a potentially valuable addition to its study.

[40] Talbert (2010), 60, 70. A full-size photograph of each of the map's eleven parchments was published (no author) in 1888, but this set remained an expensive item to be found in few libraries: ibid., 63, 68, with webpage Map B. The next full-size set to be published—after nearly another century, and in accurate color for the first time—was that of Weber (1976).

[41] On the rare occurrence of blue, see essay 13 n. 41 above. Supplementary red linework drawn unsystematically alongside some river courses in parchments 1–4 gives the impression of being a copyist's initiative that was left incomplete: Talbert (2010), 76–77, 97–98, 125.

[42] See further ibid., 124–26, 165–66, and listing on webpage.

PART III
FROM SPACE TO TIME

16

Roads Not Featured

A Roman Failure to Communicate?

The Romans are naturally to be counted among those premodern societies long admired for their road systems. In their case, moreover, there is no shortage of testimony, both material and written. So it would seem a reasonable expectation that the means exist to determine what place roads occupied in the mindset, values, and self-identity of the Roman authorities and their subjects.

Up to a point, this expectation certainly seems to hold out promise. It is possible to trace how the system developed over several centuries, penetrating to every region of Rome's empire as it expanded immensely.[1] Romans constructed their first paved road—south from the city of Rome itself—in the late fourth century BCE, possibly taking Etruscan practice as their model. Half a millennium thereafter, by the late first century CE, the empire and its highways stretched from the north of England through most of Europe and across Asia Minor to the Euphrates river, as well as along the North African coast, far up the Nile valley, and over to the Red Sea (fig. 16.1). The system remained at this astonishing extent for a further three centuries and more, until the west of the empire could no longer be held against outside invasions.[2] Many Roman highways were paved, but not all were; surfaces varied according to local conditions (fig. 16.2a, b). The courses of highways, too, were not invariably marked by milestones, nor was the milestone a Roman "invention."[3] Even so, it is an accurate impression that paving and milestones did become two distinctive features of the huge accomplishment and investment which Roman highways represented.

[1] Note, for example, Staccioli (2003); and in more depth *BNP* 12 (2008) s.v. Roads V, cols. 622–47, with Wittke (2010), 194–99, 261–62. The puzzle of how the long, straight stretches of many Roman roads were planned is addressed by Lewis (2001), 217–45. On road construction, see further Quilici (2008).

[2] For changes in Late Antiquity, see McCormick (2001), 64–82; Belke (2008); Leyerle (2009).

[3] Such markers were already to be found along Persian highways; on these, and their Assyrian forerunners, note the valuable overview by Silverstein (2007), 7–28, especially 16.

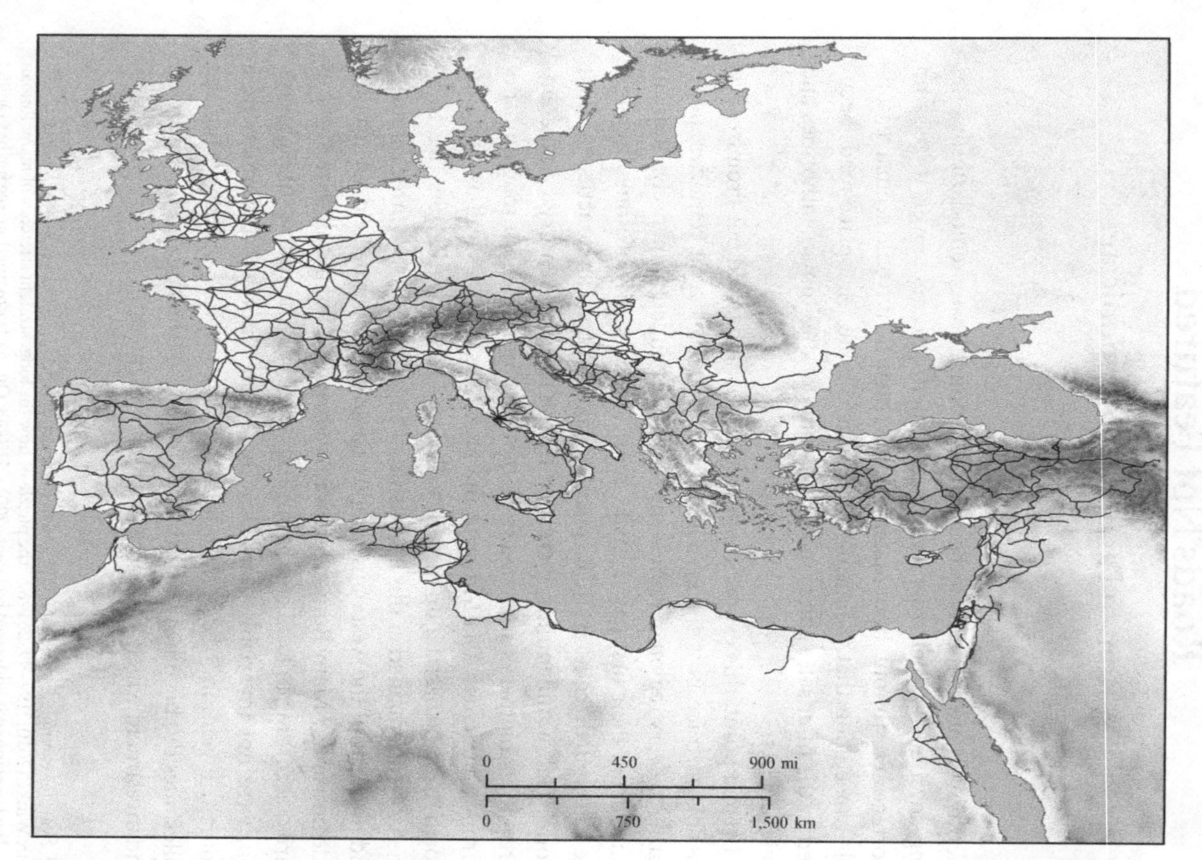

Fig. 16.1 The Roman empire's highways around 100 CE outlined on a modern map base. Map: Ancient World Mapping Center.

Fig. 16.2a, b (a) Paved Roman highway: the Via Egnatia near Philippi in northern Greece. Photo: Art Resource, NY (81486). (b) Unpaved Roman road in Egypt's Eastern Desert: a gravel surface cleared of boulders for a width of approximately 8 m. Photo: Jackson (2002), 57.

Seven to eight thousand of these milestones survive—spanning several centuries, scattered very unevenly across the empire. They, and their inscribed record, have understandably attracted intense scholarly attention.[4] As has been recognized, any modern assumption that they were designed primarily to serve as aids to travelers on their journeys may be erroneous. Instead, it would seem that the principal purpose was more often to promote the image of the official or emperor responsible for constructing a new stretch of road or for making major repairs to an existing one.[5] Take the typical instance that figure 16.3 represents. Here it is the name and many titles of the emperor (Trajan) who commissioned the construction early in the second century CE that are intended to dominate viewers' attention, together with the affirmation below in smaller lettering that *his* money funded the work all the (considerable) way from Beneventum to Brundisium.[6] In addition, the clear confirmation that at this stage travelers have reached (mile) 79 is important; but notably missing is any mention of the further distance to the road's end-point.

Of course, to learn anything from a Roman milestone, travelers needed a modest level of literacy (not widespread among the population) and working knowledge of Latin. Increasingly over time, Latin was not a source of difficulty within Italy (where the milestone illustrated was erected). However, it is liable to have remained so elsewhere, in particular throughout the east of the empire, where Greek was the shared language of educated people and where many other "pre-Roman" languages and cultures continued to be widespread, such as Syriac and Aramaic. Nonetheless, the overwhelming majority of Roman milestones throughout the east are inscribed only in Latin, relentlessly proclaiming the ruling authority's achievement in its own alien language and script, and recording distances in its alien unit of measurement.[7] To be sure, some travelers will have felt informed, proud, and safer as a result;[8] others, by contrast, will have had their uncomprehending resentment reawakened every single mile.

[4] Kolb (2004, 2007), with essays 7 above and 17 below.

[5] Witschel (2002).

[6] *BAtlas* 44G3–45G3.

[7] See further Isaac (1992), 304–309. Some milestones inscribed in both Latin and Greek are found in Asia Minor; from the late second century CE, distance figures commonly appear in both languages on milestones in Judaea, Syria, and Arabia.

[8] As both the rhetorician Quintilian in the late first century CE (*Institutio Oratoria* 4.5.22) and the poet Rutilius Namatianus in the early fifth (*De Reditu Suo* 2.7–8) reflect, the weary traveler can be soothed by reading the distances on successive milestones.

Fig. 16.3 Milestone 79 on the road from Beneventum to Brundisium in southern Italy constructed by the emperor Trajan (*CIL* IX.6021). Photo: Lawrence Keppie.

Contemporary Comment

Given the pervasiveness and undeniable importance of Roman roads, a search for contemporary commentary on them as a phenomenon or system yields surprisingly little. Next to nothing overtly hostile seems to have survived,[9] and almost all the favorable comment is articulated by Greeks

[9] Among the rebel Britons' grievances in the 80s CE was supposedly the compulsion to build roads through forest and marsh under harsh conditions, although the sole testimony is a speech composed by the Roman historian Tacitus (*Agricola* 31.2). According to the Babylonian Talmud, a

adopting an outsider's perspective on Roman civilization. The literary critic and historian Dionysius of Halicarnassus is the earliest, writing at the end of the first century BCE. He expresses in passing the personal opinion that the three most impressive forms of Roman construction, best exemplifying the greatness of Rome's hegemony, are the channeling of water, the paving of roads, and the laying of sewers.[10] Not long afterward the geographer Strabo also praises these same three as distinctively Roman, declaring: "They have constructed roads through the countryside by adding both cuts through hills and embankments over valleys, so that wagons carry as much as a boat."[11]

Early in the second century CE, the biographer Plutarch in his *Life* of the second-century BCE reformer-tribune Gaius Gracchus has special praise for his road-building program. Gaius, he says:[12]

> . . . ensured that the roads were not only functional, but also aesthetically pleasing and attractive. They were to run perfectly straight through the countryside, with a surface of quarried stone firmly bedded in compressed sand. Depressions were filled up, bridges were thrown across every watercourse or ravine which intersected the route, and from one side to the other the roads were made flat and level, so that the work presented an even and beautiful appearance throughout. He also measured every road in miles (a mile is a little less than eight stades), and set up stone columns as distance-markers. At smaller intervals he also placed further blocks of stone either side of the road, to make it easy for riders to mount their horses from them without needing a leg-up.

Plutarch's characteristically double compliment—admiring both function and form—must surely reflect his own experience of traveling along Roman roads.[13] The same may be said of the sophist Aelius Aristides in the mid-second century CE. His oration *To Rome* praises the freedom and ease of travel that Roman rule has brought (101):

second-century CE rabbi counters a fellow rabbi's praise of Roman marketplaces, baths, and bridges (and thus, by extension, roads) by declaring: "everything they have made they have made for themselves: marketplaces for whores, baths to wallow in, bridges to levy tolls"; see De Lange (1978), 268.

[10] Dion. Hal. 3.67.5.

[11] Strabo 5.3.8; cf. 4.6.6.

[12] 28.7. Translation adapted from Waterfield (1999), 105.

[13] This experience may equally have been in his mind when he describes (or imagines) the youthful Alexander hosting Persian envoys to Macedon in the absence of his father, King Philip: Alexander puts to them a stream of notably mature questions, the first of which inquires about road distances and traveling conditions for a journey into Persia (*Alex.* 5).

> On the one hand you [Romans] have surveyed the whole inhabited world [*oikoumene*], on the other you have spanned rivers with all kinds of bridges. By cutting through mountains you have made land travel feasible; you have filled the deserts with way-stations, and you have civilized everything with your lifestyle and organization.

Road-building as an integral component of the spread of Roman civilization is singled out for conspicuous praise on more than one monument in Lycia, following the Roman annexation of this mountainous region of southern Asia Minor in the mid-first century CE (among the last regions with a Mediterranean coastline to be so annexed). In particular, one of the two Greek texts on a tall rectangular pillar at the harbor of Patara first proclaims that Claudius, "emperor of the inhabited world [*oikoumene*]," has—through his governor on the spot—constructed roads throughout Lycia.[14] It then lists the distances (in stades) between at least fifty places there, as well as onward to one or two others in the neighboring province of Asia (fig. 16.4). Equally, a roadside altar in the territory of Limyra records a dedication to Claudius by the Lycians (again in Greek) expressing their gratitude for peace and road-building.[15]

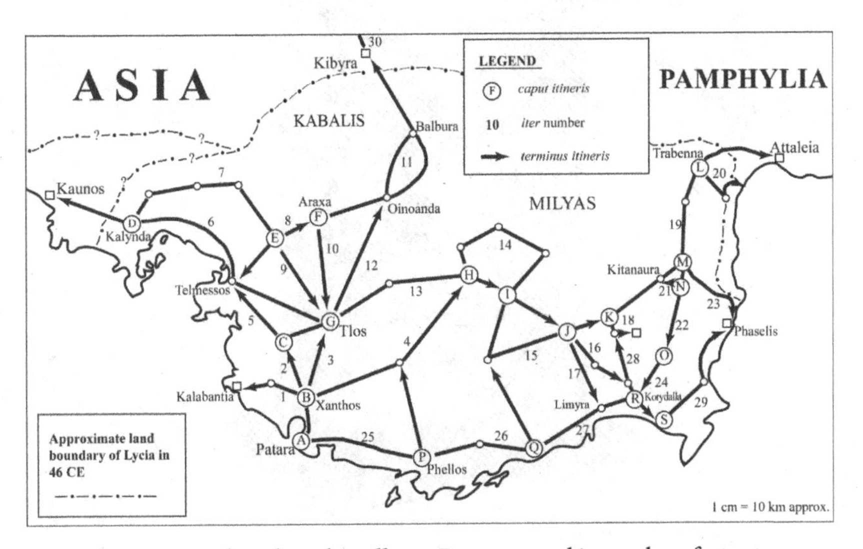

Fig. 16.4 Routes as listed on the pillar at Patara traced in modern format. Map: Salway (2012), 210 (modified).

[14] Şahin and Adak (2007); Grasshoff and Mittenhuber (2009).
[15] *SEG* 52 (2002), 1438.

Given Claudius' emulation of Rome's first emperor, Augustus, it is conceivable that the Patara pillar was meant to recall the so-called *miliarium aureum* or "golden milestone" (now lost) that Augustus set up in the forum at Rome in 20 BCE, evidently also in the form of a pillar.[16] Certainly at Patara, and perhaps at Rome earlier, too, the information inscribed on the pillar must be reckoned to fulfill more of a symbolic rôle than one genuinely useful to travelers, because much of it was set too high up for viewers to read (fig. 16.5).

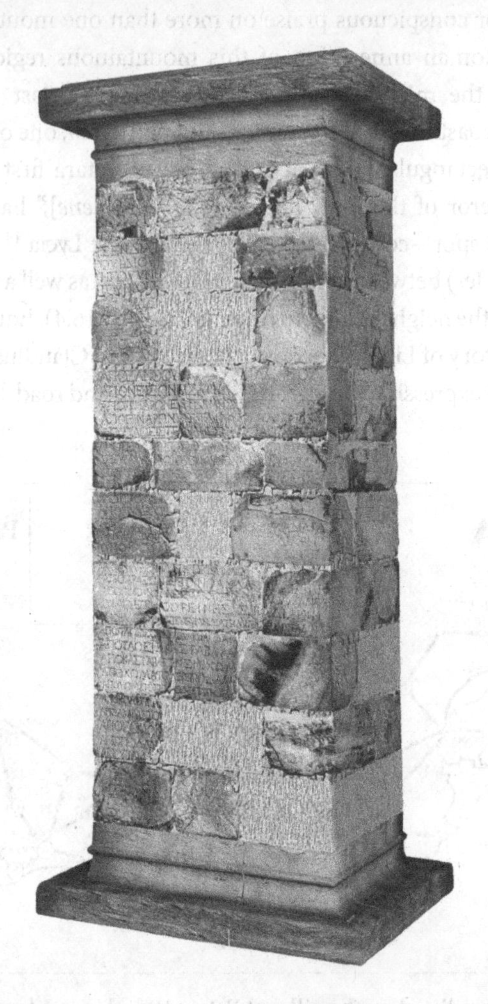

Fig. 16.5 Three-dimensional rendering by Fatih Onur of the pillar at Patara, featuring the face where the first thirty-nine route-stretches were listed. Approximate dimensions: 1.6m wide × 2.35m deep × 5.5m high. Figure: Salway (2012), 209.

[16] Plutarch, *Galba* 24, with *LTUR* 3 (1996), 250–51, s.v. *Miliarium aureum*.

Fig. 16.6 Arch erected to Augustus at Ariminum (modern Rimini)—end-point of the Via Flaminia across the Apennines from Rome—to commemorate his repair of Italy's highways. Photo: Art Resource, NY (89714).

Augustus erected the golden milestone as part of a long-term commission to restore and manage the roads of Italy.[17] We find this work of his variously recognized. In 27 BCE, the senate and Roman people erected an arch to him at Ariminum (modern Rimini), "[because thanks] to his initiative and funding, the Via Flaminia and the rest of Italy's most celebrated roads have been repaired" (*[mu]niteis*) (fig. 16.6).[18] Likewise, the legend on silver coins issued between 18 and 16 BCE runs: "Because Roads Have Been Built" (*munitae*) (fig. 16.7).[19]

[17] The modern misconception that the *miliarium aureum* was intended to serve as a nodal point for the roads of the entire empire, rather than just of Italy, is widespread. United States president Warren Harding reflected it in his speech at the dedication of the "Zero Milestone" in Washington, DC, on June 4, 1923: see www.fhwa.dot.gov/ infrastructure/zero.cfm.

[18] *ILS* 84 = Cooley (2003), 254, no. K68.

[19] Cooley (2003), 255, no. K69. The Latin verb *munire* can signify both construction and repair (as may be the intention here).

Fig. 16.7 Reverse of one of several silver coin-types issued between 18 and 16 BCE celebrating Augustus' construction and repair of roads: QVOD VIAE MVNITAE SVNT. The design recognizes the speed of travel now attainable, and the importance of bridges or viaducts—two benefits later praised by the poet Statius (*Silvae* 4.3). Photo: American Numismatic Society (1944.100.39075).

Muted Imperial Engagement and Possible Explanations

Remarkably, Augustus' work on the Via Flaminia is the only road-building or repair anywhere that he chooses to mention in the extensive record of his achievements that was set up in prominent centers across the empire after his death in 14 CE.[20] In consequence, Augustus misses what would seem to be an ideal opportunity to convey how his concern for the road system (both in Italy and far beyond), and his control of it, combine to reinforce his grip on the empire. It would appear that his successors, in turn, never articulate this point explicitly either, let alone feature it as a matter of pride or celebration on, say, inscriptions or coins.[21] Nor are Augustus' successors praised explicitly in such terms by others. It is true that around 100 CE both the court poet Statius in Latin and the orator-philosopher Dio Chrysostom in Greek each happens to affirm to the emperors Domitian and Trajan, respectively, that it is characteristic of the "good ruler" to build roads for his subjects.[22] In this way,

[20] *Res Gestae* 20.5.

[21] Such restraint is noted by Kissel (2002), 149. Indeed, it applies to public works in general funded by the imperial authorities, as observed by Harris (2003), 296, rebutting Schneider (1986).

[22] Stat. *Silvae* 4.3.20–23; Dio Chrys. *Or.* 3.127.

the poet and the philosopher are choosing to promote the so-called "beneficial ideology" for which second-century emperors gained special praise. At the same time, both authors perhaps deliberately stop short of formulating any unequivocal statement that recognizes imperial control of roads.

Such silence and restraint—if correctly suspected—seem surprising and in need of explanation. By contrast, in the case of other imperial states with a highway system, neither rulers themselves nor observers of them have been slow to express their realization of how vital the control of highways was to the maintenance of the regime. Writing in the fourth century BCE, the Greek observer Xenophon stated without hesitation that he saw the rapid delivery of comprehensive intelligence to the Great King as the primary function of Achaemenid highways.[23] Comparable insight underlay Ibn Khurradādhbih's presentation of a detailed *Book of Routes and Kingdoms* (Kitāb al-masālik wa al-mamālik) to the Abbasid caliph in the mid-ninth century. It specified among much else exactly how many stations there were and how many dinars the upkeep of the entire centralized system cost.[24] In the tenth century, lands and routes across them were a particular concern of the so-called Balkhī school of Islamic cartographers.[25] Japan's Tokugawa rulers, too, considered it vital to maintain control of the country's overland transport. In consequence, travel along the Gokaidô, or Five Highway, system with its post stations was tightly regulated.[26]

It would be a mistake to imagine that Augustus and his successors during the next three centuries somehow altogether failed to grasp the potential value of the Roman system both to themselves and to their enemies (within the empire and beyond) for the rapid delivery of intelligence and the movement of troops. According to the contemporary officer and historian Velleius Paterculus, Augustus was so shaken by the scale and effectiveness of the Pannonian rebellion on the Danube in 6 CE that he expressed to the senate his fear of the enemy reaching Rome itself within ten days.[27] Later, the biographer Suetonius explains in his *Life of Augustus* 49:

> And so that events in all the provinces could be more speedily and promptly reported and known, he first stationed young men, and later vehicles, at

[23] *Cyropaedia* 8.6.17–18.
[24] See Silverstein (2007), 90–97; De Geoje (1889), 1–208 offers a French translation.
[25] See Tibbetts (1992b); Rapoport and Savage-Smith (2008), 130–32.
[26] See Vaporis (2012), with essay 17 and fig. 17.3 below.
[27] Vell. Pat. 2.111.1. A comparable, but no more than general, reference to the potential importance of roads for effective campaigning is made around the same period by the Greek geographer

> moderate intervals along the military roads. The latter arrangement seems more convenient as it means that the men who bring letters from a particular place can also be questioned, if the need arises.

Notable here is Augustus' evident improvement of his initial arrangement, as well as Suetonius' use of the loose term *militaris via* (military road). This term is widely attested in different types of source from the 50s BCE onward to signify a road made for the army, by the army, or both.[28] It is important to understand, however, that neither the Roman army nor the courier service (*vehiculatio*) that Augustus instituted for government business ever enjoyed exclusive use of any highway. It is true that a traveler needed an official permit, or *diploma*, for the right to claim accommodation, animals, and vehicles at stopping-points; the authorities found themselves constantly struggling to prevent fraudulent claims and the exploitation of poor folk by this means.[29] Even so, travel itself was genuinely unrestricted throughout the Roman empire for almost all free civilians, male and female. There was next to no regulation or taxation of movement as such,[30] nor of the types of traffic or vehicle allowed on a road, as was clearly the case in Japan and elsewhere by comparison.[31] This said about Japan, it is also clear that such controls there were far from being very repressive in practice, let alone uniform. But still the very notion of comprehensively regulating civilians' freedom of movement represents a fundamental contrast. To the best of our knowledge, no such attempt was ever contemplated by the Roman authorities; the empire was too vast, and the Roman administrative presence too thinly spread across it. That guards were placed on Italy's roads and even paths by the emperor Maximinus' opponents in 238 CE—with a view to blocking his return there from the north, as well as the transmission of any news to him—was an exceptional precaution worthy of record.[32]

Strabo (1.1.17), who draws attention to the skill of "Germans and Celts" in concealing from Roman forces the whereabouts of such roads as existed in their otherwise trackless barbarian wasteland. Strabo again (16.4.24) stresses how the campaign into Arabia led by his friend Aelius Gallus had been hampered by the lack of roads there.

[28] See Rathmann (2003), 23–41.

[29] See Mitchell (1976), overlooked by Silverstein (2007); note also, for example, Hauken (1998), 74–137 (Skaptopara inscription, 238 CE), and, among more recent discoveries of further such instances, Hauken and Malay (2009).

[30] Some exceptions (in certain frontier zones especially) to the absence of regulation are discussed by Moatti (2000). The evident control of movement in and out of Egypt was untypical, but it hardly seems to have extended to travel within the province: see Adams (2001), 157–58.

[31] See further essay 17 below.

[32] Herodian 8.5.5.

Augustus' introduction of restrictions on the movement of the group from which he had most to fear, his 600 or so fellow senators, plainly indicates his awareness of the constant risks to his rule from rivals. Senators were banned altogether from the strategically sensitive new province of Egypt annexed in 30 BCE; and whenever any senator wished to travel elsewhere outside Italy, he was required to gain permission in advance from the emperor.[33] Cassius Dio, a senator of the early third century, states that this requirement still applied in his own day,[34] although it is impossible to say how methodically it was enforced over time. It is no coincidence that during Augustus' rule, too, exile became a standard sentence for a senator convicted of some high crime. In fact two levels of this sentence were instituted: comfortable exile (*relegatio*), potentially for a limited period only; and harsh exile (*deportatio*), in principle for life. As a further innovation, both levels required residence on an island designated by the authorities, no doubt in order to keep the exile under closer surveillance, and to limit access to him more effectively than might be feasible at any mainland location on a highway or near one.[35]

In referring to Augustus' exclusion of senators from Egypt, the historian Tacitus terms this ban as being among the *arcana dominationis*, "the secrets of his domination."[36] Silence on the part of Augustus and his successors about the road system and their control of it could be reckoned as deliberate concealment of another such *arcanum*. It is also conceivable that the system, for all its conspicuousness, was so taken for granted as an everyday feature of the Roman environment that it did not occur to the authorities to exploit its propaganda potential, except on a local basis with milestones or, in rare instances, an arch. Of course, these two possibilities need not be mutually exclusive.

We may well imagine that no emperor would be eager to raise public awareness of the overall excellence of his empire's road system in ways that might come to the attention of enemies across the Rhine, Danube, or Euphrates rivers and only serve to encourage them. Equally, the more the emperor promoted *his* control of *his* system, the greater the likelihood that the empire's communities might press for *him*, therefore, to organize

[33] See Talbert (1984), 139–41, 515. Claudius as censor gave offense by evidently trying—without advance notice—to extend the restrictions on senators' absences from Italy to *equites*, the much larger class of wealthy Roman citizens who ranked immediately below senators (Suet. *Claud.* 16).
[34] 52.42.6.
[35] Drogula (2011).
[36] *Ann.* 2.59.

(a)

(b)

Fig. 16.8a, b Even though not widely attested, the norm for representation of a road evidently became a female figure reclining on a wheel, presumably a match for the typical personification of a river. (a) This silver coin reverse celebrates Trajan's construction of the new VIA TRAIANA from Beneventum to Brundisium: for its milestone 79, see fig. 16.3. Photo: American Numismatic Society (1905.57.344). (b) This panel—originally from a triumphal arch of Marcus Aurelius, and later reused in the Arch of Constantine at Rome—depicts a "departure" ceremony (*profectio*). The female figure (lower right) who reclines on a wheel and greets the emperor is taken to personify a road, probably the Via Flaminia, along which he would travel north from Rome (cf. fig. 16.6). Photo: Art Resource, NY (383449).

and fund its day-to-day maintenance, as well as that of the official courier service. During the first three centuries CE, in fact, these were unwelcome burdens largely passed on to the communities, along with the provision of multiple other services.[37] At this period, except in remote frontier zones of strategic importance, the Roman authorities typically kept their administrative responsibilities to a bare minimum. It was in part Rome's sheer dependence that caused emperors to maintain an attitude of *civilitas* toward cities and peoples, respecting them and fostering local pride, rather than cowing and alienating them with arrogant affirmations of power. To boast of controlling roads as opposed to, say, conquering a people or region, had in any case never featured among traditional Roman means of flaunting and legitimating power.[38] In addition, it would seem that no visually compelling form of image for representing a road—an awkward challenge—was ever developed (fig. 16.8a, b).[39]

Limits of Conceptual Awareness

More fundamentally, there is cause to question how far even emperors or their better-educated subjects were in the habit of conceptualizing the empire's highways as a "system." To be sure, this is a perspective unhesitatingly adopted by modern students of the Roman empire. To them, both long-distance road travel and global cartographic awareness are routine, while centralized control of a state's infrastructure by its proactive government is to be taken for granted. The Roman mindset was quite different, however. Long-distance travel of any kind was perilous, and Augustus may not have been alone in his preference for making journeys by water whenever possible rather than overland.[40] Of like mind, evidently, were the Praetorian Guardsmen shipped out (it seems) in haste from Rome to Spain in order to escort the new emperor Galba overland through the Pyrenees and Alps

[37] Note in this connection Kissel (2002), 133–46, 153–55; Potter (2006), 186–88, 235–40, 254–55, 275–78.

[38] Notably, the Republican practice of naming a road after the magistrate responsible for its construction declined in the imperial period. Only the emperor's name could be used then, and the practice occurs relatively seldom. See *BNP* 15 (2010), s.v. *Viae publicae*, cols. 373–80.

[39] Gerhard Koeppel's interpretation (1980) of a zigzag band on Trajan's Column as a route (*itinerarium*) taken by troops in the Dacian War is imaginative, but by no means compelling.

[40] Suet. *Aug.* 82. For a contrary attitude, consider the *scholasticus* Theophanes discussed in essay 10 above: he risked Nile riverboats but was seemingly averse to a sea voyage for any part of his return journey from Egypt to Syrian Antioch.

back to Italy in 68 CE. This march by road, in armor, over a huge distance—"immensa viarum spatia" are referred to—became a ready source of complaint.[41] Moreover, in general Romans' cartographic awareness was minimal, and they appear to have had little practical use for maps, except perhaps very local ones of the landholdings within their community.[42] Nor was there any kind of "Imperial Roads Department" for either construction or maintenance; there is no clue to how such work was actually initiated.[43]

Consideration should be given to the possibility that it may not have been typical for Romans to conceive of their empire having a single connected system of roads. The one instance known to me where such a reference occurs (without recourse to any special term) is poetic. Statius (the "court poet" mentioned above), in a eulogy for a deceased ex-slave who rose to be the emperor's chief financial accountant (*a rationibus*), lists among his budgeting responsibilities the needs of "the far extended network of roads."[44] There are occasional instances in the provinces where a milestone does state the distance from faraway Rome itself: one at Savaria (modern Szombathely, Hungary) records 675 miles.[45] Even so, the strong impression from such notices and much other varied testimony is that most of the empire's inhabitants had only a local perspective, or at a stretch a regional or provincial one; the need to think more broadly than this seldom arose for them in any case.[46] At the same time, decisions by emperors or their staffs on where to invest heavily in road repair can by no means always have been governed by strategic priorities. In that event, Trajan would never have authorized such a massive repair program in Spain in the early second century CE, for example, nor would Maximinus have authorized a similar one in Africa during the 230s.[47] In neither instance was Spain or Africa a region of strategic concern at the time, whereas others were. Needless to add, the empire had no Pentagon or military staff college. The claim made by Brian Campbell in his *Rivers and the Power of Ancient Rome*—"the Romans had no empire-wide strategy or policy for using rivers consistently as some kind of military control"—applies equally to roads.[48]

[41] Tac. *Hist.* 1.23.
[42] See further essays 2 above and 18 below.
[43] See further essay 1 above.
[44] *Silvae* 3.3.102, "longe series porrecta viarum."
[45] *BAtlas* 20D2, with Kolb (2007), 172–74.
[46] See essays 7 and 8 above.
[47] Rathmann (2003), 229, 254.
[48] Campbell (2012), 198.

Comprehensive route data for the empire's highways does not seem to have been available, and quite possibly was not gathered and organized systematically.[49] A few highways happened to be named, but randomly; there was never coordinated naming of them, let alone any numbering. Nor does even the merest hint survive of a handbook comparable to the one by Ibn Khurradādhbih mentioned above. It is conceivable that emperors purposely blocked the production of such a useful reference tool; or maybe the apparent lack of one is just another illustration of the "unprofessional" approach which Romans so long maintained toward administration of their empire.[50] In the same vein, if anyone did decide to seek out a map displaying the empire's highways comprehensively as a system, it is dubious whether there was one to be found. Such few references to maps as survive never mention roads as a feature. This point applies in particular to the famous (but now lost) map of the "world" (*orbis terrarum*) sponsored by Augustus' close associate Agrippa and placed on public display in Rome around the end of the first century BCE;[51] the spread of Roman highways was in any case only just beginning at this period. Ptolemy's *Geography* of the mid-second century CE, a surviving Greek scientific reference work in the Alexandrian tradition, proves equally uninformative in this connection. Roads are not among the types of feature for which it offers coordinates.[52]

The Peutinger Map: A Creative Advance?

Against this background, the Peutinger map—as we have it, an incomplete medieval copy—merits greater recognition as an apparently creative advance than it has so far received.[53] There can be little doubt that the original map represented almost the entire world known to the Romans, from the British Isles and the Atlantic across to Persia and India. Conspicuous among the many features selected for marking are land routes, presented as red linework that fans out from Rome to forge horizontally across the entire empire and even beyond to the east. Scholars' traditional interpretation of the map is arguably too selective and literal in assuming it to be no more than

[49] See essay 9 above.
[50] See further Potter (2006), 192.
[51] See Arnaud (2007–2008).
[52] Stückelberger and Grasshoff (2006), with German translation. Equally, roads held no special interest for the Latin encyclopedist Pliny the Elder in the late first century CE.
[53] See essays 13 and 14 above.

the equivalent of a modern automobile association's atlas for drivers. The fact is, however, that Rome's central placement inevitably compounds the already severe skewing of the landmasses, while lack of concern to present routes in ways designed to be of practical help to the traveler is all too often apparent when they are subjected to close scrutiny. I think it more appropriate to see this as a map for display, therefore, designed as propaganda to reinforce Roman claims to world power, most plausibly during the period of Diocletian's Tetrarchy around 300 CE.[54] The Tetrarchy instituted a style of rule far more openly autocratic and ritualized than the Augustan model, and the appearance of the roads on the map may be regarded as an inspired reflection of that shift. Here—quite possibly for the first time—an ingenious designer exploited the opportunity to highlight the roads as an interconnected system, and their control by rulers now themselves constantly on the move as a key component of Roman imperial power. It is striking that even so, the presentation lacks menace, because no overtly military feature is marked; legionary bases, garrisons, and fortified frontier lines are all absent. Instead, the Tetrarchy's formidable military might is taken for granted.

The map's aim, rather, is to demonstrate that the empire is now restored to peace and stability, rebellions and invasions are overcome, and everyone may relax and travel freely in all directions as they please. In this respect the map skillfully recalls the "beneficial ideology" of the utopian second century, while diverting attention from the Tetrarchy's harshness and ongoing struggles. Whatever statement the map may be making about imperial control of roads, it is still a restrained one when so many other features are conspicuously visible, all of which between them serve to prompt a variety of impressions. Consequently, whether viewers of the map in antiquity would have shared traditional modern scholarship's highly focused preoccupation with its presentation of roads alone has to remain a matter of considerable doubt.

Conclusion

To regard Rome's roads as an interconnected system, and on that basis to advance broad findings about the development of the empire's economy and society, needs to be understood as a distinctly modern approach. Hence my

[54] Note essay 18 and especially fig. 18.5 below.

concern here to question whether we should assume the same conceptual awareness on the part of the Romans themselves. Ample justification for caution emerges. Even at the highest level, it seems, the mindset, motivation, reference tools, and administrative resources to recognize the empire's highways as an interconnected system and to exploit their potential as such were all absent. Disappointing these shortcomings may well be to us today, given our admiration for the system and our recognition of its extraordinary potential. However, to judge by the patent and persistent limitations of Roman imperial government in many other respects, such lack of more active appreciation for roads among much else should hardly surprise us.

17

Roads in the Roman World

Strategy for the Way Forward

The kind invitation to give the keynote lecture at this conference *Roman Roads: New Evidence—New Perspectives* is both an honor and a daunting challenge. A breadth of perspective is called for, as well as more general reflection than might prove practical or appropriate within the scope of a single paper. Permit me to use this opportunity, therefore, to draw attention tactfully to two concerns, and to present some recommendations about them. My aim is no more than to stimulate productive discussion. There is no intent here to expose faults, or to win arguments, or to press for radical changes to current approaches.

The first concern relates to my reactions when I seek to gain (so far as is possible) an informed overview of the work at present being pursued on Roman roads, milestones, and associated matters. There is every reason to find current work sound and useful, not to say vital. Our understanding of the courses of many roads certainly needs improvement, for example. Undiscovered milestones, documents, and archaeological finds are always emerging; naturally, the entire scholarly community benefits from their methodical study and publication. However, just because there continues to be so much to preoccupy us in these typically very focused basic respects, other approaches are liable to remain relatively neglected by contrast, and so still await realization of their latent potential. This observation applies in particular, I suspect, to the wider roles of Roman roads in the cultures and mindsets of the empire's peoples. To be sure, far be it from me to overlook or to devalue the explorations that have been made of such roles, and that continue to be made—for example, in the entire substantial volume 24 (2016) of *Antiquité Tardive*, just published, devoted to *Le voyage dans l'Antiquité tardive: réalités et images*.[1] I merely urge the cultivation of such broader approaches as an aspect of our own work if it is not already present there. We should also

[1] Talbert (2016) is among the contributions; cf. the broader overview by Cioffi (2016).

welcome interest in this level of approach on the part of colleagues who make no claim to be experts on roads, but who recognize that their place in the Roman physical, cultural, and mental landscape repays attention.

As in the study of antiquity from any perspective, it is essential to keep in mind the fundamental point that Romans' vision of their roads did not necessarily match our vision of them today. At an earlier conference entitled *Highways, Byways, and Road Systems in the Pre-modern World*, I dared to question provocatively "how far even [Roman] emperors or their better-educated subjects were in the habit of conceptualizing the empire's highways as a 'system.'"[2] This is a perspective which comes so naturally to us today—with our scientific and cartographic grasp, and our planned national highway systems—that we may too readily project it back into antiquity. I concluded with reference to Rome:[3]

> Even at the highest level, it seems, the mindset, motivation, reference tools, and administrative resources to recognize the empire's highways as an interconnected system and to exploit their potential as such were all absent.

Clearly, there is ample scope for differences of opinion here. Major issues arising include, for instance, the degree to which the Roman authorities sought to control and exploit their subjects, as well as the authorities' ability to produce maps at a reasonable scale that marked the course of roads, and then their inclination to use such maps. Specifically, how typical is the featuring of roads on the Peutinger map? Was this map intended to have some practical use? Would it have prompted viewers to think of the roads as a network? My altogether skeptical conclusions failed to convince at least one reviewer, Ray Laurence, whose discussion of all the papers emerging from the conference, and mine in particular, was notably generous and thoughtful.[4] However, as so often in scholarship, the new questions opened up by a book or a paper may prove more influential in the long term than the conclusions advanced in the first instance.

One approach that the *Highways* conference aimed to encourage specifically was comparison of the experiences of premodern states worldwide noted for their highways. The initial impetus came one morning during a still earlier conference that Kurt Raaflaub and I organized on ideas about

[2] Essay 16 above, p. 217.
[3] Ibid. p. 221.
[4] Laurence (2016) and further (2020).

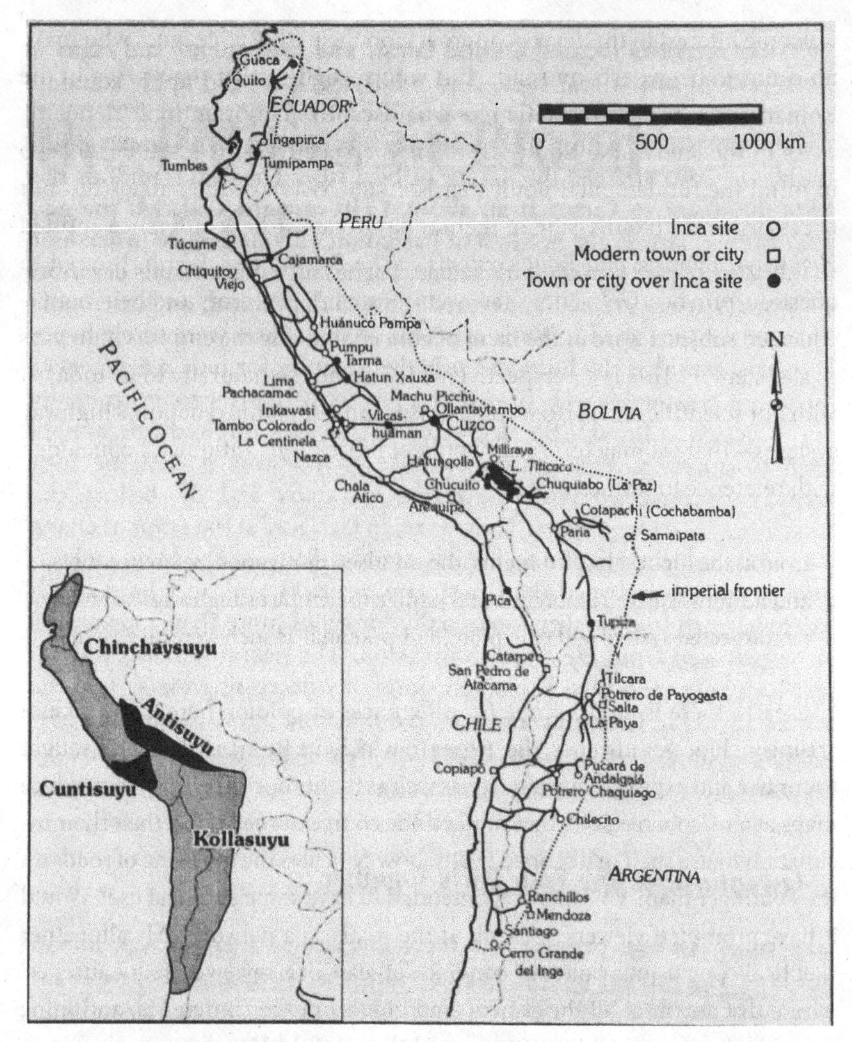

Fig. 17.1 The Inca empire (Tawantinsuyu) in the early 1500s, featuring its highways. Map: Julien (2010), 129.

geography and ethnography in premodern societies.[5] In the course of our breakfast Catherine Julien, an Inca expert, referred in passing to how the sixteenth-century Spanish conquerors of Peru often made comparisons between Roman and Inca roads (fig. 17.1). It struck me that such comparisons could usefully be extended and deepened worldwide.

During the planning stage of the *Highways* conference, two memorable surprises emerged. The first was that an effort to make such comparisons on

[5] Raaflaub and Talbert (2010) was the outcome.

the global scale seemed almost unprecedented. We could find no published record of an earlier such initiative, although we were delighted to discover that one of our contributors, James Snead, was coediting a volume (the outcome of a conference like ours), *Landscapes of Movement: Trails, Paths, and Roads in Anthropological Perspective,* that was to be published in 2009. In fact almost all its twelve chapters relate to the American continent only, but the broad premise on which they are based clearly invites application elsewhere. The coeditors state in their preface:[6]

> Our discussion began from the shared premise that trails, paths, and roads are the manifestation of human movement through the landscape and are central to an understanding of that movement at multiple scales. The study of these features connects with many intellectual domains, engaging history, geography, environmental studies, and, in particular, anthropology and its subfields. In the process we are developing a better understanding of infrastructure, social, political, and economic organization, cultural expressions of patterned movement, and the ways that trails, paths, and roads materialize traditional knowledge and engineering, world view, memory, and identity.

The second surprise was that in several instances colleagues with the relevant expertise were very hard to find; and of those that we did find eventually, none (we learned) had ever been involved in a comparative initiative. Securing an expert for the Classical Era (late fourth century BCE to early fourth century CE) in China was an exceptional challenge. Michael Nylan at Berkeley was highly recommended to us, but her initial response when approached was discouraging. She was unable to help, she replied, and in any case (she assured us) there was next to nothing to say about highways in premodern China. In desperation we begged her to contribute whatever she could all the same. She kindly agreed, and then three months later reported in a very different mood that she had found plenty to say in fact. For this reason, we accepted from her a contribution double the length of any other in view of the importance and range of her findings. These extended to the road as metaphor, as well as to road deities, cults, and rituals.[7]

To be sure, the Roman case is different insofar as specialist interest in Rome's roads is long established. Even so, to compare Rome's roads with those of other premodern states still remains a novel perspective where there

[6] Snead, Erickson, and Darling (2009), xv.
[7] Nylan (2012).

is surely unrealized potential to be tapped. My own initial efforts have at least acted to highlight formative characteristics that I suspect go unappreciated, or at best underappreciated, by those of us in today's first-world Western societies who limit our attention to Roman roads. Let me identify, and briefly comment upon, three such characteristics.

The first characteristic is the quality and empire-wide range of Rome's highways or "public" roads. The level of organization, along with the deployment of immense manpower and material resources, that were required to construct and maintain these roads for centuries merit recognition as an astonishing achievement and investment for any state worldwide at any period, especially in a premodern era. Specialists on Roman roads are well aware of the achievement, naturally; but they have hardly been concerned to set it in the global context of time and space. It may have slipped their notice, too, that the "premodern" era can extend to less than a century ago. In a major geographic overview of China, George Cressey (who had traveled very widely there during the 1920s) could state:[8]

> Inaccessibility and poor communications have handicapped China for centuries. Except where railroads or modern automobile service is available, travel is on foot, by sedan chair, on muleback, in two-wheeled carts, or by boat. Twenty miles a day is a good average, and in place of a journey of a few hours as by rail one spends days jolting along in a two-wheeled cart.

Within former Roman territory, conditions in Asia Minor toward the end of Ottoman rule a century ago were similar to those in China. The *Handbook of Asia Minor* issued by the British Naval Staff Intelligence Department around the end of World War I gives a vivid sense of them, and of the sketchiness of available data. The header to the itinerary from Adalia (Antalya) to Selefke (Silifke), for example, draws attention to the kinds of difficulties routinely encountered, as well as warning users that distance-figures are liable to be estimates derived from the time spent on the journey in the absence of accurate measurement on the ground (fig. 17.2):[9]

[8] Cressey (1934), 24–25, quoted by Kim (2012), 66. A military convoy did successfully drive across the United States from Washington, DC, to San Francisco, CA, for the first time in 1919, but it proved a fifty-eight-day struggle: see Winchester (2013), 280–93.

[9] Naval Staff Intelligence III.3 (1919), 92. This *Handbook* was planned to appear in four volumes, with vol. III in three parts, and vol. IV in two. In the event, vol. III part 1 (to cover north of the Bosphorus to the Halys) and vol. IV part 2 (to cover the area between the Black Sea and Kayseri) never appeared.

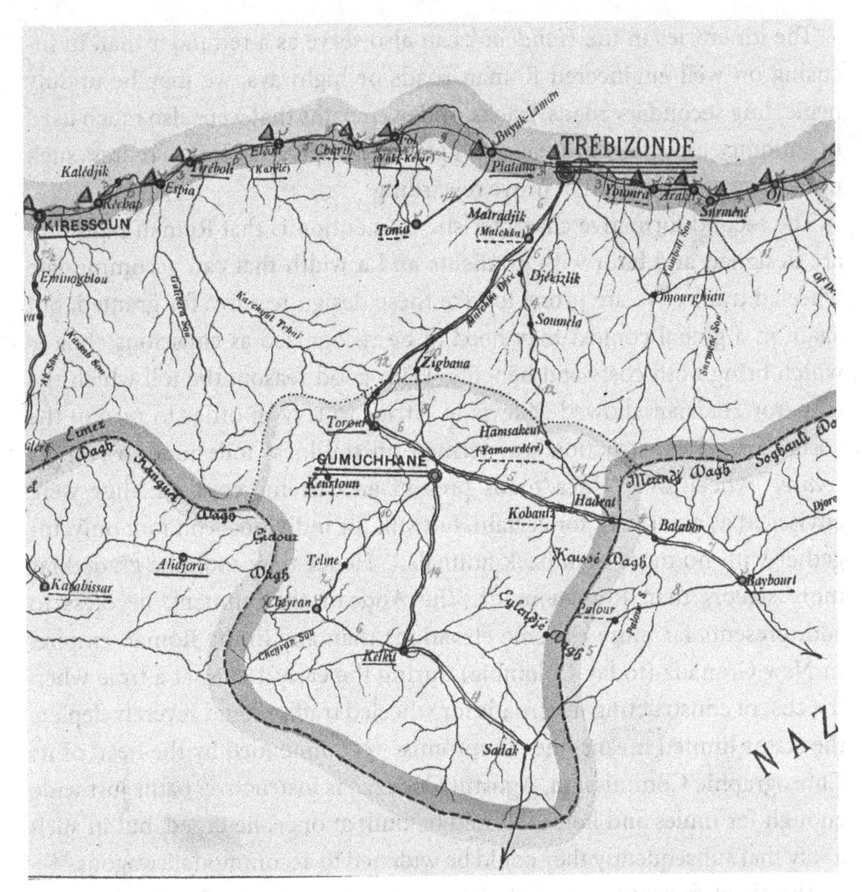

Fig. 17.2 Unlike the British Naval Staff Intelligence, which estimated distances in Turkey, other mapmakers instead stated the number of hours to reckon for a journey from one place to the next. The detail here of the Trabzon region in the northeast is from an undated (around 1900?) 1:750:000 scale map made—without showing any elevations—by the Ottoman Public Debt Administration, probably for its (European) governing council. The hour-figures (some with a fraction, and elsewhere even a question mark) were marked in red. For the Administration's operations, see Birdal (2010). Map: Bavarian State Library, Munich, Mapp. XVII, 6 db-2.

The greater part of this route is merely a track. It does not appear to be passable for wheels, except possibly over short stages, and is not in regular use. The main authorities used for the itinerary are archaeological reports, and the route described may in places not be the track most commonly followed. The distances are calculated from the times noted by the various travellers whose reports have been consulted, and can only be regarded as approximate.

The itineraries in the *Handbook* can also serve as a reminder that, in focusing on well-engineered Roman roads or highways, we may be unduly neglecting secondary roads, tracks, and even paths that were also much used in antiquity, especially in areas of rugged landscape. An effort to redress such imbalance would no doubt prove rewarding.

The second formative characteristic to mention is that Roman highways are designed and built with gradients and a width that can accommodate wheeled traffic. We are prone to take these design features for granted, but again in a global context they need to be recognized as conscious choices which bring both costs and benefits. With good reason, the toll which the emperor Hadrian allowed Milevis in Africa to levy in order to recoup the costs of road construction was termed on the local milestones a *vectigal rotare*, "wheel-toll."[10] Inca roads present an extreme contrast: they were envisaged as a network for certain, but still for individuals on foot only, together with no more than pack-animals.[11] Hence they could be graded far more steeply than Roman roads. The Andean environment, needless to add, presents far more extreme elevations than any in the Roman empire. In New Granada (today Colombia) during the early 1850s, at a time when the cost of constructing new roads for wheeled traffic would severely deplete the state's limited means, the compromise recommended by the head of its Chorographic Commission, Agustín Codazzi, is instructive: paths just wide enough for mules and horses should be built at once, he urged, but in such a way that subsequently they could be widened to accommodate wagons.[12]

The third formative characteristic is the openness of Roman roads to all comers and to vehicles of all kinds. Slaves needed their owners' permission for any movement, of course; exiles were restricted; and Roman senators' private travel beyond Italy, Sicily, or Narbonese Gaul was subject to the emperor's approval. But otherwise, generally speaking, anyone could use these roads without requiring prior authorization or the equivalent of a modern passport. While freedom of movement was indeed a prized feature of the Roman citizen's *libertas*, throughout the empire this particular freedom was by no means limited to citizens. Moreover, tolls were the exception, not the rule; individuals and their vehicles could usually travel along a

[10] *CIL* VIII.10327–10328.

[11] See Julien (2012).

[12] Appelbaum (2016), 112. Notoriously, in some regions of New Granada the trails were so steep and eroded that it was safest for a traveler to submit to being carried in a chair on the back of a local porter (ibid., 81–83).

road without charge. All this is very different from the arrangements in both Japan and China, for instance. Here the authorities exercised tight control on who could take which roads, for what fee, with what documentation, and even at what date during the year. In early modern Japan there was added discrimination by gender on the highways of the Gokaidô network (fig. 17.3). Not only Japan's authorities (until 1867!) but also family pressure especially discouraged travel by females.[13] Nylan says of China in its Classical Era:[14]

> Obviously, the dynasty hoped to control the flow of people, things, and ideas as much as possible, lest too much commerce and too much movement disrupt subject populations engaged in sedentary agriculture, the basis of stable rule within civil society.

If our focus remains confined to Rome's world, there is the risk that we shall take almost for granted the empire's "connectivity" (as we like to call it), without maybe acknowledging quite how remarkable a phenomenon this was. Connectivity overland, after all, is a product not just of the construction and maintenance of so many roads. As noted with reference to Japan and China, a state where the authorities place severe limits on who can travel creates a connectivity inferior in quality to Rome's. A further contributing

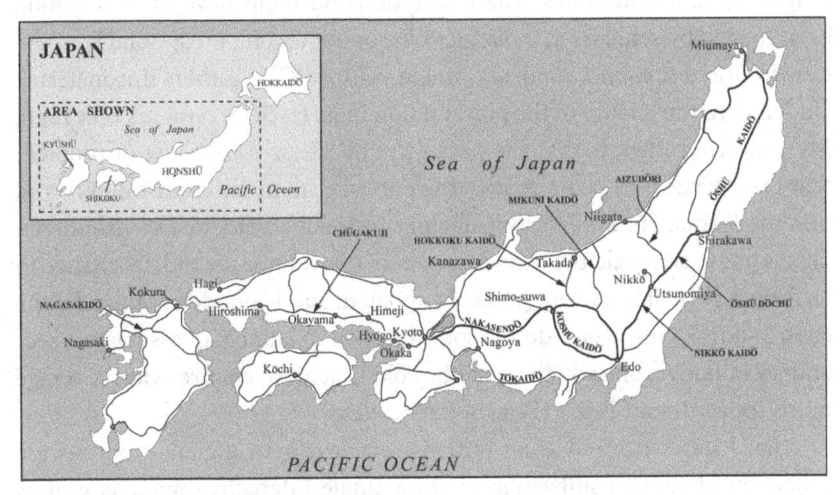

Fig. 17.3 Gokaidô highway network. Map: Vaporis (2012), 91.

[13] Vaporis (2012), 93, 95–98.
[14] Nylan (2012), 42.

factor easily overlooked is Rome's lack of restrictions on vehicles. In China, by contrast, who could use what vehicles was closely regulated; in Japan the public road network could accommodate wheeled vehicles, but their use on it was banned.[15]

By the same token, there may be marked variation in the components of a state's highways and its investment in them, with serious consequences for travelers and their rate of progress. In particular, the provision of bridges built in stone—perhaps the most challenging, expensive, and impressive single component of a Roman highway—is not to be taken for granted. Inca suspension bridges, by contrast, were made of straw and other plant materials.[16] In Japan, for a variety of reasons (rather than merely cost), the Tokugawa authorities mostly opted not to build bridges. So even on Japan's most developed highway—the Tokaido, running for 500 km between Kyoto and Edo (modern Tokyo)—travelers were compelled to cross eight major rivers by ferry and to ford four others; only over two further rivers were bridges built.[17] Considered in this light, the smoothness of Roman connectivity overland for travelers and their vehicles is all the more striking. Thus to place this Roman achievement within a global context seems a perspective ripe for further exploitation.

****

The second of my two concerns is different and more delicate: the publication of Roman milestones. The inscriptions on them have of course long been treated by scholars as a distinct type of document, most notably in the *Corpus Inscriptionum Latinarum*, where volume XVII gathers this material. Undeniably, the quality of the presentation in its fascicles remains very high. My concern, rather, is that the pace of publication is too slow (a problem that I recognize may lie more with the publisher than the contributors), and that the format is no longer satisfactory in today's academic environment. This is not to advocate the abandonment of the volume, or radical change to its format (which has been modestly updated, and is not determined by the contributors). However, I do see sound reason to launch an associated, secondary initiative, one which would be able to achieve empire-wide coverage much sooner than *CIL* XVII can be expected to.

What I have in mind is an initiative to develop a digital, empire-wide collection of Roman milestones, with a single interactive map, as well as

[15] Ibid., 39–40; Vaporis (2012), 101.
[16] Julien (2012), 151.
[17] Vaporis (2012), 102.

photographs and drawings, all accessible without charge. The initiative should be organized so as not to impede working across national boundaries. The record of each milestone's inscription should conform to a standard format, a concise one, limited to presenting and translating its text and explaining the important features, with a minimum of bibliography. Normally, therefore, there would be no account of an inscription's publication history such as *CIL* XVII offers, often at length. Even though scholarly in character, the record would be written to render it accessible to nonspecialists. Consequently, it would not be written in Latin (like *CIL*) but rather in a widely understood modern language.[18]

Looking further into the future, the standard format developed for each milestone's record might also serve as a useful guide for the compilers of further collections to be published.[19] Moreover, with adoption of a digital format rather than a print one, it is a matter of no great difficulty to introduce corrections, modifications, and additions. *CIL* XVII suffers in this important respect because it is issued only in print, and inevitably its fascicles have become more and more expensive. As a result, access to them is reduced, since fewer libraries—let alone individual purchasers—can afford such costly items. That is an unwelcome development at just the time when, in my view, we should be striving to broaden the audience for our work, to engage nonspecialists, and to facilitate a global perspective.

To be sure, both the preparation and the long-term maintenance of a digital resource which is accessible free of charge will still incur substantial costs. But a collaborative initiative to present Roman milestones empire-wide, including a seamless map and photographs, is surely the type of international project that ought to be competitive for funding.[20] While the entire undertaking would indeed prove far from straightforward, it is a feasible one nonetheless, and certain to be invaluable for the long term. Altogether, then, here is a sound strategy for enhancing and linking all the valuable work currently being done on Roman roads and milestones.

[18] Disappointment that Galician was used for one substantial record—Rodríguez Colmenero (2004)—is understandable; see Kolb (2006), 582.

[19] For another instance of a distinctive format for the publication of an important collection, note French (1981–2016).

[20] A suitable geo-referenced map base could be created using the *Map Tiles* tool offered free of charge by the Ancient World Mapping Center (awmc.unc.edu). Compare the map designed to accompany the translation of Pliny the Elder's geographical books discussed in essay 5 above.

18
Communicating Through Maps
The Roman Case

For certain, any reasonably wide-ranging consideration of communications in classical antiquity would be incomplete if it were to overlook maps. However, to anyone today with a typical Western education and intellectual background the findings are liable to prove a source of puzzlement, as well as a sharp reminder of just how foreign a country the past can prove to be. Incredible though it may seem to us, throughout classical antiquity maps as we know them seldom attained more than marginal importance, and their potential value went largely unexploited.[1] A great variety was developed, to be sure, although the number of surviving specimens is frustratingly small. Even so, it is clear that standard conventions for the presentation of a map hardly came to be established. Thus for orientation, scale, and symbology (among much else), common practice was lacking. Hence in the notorious case of a recent discovery on papyrus—a map that admittedly was never finished—scholars' interpretations of what is being represented have ranged from a single estate to the Iberian peninsula, or even the entire Mediterranean.[2]

In neither Greek nor Latin vocabulary was there ever a term in use that unequivocally signifies "map," nor were the very concepts "atlas" and "cartography" articulated; in fact, the former only dates to the sixteenth century, the latter to the nineteenth.[3] There was seemingly no profession of "mapmaker," let alone much in the way of widespread, recognized uses for the range of materials that we today would broadly categorize as maps. If there was a role for maps in formal education (where literature and rhetoric dominated the curriculum), it is all but invisible to us.[4] In any case, nowhere in classical antiquity was instruction organized on the scale of the mandatory, publicly

[1] Note in this connection Dueck (2013), and contributions in Talbert (2012).
[2] See essay 11 above.
[3] Grafton et al. (2010), 103; Harley (1987), 12.
[4] Gautier Dalché (2014); and, more fully, Racine (2009); Johnson (2015).

funded programs for children instituted in the West in the late nineteenth century; nor was there the mass circulation of materials that the invention of printing made possible from the sixteenth century. By its very nature, any map more complex than a mere sketch is liable to prove a severe challenge to reproduce accurately by hand, a far more taxing task than for a text.[5]

Some use of maps in administration can be detected, but even at best (it seems) this use remained minimal and highly localized. Harder still to detect is the use of maps in the conduct of a state's foreign affairs, or of its military campaigns by land or sea.[6] Private travelers and mariners appear not to have used them much either; their recourse was rather to itineraries in the form of lists.[7] Geographical and ethnographic knowledge was successively expanded by such developments as Greek colonization, Alexander's eastern campaigns, and Augustus' expansion of the Roman empire, northward especially.[8] Ironically, a succession of Greek scientific thinkers at Alexandria—from Eratosthenes in the third century BCE to Ptolemy in the second century CE—did successfully formulate principles (which remain standard today) for representing the earth's curved surface on a two-dimensional plane, and for attempting to fix and record specific locations by means of latitude and longitude coordinates.[9] On a technical level, however, not even these scientists were able to measure either distance or time with precision; in consequence, accurate calculation of longitude in particular was beyond them. For several centuries, dissemination and exploitation of these methods and results barely extended beyond the scientists' own very restricted circle, not least because widespread communication of their learning was of minimal concern to them. In a surprising development, their work may have become somewhat better known during the second century CE after Ptolemy developed a more concise, "user friendly" style for expressing coordinate figures. Even so, as will emerge below from a curious test case, this spread of knowledge was not enough to stimulate keener recognition of the value of maps.

[5] See further essay 15 above.

[6] Mattern (1999), 24–80, with reference to Rome. Although Gautier Dalché (2015) focuses on the Middle Ages, his approach and his comments on Vegetius, *Mil.* 3.6 merit notice (passage quoted in essay 9 n. 80 above).

[7] See further essays 7 and 9 above. It is sobering to reflect that itineraries of variable accuracy continued to form the basis of many travelers' maps into the twentieth century: see, for example, the detailed appraisal of the first edition of Richard Kiepert's *Karte von Kleinasien* (1901–1907) by Guillaume de Jerphanion (1909), with Talbert (2019), 110–12.

[8] Irby (2012) and Roller (2015) offer overviews.

[9] For Eratosthenes, see Roller (2010); for Ptolemy's *Geography*, see Berggren and Jones (2000); Stückelberger and Grasshoff (2006; 2009); Jones (2012).

From today's Western perspective, therefore—in our highly literate cultural environment where maps are a well-defined genre with familiar standard characteristics, and are taken for granted as invaluable sources of information, through digital media especially—this indisputably limited exploitation of maps is sure to seem a huge missed opportunity. The sense of disappointment (insofar as that is an appropriate attitude in this context) is only deepened by awareness that at least one other major ancient civilization did develop a remarkable map consciousness: namely China, although its creative reliance upon maps was never matched by the establishment of fundamentals that Eratosthenes and his successors achieved.[10] Meantime, in classical antiquity, despite that achievement, whatever further conditions or mindset might have empowered a breakthrough to a more engaged level with maps never emerged. Such a breakthrough could have occurred, but it did not.[11]

We should bear in mind that similar inconsistency has long continued to occur elsewhere, too, even under otherwise favorable circumstances. For example, although Arabic scholars demonstrated immense enthusiasm for the methodology and coordinates in Ptolemy's *Geography*—a work known to them from the ninth century, it seems—this knowledge never led to latitude and longitude being made the basis for Arabic mapmaking.[12] Equally, as late as the start of the twentieth century, Britain's political agent and consul at Muscat could be cautioned against undue exertion to gather intelligence on his surroundings in the Persian Gulf. His informant wrote privately: "I know from experience that the FO [Foreign Office] has a distinct distaste for acquiring geographical knowledge."[13] Despite sporadic efforts by the War Office and the Admiralty to remedy the deficiency, lack of maps (or disregard for them in certain instances) had contributed to serious defeats suffered by British forces in the Crimean and Boer wars. On the outbreak of war with Germany in 1914, a mere century ago, the Geographical Section of the General Staff (formed in 1906) was able to supply large-scale maps of Belgium and France, but there was no index of what *The Times* later termed

[10] See, for example, Henderson (1994; 2010); Yee (1994); Hsu (2010). For comparison of Rome and China, essay 2 above; Brodersen (2004), 183–84.

[11] Contrast the shift to concern for knowing the hour from the mid-third century BCE onward, discussed in essay 19 below.

[12] Tibbetts (1992b), 93–101.

[13] H. Whigham to P. Cox in 1902, quoted by Hamm (2014), 896; see 886 for a similar complaint in 1904 by Captain Francis Maunsell, British military attaché in Constantinople.

their "horribly unpronounceable place names," and no overall map of Europe available even at as modest a scale as 1:1,000,000.[14]

A striking feature of Roman mapmaking as we know it from what survives is the ingenious creation of maps whose varied character consciously transcends the scientific and factual. Such maps represent initiatives by anonymous artists who developed dynamic cartography of a type that perhaps had no counterpart earlier in the Mediterranean world except to a limited degree in Egypt.[15] These artists astutely perceived the appeal and potential of maps in the range of media that could be exploited to justify and celebrate the spread of Roman rule. The concern to communicate to Roman society through maps in this way is especially thought-provoking. While little enough relevant material survives, there happens to be more than for any other type of Greek or Roman mapping; hence my choice to focus here on the Roman case.

A further reason for the choice is the relative novelty of an approach that considers these maps primarily in relation to their ancient makers and viewers, rather than preferring to pursue perspectives and questions that occur to modern viewers. They naturally enough expect maps to be factual, practical resources; and until the 1980s there was scant concern even among scholars to question whether or not this was the assumption in antiquity too. Only then did a decisive shift ensue in how mapping among premodern societies generally should be approached and evaluated. The shift was set in motion by Brian Harley and David Woodward with their launch of the transformative, ongoing *History of Cartography* project.[16] Inevitably, application of their approach has taken time to gain momentum, so that the claims and conclusions stemming from it remain controversial and still in process of formation.[17] Issues of communication are central to this approach—both what mapmakers were meaning to convey and the range of reactions we may fairly imagine to have been forthcoming from viewers or readers.

Issues of communication are plainly evident in a passage from a Latin panegyric delivered in the late 290s CE that refers to one or more maps (all

[14] The Royal Geographical Society was hurriedly ordered to produce the index and volunteered to begin work on the map at once: see Heffernan (1996), 508. Whether names were easily pronounceable by a Latin speaker was a matter remarked on by Pliny: note *NH* 3.7, 139; 5.1.

[15] O'Connor (2012), 55–58.

[16] See Edney (2005a; 2005b). Publication by the University of Chicago Press to be completed in six (mostly multipart) volumes: www.geography.wisc.edu/histcart.

[17] For consideration of the challenges that had to be overcome in stimulating reassessment of Greek and Roman mapping, note Talbert (2008), 9–15, with essay 1 above.

now lost). The speaker, Eumenius, is the new head of a school of rhetoric at Augustodunum (modern Autun) in Gaul, which has suffered damage. He seeks the provincial governor's permission to rebuild it at his own expense. A feature of the school that is already in place, apparently intact, is a large map of the *orbis terrarum* (and possibly some regional maps, too); the governor has even seen it himself. In the climax to the speech, Eumenius expands upon the potential value of this map:[18]

> In [the school's] porticoes let the young men see and examine daily every land and all the seas and whatever cities, peoples, nations, our most invincible rulers either restore by affection or conquer by valor or restrain by fear. Since for the purpose of instructing the youth, to have them learn more clearly with their eyes what they comprehend less readily by their ears, there are pictured in that spot—as I believe you have seen yourself—the sites of all locations with their names, their extent and the distance between them, the sources and mouths of rivers everywhere, likewise the curves of the coastline's indentations, and the Ocean, both where its circuit girds the earth and where its pressure breaks into it.
>
> There let the finest accomplishments of the bravest emperors be recalled through different representations of regions, while the twin rivers of Persia and the thirsty fields of Libya and the convex bends of the Rhine and the fragmented mouths of the Nile are seen again as eager messengers constantly arrive. Meanwhile the minds of those who gaze upon each of these places will imagine Egypt, its madness set aside, peacefully subject to your clemency, Diocletian Augustus, or you, unconquered Maximian, hurling lightning upon the smitten hordes of the Moors, or beneath your right hand, Constantius, Batavia and Britannia raising up their grimy heads from woods and waves, or you, Maximian Caesar [Galerius], trampling upon Persian bows and quivers. For now, now at last it is a delight to examine a picture of the world [*orbem depictum*], since we see nothing in it which is not ours.

For all his rhetoric, Eumenius' description of the map leaves no doubt that its maker had aimed to make it a large creation, geographically accurate and comprehensive, extending—with its orientation unknown—north

[18] 9(4).20.2–21.3 Mynors (1964); for commentary, Nixon and Rodgers (1994), 171–77.

to Britain, south far up the river Nile, and at least as far east as Mesopotamia (the land of the "twin rivers of Persia"). There is no knowing the map's origin: whether it was in fact a copy of a map already to be found elsewhere, or whether it was a product tailor-made for the prescribed needs of the Augustodunum school and for the specific location where it was displayed there. The natural inference is that its representation of physical geography derives from the Alexandrian cartography instituted by Eratosthenes. This cartography, as we see it reflected most fully in Ptolemy's *Geography*, maintained a scientific, objective approach. It aimed to span the world, and avoided close linkage with any political power or specific period of time. In consequence, a user of the *Geography* is barely made aware of the existence of the Roman empire, or of the relative size and importance of the principal cities within it, including Ptolemy's own Alexandria. The feature that happens to be central in Ptolemy's rendering of the world's geography is the Persian Gulf.[19]

Others, however, had grasped the potential for them to derive added or alternative significance from a map of this accurate, comprehensive type, because it offered an ideal medium through which to encapsulate the Roman imperial achievement. Seemingly, the earliest Roman patron to realize such potential was Augustus' close associate Agrippa. He commissioned a now-lost world-map, whose design has attracted endless scholarly speculation.[20] At least there is no cause to doubt that it was large in size and scope, as well as geographically accurate in character, and that it remained on permanent display to the public in the Porticus Vipsania at Rome as "the world for the world to see" (*orbis orbi spectandus*) in Pliny the Elder's phrase.[21] As envisaged by the panegyrist Eumenius, the map in his school was to serve a similar dual communicatory role—to be informative about physical and cultural geography, and to raise pupils' awareness of Rome's imperial achievement and their pride in its revival by Diocletian's Tetrarchy.

In terms of communication, the large display-maps of Agrippa and Eumenius reinforced a traditional Roman taste (extending far back into the Republic) for publicly displaying objects or documents or images that both informed Romans and boosted their pride. The variety of expressions

[19] Points stressed by Jones (2012), 125–27.
[20] Arnaud (2007–2008).
[21] Plin. *NH* 3.17, with Boatwright (2015). Some manuscripts read *urbi* instead of *orbi*: "the world for the city [of Rome] to see."

developed for these displays expanded with the consolidation of the Principate and the conscious sense of empire fostered by Augustus and so confidently projected in his *Res Gestae*.[22] The duration of the displays, too, expanded from ephemeral (as in the case of many objects and images carried in triumphal processions) to long-term or permanent (as most obviously with texts inscribed on metal or stone).[23] Among display-maps, variation in the balance between the "informative" and "boastful" elements is to be expected. The large-scale bronze or stone map of its "centuriated" land that each Roman community was expected to keep on public display doubtless had the capacity to boost the pride of, say, a Roman colony planted in a newly subdued region; but still it is appropriate to regard these land-maps as designed primarily to serve legal and fiscal purposes (fig. 18.1).[24] As its primary purpose, Eumenius' map was evidently intended to be a resource for fostering awareness of geography on an expansive scale, although the associated prospect of boosting Roman pride in the process was far from being a negligible secondary aim.

Two large Roman display-maps, substantial parts of which survive, each offer in their own different ways powerful instances where it can be argued that the makers have been sufficiently bold and creative to swing the balance decisively in the opposite direction, rendering the communication of Roman pride the primary purpose and relegating the map's informative element to a subordinate role. We lack testimony for how either map would have been referred to in antiquity. Today they are typically called the "Forma Urbis," or "Rome's Marble Plan," and the "Peutinger map." In each case, the emphasis adopted is a deft accomplishment, insofar as the informative element remains very substantial. As a result, only when the communicatory impact of each map is considered within the context intended for its display does the maker's priority become clear. In each instance, it may be said, failure to attach sufficient importance to the matter of intended context has been a serious flaw in modern scholars' interpretations.

[22] Especially its sects. 25–33; see Cooley (2009), 36–37, 213–56, for comment.

[23] Östenberg (2009). The map evidently displayed by the emperor Claudius at a meeting of the Roman senate was also perhaps only for short-term use: see essay 12 above. For Pliny's *Natural History* approached as a literary "display text," see essay 4 above.

[24] For the "centuriation" of cultivable land into square or rectangular divisions by professional surveyors (*agrimensores, gromatici*), see *OCD*[4] ss.vv. *centuriation, gromatici*; Roby (2014); Morris (2018); and figs. 2.5a and b above.

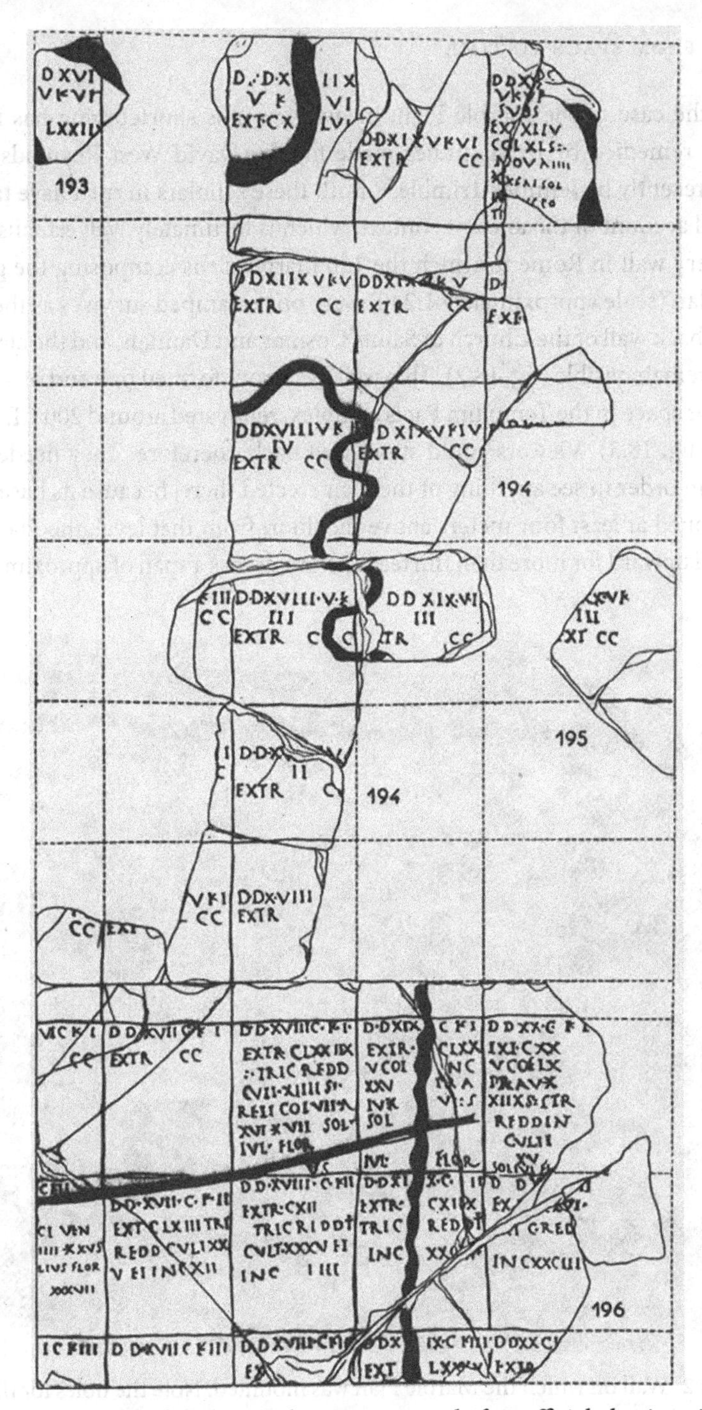

Fig. 18.1 Drawing of all that can be reconstructed of an official plan inscribed on marble to record the centuriated subdivisions of part of the territory of Arausio (modern Orange, France). This plan was made around 100 CE at a scale of approximately 1:6,000. Fig. 2.5b shows the fragments placed lowest. Figure: Piganiol (1962), planche XXI.

In the case of the Marble Plan, in my view this shortcoming has been amply remedied by fresh studies made first by David West Reynolds and more recently by Jennifer Trimble.[25] Both these scholars in turn have taken careful account of the ancient context, which is fortunately well established. The very wall in Rome to which the 150 marble slabs composing the giant city-plan (scale approximately 1:240) were once clamped survives as the exterior back wall of the Church of Saints Cosmas and Damian, and the clamp-holes remain visible (fig. 18.2). This wall, we know, formed one end of a long interior space in the Templum Pacis complex, renovated around 200 CE after a fire (fig. 18.3). Viewers could stand well back, therefore. They needed to do so in order to see anything of the Plan erected there, because its base was positioned at least four meters above the floor; from that level, the Plan extended upward for more than thirteen meters, across a span of approximately

Fig. 18.2 Wall on which the Marble Plan was mounted. Note the holes for the clamps which kept the slabs in place. Photo: Elizabeth Robinson.

[25] West Reynolds (1996); Trimble (2007; 2008). See further now Wolfram Thill (2019).

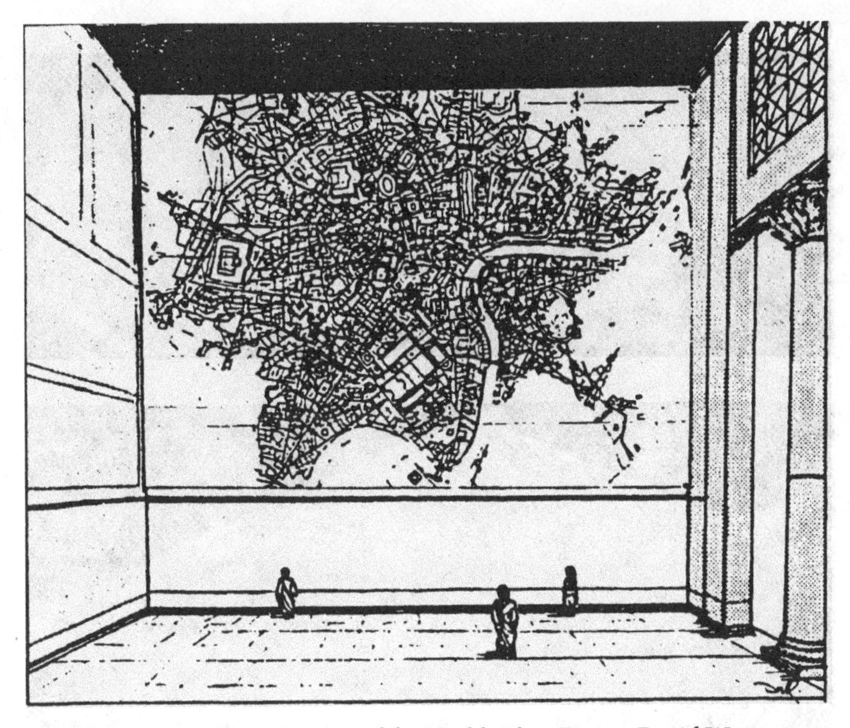

Fig. 18.3 Imagined presentation of the Marble Plan. Figure: David West Reynolds (1996), 296.

eighteen meters. Altogether, therefore, this immense inscribed monument covered about 235 square meters, stretching as high as a modern building of four to five stories. Lighting conditions within the interior space are unknown. All the same, there can be no question that most of the astonishingly rich detail shown of the city at ground level, which we can easily marvel at today from viewing the fragments close up—noting even individual columns and small rooms with their doorways, as well as flaws in presentation—could seldom, if ever, have been appreciated by ancient viewers (fig. 18.4a, b).[26]

Thus the Plan's placement rendered it impossible to communicate this detail adequately to the viewers at floor level. Without doubt, its makers were fully aware of that limitation from the outset. However, they never intended the Plan to serve any practical function, although its data must have

[26] To date, approximately 1,200 fragments can be documented, representing around 12% of the entire Plan.

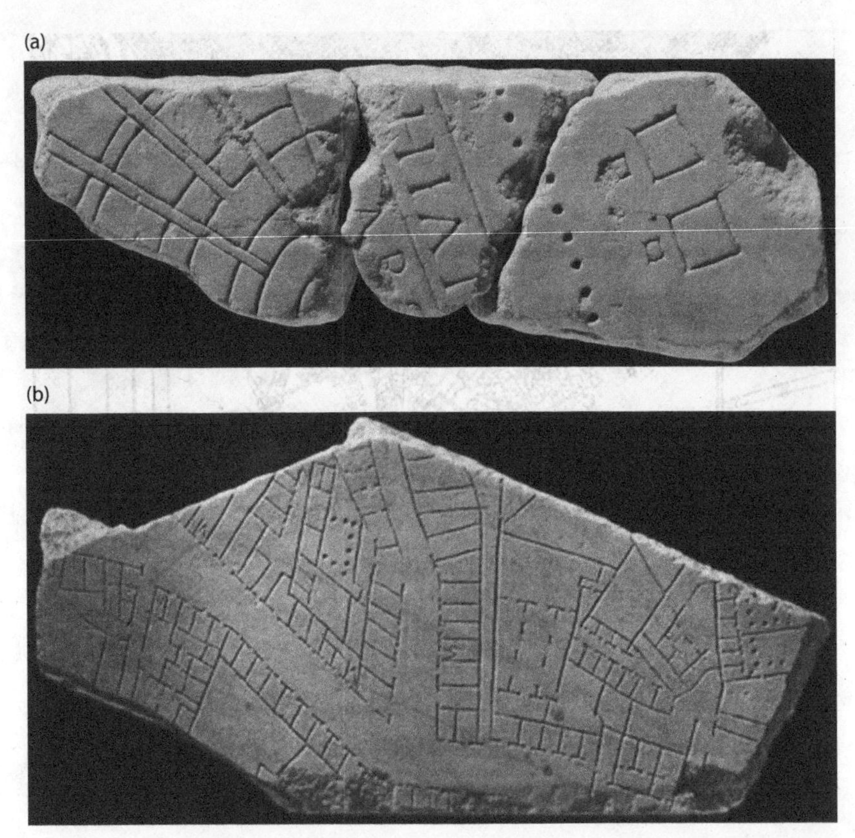

Fig. 18.4a, b Marble Plan fragments: (a) Theater of Marcellus in bird's-eye view perspective; (b) buildings and rooms squashed or stretched. Photos: Digital Forma Urbis Romae Project, Stanford University.

been derived (with some simplification) from painstaking official surveys of the city presumably preserved on papyrus and never intended for circulation. Rather, the makers' main intention was to communicate messages of a broader nature, and above all to fire Roman pride. At the same time, it may not have escaped them or their high-ranking sponsor (conceivably the emperor Septimius Severus himself) that quite the opposite responses might be stirred, too: for example, fear and loathing at such arrogant Roman control of the environment, both built and physical; the command of extensive resources, both human and natural; and the extraordinary level of urban so-called civilization. But possible rejection of this type was likely to be dismissed as merely irrelevant mis-communication of no concern to the Plan's makers.

The case of the Peutinger map is more awkward and delicate.[27] It is awkward because not just the left-hand end is lost (leaving the extent of that loss a matter of conjecture); lost, too, is any trace of the original Late Roman map itself. All that survives is an incomplete medieval copy made on parchment around 1200. Inevitably, therefore, uncertainty and argument persist about the extent to which the map in this form has been "improved," as well as miscopied, by an irrecoverable succession of alternately well-meaning or careless scribes over almost a millennium. To add to all this awkwardness, our copy offers no pointer to the context for which the original map was produced, let alone to the nature of the surface on which it was presented; and clues to determining at all precisely the date of the map's original production are minimal at best.

The case is made delicate by the fact that, although lively scholarly interest in the Peutinger map has been maintained ever since its undocumented "rediscovery" around 1500, this interest has remained narrowly fixated on the land routes shown—a distinctively prominent and colorful feature—and the factual accuracy of their presentation. One of my own recent concerns has been to widen the focus by addressing fundamental issues relating to the map's design; traditionally, these have been taken for granted or accorded only minimal attention. It is also vital in my view to imagine the original context in which the map was to be presented, along with the impression that it was intended to communicate there to its viewers.

Interpretation is called for to account for the shape and content of the map: in particular, its extreme length (as we have it, a little under seven meters) contrasted with marked squatness (about thirty-three centimeters tall); its spanning of the entire known world as a seamless whole without boundaries anywhere; the privileging of land over sea, with much open water removed; and (I think) the placement of the city of Rome at the map's center—thus a most conspicuous placement, but also a very disruptive one for the mapmaker. It calls for equalizing coverage westward from Rome to the Atlantic (presumably, at the now-lost left-hand end), with the same length eastward from Rome for the much greater distance (on the ground) to Sri Lanka. The mapmaker's ingenious solutions to these challenges are first to present Italy as uniquely large, and then to subject the presentation of Persia and India to severe compression. A notable consequence is that the Mediterranean dominates the map. Moreover, the tracing of land routes everywhere, while it

[27] See essays 13–15 above.

may indeed appear informative and without doubt must derive from factual sources, is at best of limited practical value, given the virtual elimination of a North–South dimension and the extraordinary distortions required to accommodate the placement of Rome. Needless to emphasize, the map lacks a consistent scale; in consequence, the length noted for any stage of a route bears no relation to the distance to be traversed on the ground for that stage.

In my estimation, it follows that the traditional literal-minded reading of the map as a practical route guide for land travelers, and nothing more, is a mistaken one; the map simply does not fulfill scholars' eager wish to recover such a document from the Roman empire. Rather, the map is to be regarded as closer to the Marble Plan in its nature and purpose, although the map's design demanded far bolder cartographic creativity and adaptation than did that of the Marble Plan. The Peutinger map was meant to reinforce claims—likewise advanced in literature and other artforms, including coinage—promoting Rome's rule of the world and even of the cosmos.[28] For communicating that aim, general impression mattered most, and the detail (as on the Plan and on monuments like Trajan's Column) had no more than secondary importance; it does not even relate consistently to the same period. This is certainly not to deny that care was taken over the gathering and presentation of detail, or that it could make an impact. Viewers able to inspect it and understand it on the map, for example, could marvel at the remarkable inclusiveness generated by the marking of well over 1,000 settlements too minor to attract the scientific attention of Ptolemy.

The date and context of the original Peutinger map can only be matters for speculation. I consider it most appropriate to associate them with the recovery of the empire by Diocletian's Tetrarchy around 300 CE (the same period in which Eumenius stressed the value of the map in his school).[29] The map's extraordinary shape seems our best clue to the context for which it was designed. It may conceivably have formed only part of a larger artwork now otherwise lost—the surviving "landmap," for example, being one of a set of three that also included counterparts for the sky and the sea. Or perhaps what survives is the *oikoumene* part of a tall globe image divided (according to traditional Greek thought) horizontally into "zones." The representation of this habitable part (*oikoumene*) of the northern hemisphere between the

[28] See, for example, Nicolet (1991), 29–56.

[29] However, a production date somewhat later in the fourth century—in 336, for example, when Constantine celebrated thirty years of rule—is not to be ruled out: see Barnes (2011), 378.

Fig. 18.5 Globe-map image imagined within the apse of a late Roman *aula* where a ruler sits enthroned. Figure: Daniel Talbert.

frigid Arctic and the torrid equatorial zone (both barely habitable) would act to highlight the thriving peace, civilized urbanism, and secure connectivity maintained by Roman rule here, in stark contrast to impoverished barbarism, isolation, and conflict elsewhere.[30] The formal court procedure newly instituted by Diocletian's regime required payment of groveling homage to a Tetrarch. Ideally, the ceremony occurred in a hall (*aula*) where his throne was set in an apse at one end. A tall globe image of the type just envisaged (painted on panels, say) would loom as a powerful backdrop in such an apse, especially when the city of Rome at the center of the *oikoumene* would then appear most prominent directly above where the Tetrarch sat (fig. 18.5).

Matthew Canepa's study *The Two Eyes of the Earth: Art and Ritual of Kingship between Rome and Sasanian Iran* (2009) gives reason to suspect a further possible dimension to the Peutinger map's all-encompassing communication of Rome's claim to world rule. He draws attention to assertions by the Sasanian dynasty that its empire, too, conceived of itself as "a universal domain that

[30] Cf. Plutarch's reference (*Theseus* 1.1) to the habit of filling out the remotest parts of maps with such notices as "Beyond are waterless deserts infested by wild animals," "Murky bog," "Scythian cold," "Frozen sea." Appian (*Roman History*, Praef. 7) claims to have witnessed envoys from impoverished, unproductive barbarian regions begging the emperor—in vain—to bring them under Roman rule.

ruled the entire civilized world under a divine mandate," with "the Sasanian king of kings reigning at the center of Iran, Iran at the center of the empire, and the Sasanian empire at the center of the earth."[31] Given that the Tetrarchy was a period of active diplomacy and warfare between Rome and Persia, with Rome decisively gaining the upper hand, the map could be regarded as an item in the "agonistic exchange" (Canepa's phrase) between Roman and Sasanian rulers. For certain, Sasanian envoys who saw it would only be provoked to find Persia diminished in size and marginalized, while Rome occupied the center and dominated the world. Altogether, the map communicated Roman imperial reach, power, and values even more ambitiously than the Marble Plan.

*****

The more or less severely distorted forms in which the shape of the Peutinger map required the known world's landmasses to appear can hardly have been how its informed makers regularly envisaged them. Rather, the makers must have adapted representations that reflected the scientific Alexandrian ideas and methods initiated by Eratosthenes. These ideas and methods unquestionably also underpin a neglected group of objects that communicate both geographical knowledge and Roman values, one that might even have stimulated wider appreciation of maps, but evidently did not. This group is a type of portable sundial. The optimal functioning of any sundial demands some grasp of the concept of "latitudes" or parallel lines imagined by Eratosthenes as encircling the globe; each such line is situated at its own distinct angle in relation to the sun, with the angle varying according to the time of year.[32] So, for its satisfactory operation, a fixed sundial's design takes into account the latitude at which it is to be installed. Without doubt, by the first century CE, fixed sundials were commonly to be found across the Roman empire in both public and private settings, and they served as the main instruments in use for telling the time. Portable Roman sundials were also developed. These were suspended in order to function, and some could even be adjusted to tell the time at whatever latitude the owner happened to be, over a considerable range.[33]

It has so far escaped scholars' attention that one group of such portable sundials offers us the prospect of gaining insight into the worldview

[31] Canepa (2009), 101, 102. According to Iranian cosmology, the earth was divided into seven continental sections, of which only the central one (the largest) was originally inhabited by humans.

[32] Hannah (2009), 116–44; Houston (2015).

[33] Winter (2013), 77–84; Talbert (2017), 4–9; and essay 19 below. Even so, to tell the time by means of such sundials—fixed or portable—remained an inexact exercise. The hours recorded were merely twelve equal divisions of the period of daylight, which varies according to the latitude and the season. At Rome itself, for example, in late December one such "hour" is no more than three-quarters of a modern fixed hour, but in late June it extends to one and a quarter modern fixed hours.

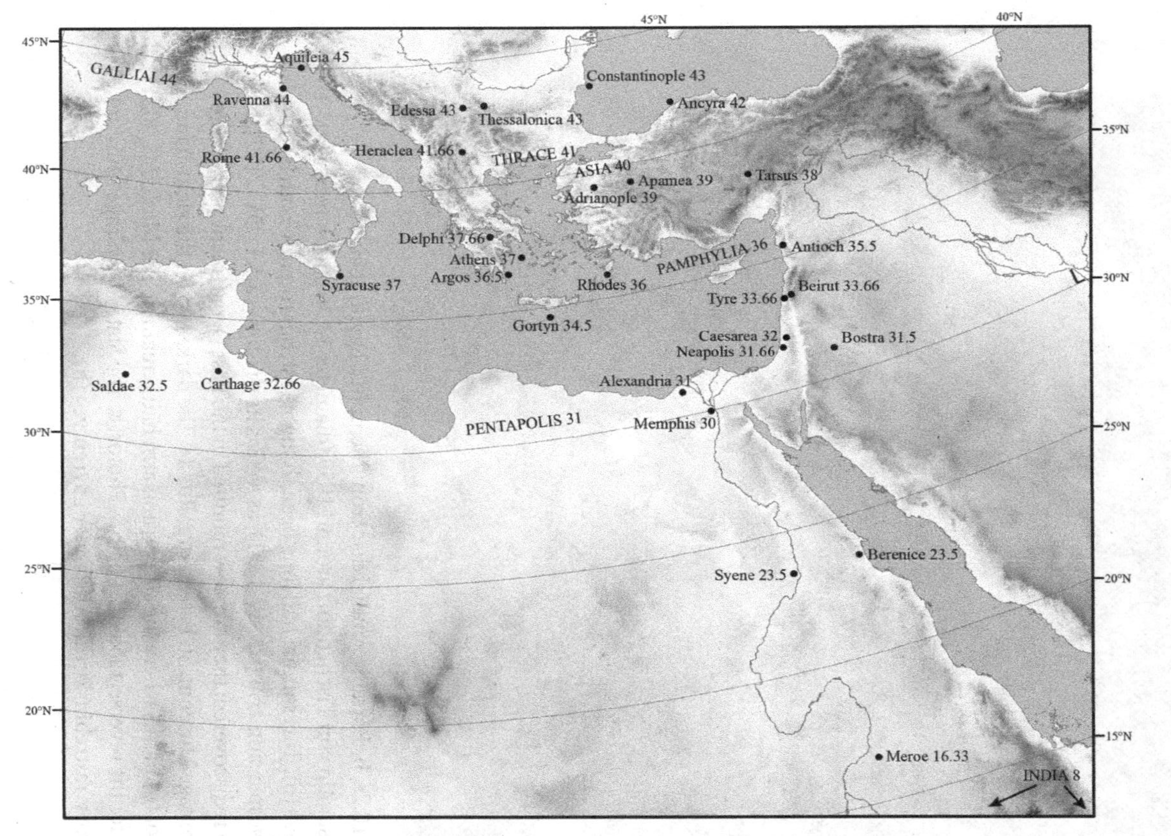

Fig. 18.6 Names on the portable sundial found at Memphis, Egypt, marked on a modern map base, each with the latitude stated for it (fig. 18.7a, b). Accurate longitude is assumed. With this sundial's rediscovery, three further city-names now emerge – Saldae, Tyre, Beirut – none of them in any other sundial's list to date. Map: Ancient World Mapping Center.

(a)

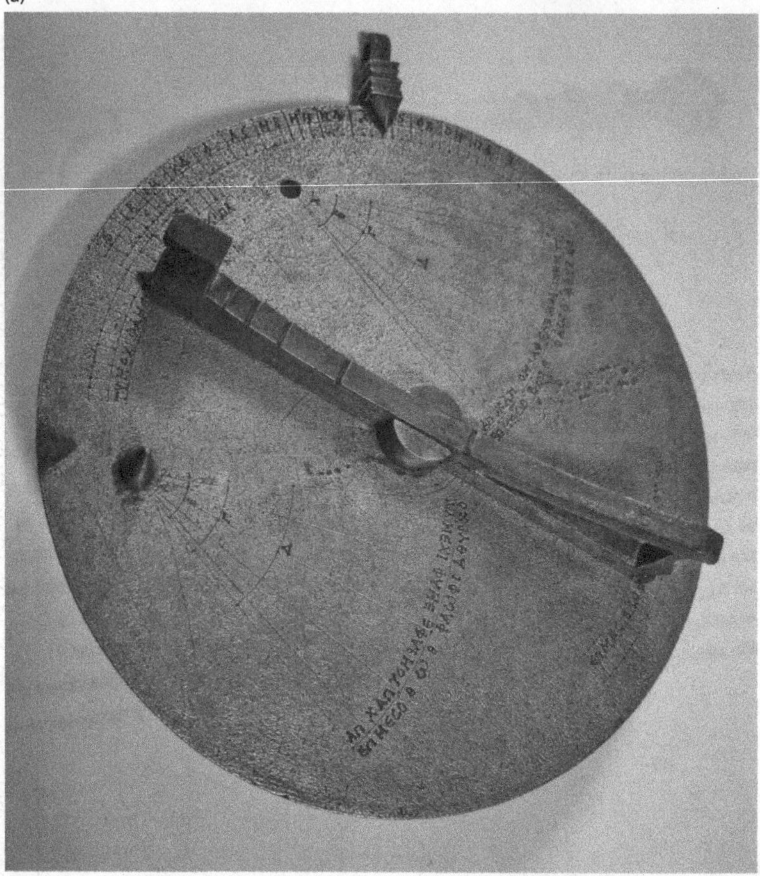

Fig. 18.7a, b Portable sundial disc from Memphis, Egypt, acquired by Constantin von Tischendorf around 1859 (diameter 13.36 cm): obverse (a), reverse (b). This sundial had long been thought lost, but recently both its disc and suspension arm (inv. W-1531), as well as its shadow-caster (DV-2177), have been rediscovered; they are now reunited in the East Department of the State Hermitage Museum, St. Petersburg, Russia. The dark strip on the right-hand side of (b) marks where the suspension arm had remained. See Maslikov (2021) and, for the reverse, table 18.1 below. Photos: © The State Hermitage Museum, Grigory Yastrebinsky (a), Alexander Lavrentyev (b).

(b)

Fig. 18.7a, b Continued

of individual Romans, because on the reverse of these sundials is inscribed a list of city- or region-names (up to as many as thirty-six), each with its latitude figure (figs. 18.6 and 7a, b). Hence, in principle, when the owner wants to tell the time in any of the locations listed, he or she will at once be informed of the latitude to which the sundial should be set, without the need for further reference or inquiry. One dozen or so portable sundials with lists of this type are known. As representative examples, tables 18.1 and 18.2 offer four lists inscribed in Greek and five inscribed in Latin; in the absence of any designated starting-point, I begin each at the name with the lowest latitude figure.

Table 18.1 Names inscribed in Greek on four portable sundials.

British Museum, London		(ex-)Time Museum, Rockford		Memphis		Aphrodisias	
ΜΕΡΟΗϹ Ιϛ	Meroe 16	ΜΕΡΟΗϹ Ιϛ	Meroe 16	ΙΝΔΙΑ Η	India 8	ϹΟΗΝΗϹ ΚΓϹ	Syene 23.5
ϹΟΗΝΗϹ ΚΔ	Syene 24	ϹΟΗΝΗϹ ΚΔ	Syene 24	ΜΕΡΟΗ ΙϛΓ	Meroe 16.33	ΜΕΜΦΙϹ ΚΘϹ	Memphis 29.5
ΘΗΒΑΤΔ ΚΔ	Thebaid 24	ΘΗΒΑΙΔΟϹ ΚΗ	Thebaid 28	ϹΟΗΝΗ ΚΓ<	Syene 23.5	ΠΕΛΟΥϹΙΟΝ ΛϹ	Pelusium 30.5
ΛΙΒΥΗϹ ΚΔ	Libya 24	[Α]ΙΓΥΠΤΟΥ ΛΑ	Egypt 31	ΒΕΡΟΝΙΚΗ ΚΓ<	Berenice 23.5	ΑΛΕΞΑΝΔΡϛ ΛΑ	Alexandria 31
ΑΙΓΠΤΟΥ ΛΑ	Egypt 31	[ΠΕΝ]ΤΑΠΟΛϹ ΛΑ	Pentapol-is/-eis 31	ΜΕΜΦΙϹ Λ	Memphis 30	ΠΕΝΤΑΠΟΛΕϛ ΛΒ	Pentapoleis 32
ΑΛΕΞΑΝΔ ΛΑ	Alexandria 31	·ΑΦΡΙΚΗϹ ΛΒ	Africa 32	ΑΛΕΞΑΝΔΡΙ ΛΑ	Alexandria 31	ΧΑΡΤΑΓΩΝ ΛΓ	Carthage 33
ΠΑΛΑΙϹΤ ΛΒ	Palestine 32	ΠΑΛΑΙϹΤΙΝ ΛΒ	Palestine 32	ΠΕΝΤΑΠΟΛΙϹ ΛΑ	Pentapolis 31	ΚΡΗΤΗϹ ΛΕ	Crete 35
ΠΕΝΤΑΠ. ΛΒ	Pentap-olis/-oleis 32	ΜΑΥΡΙΤΑΝΙΑ ΛΔ	Mauretania 34	ΒΟϹΤΡΑ ΛΑ<	Bostra 31.5	ΑΘΗΝΩΝ ΛΖ	Athens 37
ΑΦΡΙΚΗ ΛΔ	Africa 34	ΚΥΠΡΟΥ ΛΕ	Cyprus 35	ΝΕΑΠΟΛΙϹ ΛΑΓ$_0$	Neapolis 31.66	ΘΕϹϹΑΛΟΝΙΚ` ΜΓ	Thessalonica 43
ΚΡΗΤΗϹ ΛΔ	Crete 34	ϹΙΚΕΛΙΑϹ ΛΕ	Sicily 35	ΚΕϹΑΡΙΑ ΛΒ	Caesarea 32	ΚΥΖΙΚΟΥ ΜΑ	Cyzicus 41
ΚΥΠΡΟΥ ΛΕ	Cyprus 35	[illegible] Λϛ	36	ΚΑΡΧΗΔΩΝ ΛΒΓ$_0$	Carthage 32.66	ΝΙΚΟΜΗΔΙΑ ΜΒ	Nicomedia 42
ΚΟΥΛΗϹ Λϛ	Coele (Syria) 36	ΠΑΝ[ΦΥ]ΛΙΑϹ ΛΖ	Pamphylia 37	ϹΑΛΔΑΙ ΛΒ<	Saldae 32.5	ΚΩΝϹΤΑΝΤΙϛ ΜΑ	Constantinople 41
ϹΙΚΕΛΙΑ Λϛ	Sicily 36	ΕΛ[ΛΑ]ΔΟϹ ΛΖ	Hellas 37	ΤΥΡΟϹ ΛΓΓ$_0$	Tyre 33.66	ΓΑΛΑΤΙΑ ΜΒ	Galatia 42
ΠΑΜΦΥΛ Λϛ	Pamphylia 36	ϹΠΑΝ[ΙΑ]Ϲ ΛΖ	Spain 37	ΒΗΡΥΤΟϹ ΛΓΓ$_0$	Beirut 33.66	ΚΑΠΠΑΔΟΚΙΑ ΛΘϹ	Cappadocia 39.5
ΑΧΑΕΙΑϹ ΛΖ	Achaia 37	ΤΑ[Ρ]ϹΟΥ ΛΗ	Tarsus 38	ΓΟΡΤΥΝΑ ΛΔ<	Gortyn 34.5	ΤΑΡϹΟΥ ΛϛϹ	Tarsus 36.5
ΤΑΡϹΟΥϹ ΛΖ	Tarsus 37	ΑΝΤΙΟΧΙΑ ΛΘ	Antioch 39	ΑΝΤΙΟΧΕΙΑ ΛΕ<	Antioch 35.5	ΑΝΤΙΟΧΙΑ ΛΕΓ	Antioch 35.33
ϹΠΑΝΙΑϹ ΛΗ	Spain 38	ΜΑΚΕΔΟΝΙΑ Μ	Macedonia 40	ΡΟΔΟϹ Λϛ	Rhodes 36	ΦΟΙΝΙΚΗϹ ΛΓΓ	Phoenice 33.33
ΑΝΤΙΟΧΟ ΛΘ	Antioch 39	ΓΑΛΑΤΙΑϹ Μ	Galatia 40	ΠΑΜΦΥΛΙΑ Λϛ	Pamphylia 36	ΠΑΛΕϹΤΙΝΗ Λϛ	Palestine 36
ΠΕΛΟΠΠΟ [??]	Peloponnese [missing]	ΘΕϹϹΑΛΟΝΙ Μ	Thessalonica 40	ΑΡΓΟϹ Λϛ<	Argos 36.5	ΚΥΠΡΟΥ ΛΔ	Cyprus 34
ΘΕϹϹΑΛΟ Μ[	Thessalonica 40? 41?	Θ[Ρ]ΑΚΗϹ ΜΑ	Thrace 41	ϹΟΡΑΚΟΥϹΑ ΛΖ	Syracuse 37	ΛΥΚΙΑϹ ΛΕΓ	Lycia 35.33
ΡΩΜΗϹ ΜΑ	Rome 41	ΡΩΜΗϹ ΜΒ	Rome 42	ΑΘΗΝΑΙ ΛΖ	Athens 37	ΠΑΜΦΥΛΙΑ ΛϛϹ	Pamphylia 36.5
ΘΡΑΚΗϹ ΜΑ	Thrace 41	ΙΤΑΛΙΑϹ ΜΒ	Italy 42	ΔΕΛΦΟΙ ΛΖΓ$_0$	Delphi 37.66	ΡΟΔΟΥ Λϛ	Rhodes 36
ΒΙΘΥΝΙΑ ΜΑ	Bithynia 41	[Δ]ΑΛΜΑΤΙΑϹ ΜΒ	Dalmatia 42	ΤΑΡϹΟϹ ΛΗ	Tarsus 38	ϹΙΚΕΛΙΑϹ ΛΗ	Sicily 38
ΑΒΥΔΟϹ ΜΑ	Abydos 41	ΓΑΛΛΙΑϹ ΜΒ	Gallia 42	ΑΔΡΙΑΝΟΥΠΟΛΙϹ ΛΘ	Adrianople 39	ΡΩΜΗ ΜΑϹ	Rome 41.5
ΔΑΛΜΑΤ ΜΒ	Dalmatia 42	ΚΑΠΠΑΔΟΚΙ ΜΓ	Cappadocia 43	ΑϹΙΑ Μ	Asia 40	ΓΑΛΛΙΩΝ Μϛ	Galliai 46
ΚΑΠΠΑΔ ΜΓ	Cappadocia 43	Κ[Ω]ΝϹΤΑΝΤΙ ΜΓ	Constantinople 43	ΗΡΑΚΛΕΙΑ ΜΑΓ$_0$	Heraclea 41.66	ΒΟΥΡΔΙΓΑΛΑ ΜΕ	Burdigala 45
ΙΤΑΛΙΑϹ ΜΓ	Italy 43	[Α]ΡΜΕΝΙΑϹ ΜΔ	Armenia 44	ΡΩΜΗ ΜΑΓ$_0$	Rome 41.66	ϹΠΑΝΙΑϹ ΜΒ	Spain 42
ΚΩΝϹΤΑ ΜΓ	Constantinople 43	Π[Α]ΝΝΟΝΙΑϹ _Δ	Pannonia 44?	ΑΓΚΥΡΑ ΜΒ	Ancyra 42	ΗΜΕΡΙΤΑ ΛΘϹ	Emerita 39.5
ΓΑΛΛΙΑϹ ΜΓ	Gallia 43	[ΒΙ]ΘΥΝΙΑϹ ΜΔ	Bithynia 44	ΘΕϹϹΑΛΟΝΙΚΗ ΜΓ	Thessalonica 43		
ΑϹΙΑ ΜΓ	Asia 43	ΓΕΡΜΑΝΙΑϹ Ν[_?]	Germany 50 (+?)	ΑΠΑΜΙΑ ΛΘ	Apamea 39		
ΝΕΟΚΑΙϹΟ ΜΔ	Neocaesarea 44			ΕΔΕϹΑ ΜΓ	Edessa 43		
ΑΡΜΕΝΙΑ ΜΔ	Armenia 44			ΚΩΝϹΤΑ`ΤΙΝΟΥΠΙ ΜΓ	Constantinople 43		
ϹΑΡΜΑΤΙΑ ΜΕ	Sarmatia 45			ΓΑΛΛΙΑΙ ΜΔ	Galliai 44		
ΠΑΝΝΟΝΙ ΜΕ	Pannonia 45			ΑΡΑΒΕΝΝΑ ΜΔ	Ravenna 44		
ΜΕΔΙΟΛΑΝ Μϛ	Mediolanum 46			ΘΡΑΚΗ ΜΑ	Thrace 41		
ΒΡΕΤΤΑΝ Ν[_?]	Britain 50 (+?)			ΑΚΥΛΗΙΑ ΜΕ	Aquileia 45		

Table 18.2 Names inscribed in Latin on five portable sundials.

Oxford	
AEGYPT XXX	Egypt 30
LYCIAE XXXI	Lycia 31
CILICIA XXXI	Cilicia 31
ASIAE XXXI	Asia 31
GAL XLVIII	Gal 48
CAPPAD XXXI	Cappadocia 31
INSVRIA XXXVI	(As?)Syria 36
QVIRINE XXXVIIII	Cyrene? 39
SARDIN XL	Sardinia 40
NARB XLIII	Narb-o/-ensis 43
SICILI XLI	Sicily 41
AFRCA XLI	Africa 41
MAVRT XL	Mauretania 40
ISPAN XLII	Spain 42
BRIT LV	Britannia 55
ITAL XLIII	Italy 43
NARB XLIIII	Narb-o/-ensis 44
BELG XLVIII	Belgica 48
GERS XLVIIII	Germania Superior 49
GERI LI	Germania Inferior 51
ROMA XLII	Rome 42
DACIA LII	Dacia 52
PANONI XLVI	Pannonia 46
MACEDO XL	Macedonia 40
LVGD XLV	Lugdun-um/-ensis 45
TRACIA XLI	Thrace 41
GALATI XLV	Galatia 45
PHRYGI XXXVI	Phrygia 36
BITVNI XXXV	Bithynia 35
ITAL XLII	Italy 42

Vignacourt/Berteaucourt-les-Dames	
ALEXAND XXX	Alexandria 30
ASIAE XXXIII	Asia 33
CAPPAD XXXIIII	Cappadocia 34
IVDEAE XXXV	Judaea 35
SYRIAE XXXVII	Syria 37
AFRICAE XL	Africa 40
MAVRETA XLI	Mauretania 41
ITALIAE XLII	Italy 42
SANIAE XLII	Spain 42
NARBON XLIII	Narbon-is/-ensis 43
CALLECIA XLIIII	Callecia 44
AQVITAN XLV	Aquitania 45
LVGDVN XLVI	Lugdun-um/-ensis 46
BELGIC XLVIII	Belgica 48
NORICI XLVI	Noricum 46
RETIAE XLVI	Raetia 46
ILLYRICI XLVI	Illyricum 46
PANNON XLIX	Pannonia 49
GERMAN L	Germania 50
MOESIAE LI	Moesia 51
DACIAE LI	Dacia 51
SARMATIA LIII	Sarmatia 53
BRITANN LV	Britannia 55

Crêt-Châtelard, Loire	
AETHIOPIAE XXX	Ethiopia 30
AEGYPTI XXXIII	Egypt 33
HISPANIAE XXXV	Spain 35
BABYLONIAE XXXV	Babylonia 35
ILLYRICI XXXVII	Illyricum 37
SYRIAE XXXVIII	Syria 38
ARABIAE XXVIIII	Arabia 29
AFRICAE XLII	Africa 42
MAVRETAN XLS	Mauretania 40.5
BITHYNIAE XLI	Bithynia 41
ITALIAE XLII	Italy 42
NASVMIEN XXIIII	=Nemausus? 24
ANCONIS XLIV	Ancona 44
GALLIAE XLVIII	Gallia 48
GERMANIAE L	Germania 50
BRITANNIAE LVI	Britannia 56

Rome	
AETIOPI XXX	Ethiopia 30
AEGYPTI XXXIII	Egypt 33
HISPAN XXXX	Spain 40
BABYLON XXX	Babylon(ia?) 30
ILLYRI XXXVII	Illyricum 37
SYRIAE XXXVIII	Syria 38
ARABIA XXVIIII	Arabia 29
APHRICE XXXX	Africa 40
MAVRE XXX	Mauretania 30
BITHYNI XLI	Bithynia 41
ITALIAE XLII	Italy 42
NARBON XLIII	Narbon-is/-ensis 43
ANCON XLV	Ancona 45
GALLIAE XLVIII	Gallia 48
GERMA L	Germania 50
BRITAN LVII	Britannia 57

Mérida, Spain	
ALEXAND XXX	Alexandria 30
BETICA XXX IIX	Baetica 38
LVSITAN XL II	Lusitania 42
ILATIA XL II	=Italy? 42
MOESIA XL III	Moesia 43
SARMAT XL IIIS	Sarmatia 43.5
TARRAC XL IIII	Tarrac-o/-onensis 44
NARBON XLIII S	Narbonensis 43.5
AQVITAN XL V	Aquitania 45
PANN I XL V	Pannonia Inferior 45
PANN S XL VS	Pannonia Superior 45.5
RETIA XL VI	Raetia 46
LVGDVN XL VI	Lugdun-um/-ensis 46
NORICV XL VII	Noricum 47
DA IA XL VII	Dacia 47
BELGI XL IIX	Belgica 48
GERMAN L	Germania 50
SVPERIOR LIIII	[Britannia ?] Superior 54
BRITANN INF LVII	Britannia Inferior 57

It is immediately evident that the names chosen are an eclectic mix. On the one hand, they include major, prominent cities, regions, and provinces—Alexandria, Constantinople (former Byzantium), Italy, Gaul, and Spain, for example—that seem predictable choices when coverage of a wide span is intended. On the other hand, some less predictable choices occur that most probably reflect the particular movements or links of the individual who compiled or commissioned the selection, presumably for personal use. Consider, for example, Neocaesarea in the British Museum, London (Greek) list, or the many cities in the Aegean area in the Memphis (Greek) list.[34] Equally, inclusion of both Tyre and Beirut in the latter list, and of both Pannonia Inferior and Pannonia Superior in the Mérida (Latin) list,[35] suggests a deliberate personal preference; when the same latitude is listed for these two cities and minimally different ones for the two provinces, the inclusion of both names in each instance might be regarded as redundant.

Did whoever compiled such a list of names and figures have some awareness of the geographical relationship of the locations chosen for inclusion? Yes, judging by two lists at least. First, the names on the Aphrodisias (Greek) list—presented as a continuous round—may give the initial impression of being just a random jumble without even a designated starting-point (fig. 18.8).[36] The order here really is deliberate, however, because the names can be visualized to form an outline *periplous* or *periegesis* of the Mediterranean, including some of its greatest cities (fig. 18.9). Second, the placement of one particular name in the Vignacourt/Berteaucourt-les-Dames (Latin) list is revealing (fig. 18.10).[37] Here, as in several other instances (the Memphis list, for example), the order of the names is determined by the numerical sequence of their latitude figures. It follows, therefore, that the compilers of these particular lists had both the wish and the capacity to create for themselves a "mental map" reliant upon latitude as its organizing principle.

Strikingly, however, the compiler of the Vignacourt list did not rely upon latitude alone, because there is a glaring departure from the latitudinal sequence to be noted here: Belgica 48, with Lugdunum/-ensis 46 and Aquitania 45 to follow in one direction, and Noricum, Raetia, Illyricum (all 46) in the other. This compiler's geographical "mental map" evidently envisages the Gallic provinces

[34] Talbert (2017), 99–103, 29–34; and now, for the rediscovered Memphis sundial, Maslikov (2021), 316–17.
[35] Talbert (2017), 93–99.
[36] Ibid., 60–64.
[37] Ibid., 88–92.

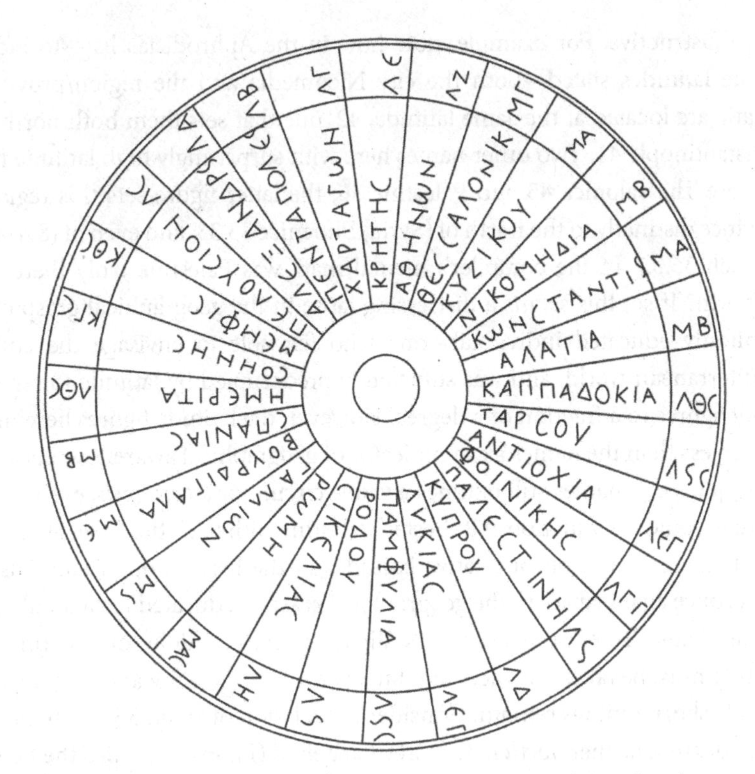

Fig. 18.8 Drawing of portable sundial disc (reverse) found at Aphrodisias, Turkey, in 1963. Figure: Ancient World Mapping Center.

Aquitania, Lugdunensis, and Belgica as three adjacent blocs. Hence the wish not to separate Belgica from the other two here, even though its name must occur out of latitude order in consequence. This said, it is plain that, by the same token, the compiler could also have switched the order of Narbon-is/-ensis and Callecia, so that Callecia (the northwest region of Spain) would then come next to Spain in the list, and Aquitania, Lugdunensis, and Belgica next to Narbonensis. Why the compiler did not switch the order of Narbon-is/-ensis and Callecia is impossible to say. Possibly he (or she) was disinclined to depart from latitude order, but was nonetheless tempted to indulge just a single exception for what may have been his (or her) own "home" province. The villa site at Vignacourt where this sundial was found is near Samarobriva (modern Amiens) in Belgica.[38]

If we may infer that the compilers of all these lists had an awareness of the geographical relationship of the locations they selected, then, in turn, some comparison of the relative latitudes they assign to them promises to

[38] *BAtlas* 11C3. For Rome's provinces as a framework for worldview, see essay 7 above.

prove instructive. For example, note how in the Aphrodisias list—to judge by the latitudes stated—both the city Nicomedia and the region/province Galatia are located at the same latitude, 42, one that sets them both north of Constantinople 41. Two other names here with surprisingly high latitude figures are Thessalonica 43 and Palestine 36; the latter figure sets this region/province distinctly to the north of (Syria) Phoenice 33.33, and even of (Syrian) Antioch 35.33. In the compiler's mental map, was Palestine truly there, so far north? If so, this seems a distressing lapse in the geographical grasp of a manifestly educated individual—one who felt able to envisage the entire Mediterranean world, and was sufficiently preoccupied by latitude to record many figures to a fraction of a degree. However, the latitude figures here may matter less than the names as an indicator of geographical awareness, because the sequence of names still outlines a viable circuit for a *periegesis*, even if the sundial would not function at its best in Palestine with the latitude set at 36.

When a comparison of relative latitudes in the lists on other sundials is made, once again lapses in the geographical grasp of educated individuals are unmistakable. To be sure, engravers' slips as well as muddles of one kind or another must be taken into account, but even after such allowances are made, notable shortcomings remain. Consider the relation of Spain 35 to Africa 42, seven degrees farther *north* in the Crêt-Châtelard (Latin) list;[39] also the figure 37 for Illyricum (at the actual latitude of Syracuse, therefore!) both here and in the very similar Rome (Latin) list.[40] Consider Bithynia 44, Cappadocia 43, and (as many as three degrees farther *south*) Galatia 40 in the Time Museum (Greek) list.[41] In the same vein, the latitude figures in the Mérida list place Tarrac-o/-nensis 44 to the north of Narbonensis 43.5. Equally, note Neocaesarea 44; Cappadocia, Asia, and Constantinople all at 43; and Bithynia 41 in relation to one another in the British Museum, London, list. Most strange are Galatia 45, Phrygia 36, and Bithynia 35 in relation to one another as well as to Lycia, Cilicia, Asia, and Cappadocia (all four at 31) in the Oxford (Latin) list.[42] Also strange here is the wide interval between the adjacent figures 46 for Pannonia and 52 for Dacia; compare too the latitudes for Spain 42, Africa 41, and Sicily 41, all of them to the north of Sardinia 40. A final such puzzle is the Memphis list's figures for Carthage (32.66) and Saldae (32.5), almost identical (as they should be) but as many as four degrees or so too far south, even when Syracuse 37 here is correct.

[39] Talbert (2017), 48–52.
[40] Ibid., 24–27.
[41] Ibid., 70–75.
[42] Ibid., 52–59.

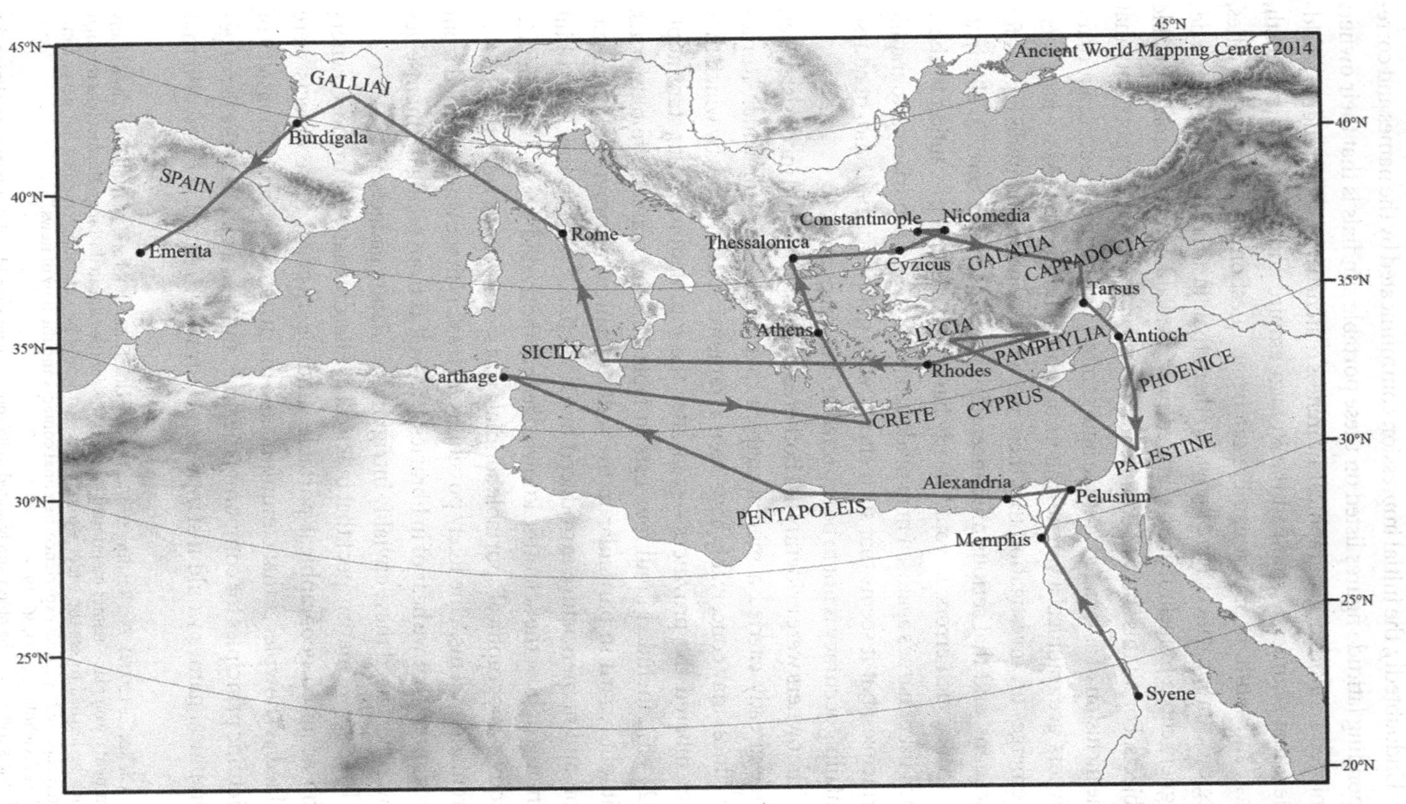

Fig. 18.9 Names on the portable sundial found at Aphrodisias marked on a modern map base at their *correct* latitudes, with a route added. Accurate longitude is assumed. Map: Ancient World Mapping Center.

Undoubtedly, the initial impression communicated by the names and corresponding latitude figures listed on these portable sundials is that their owners enjoyed a confident geographical awareness matched by a wide-ranging worldview that seamlessly spanned the Roman empire and beyond to the north, east, and south. In addition, there can be no question that many more cities, regions, and provinces were known to them than just those that they (or a designer) chose to include in the limited space available on these small portable objects. For all this show of confidence, however, the unique opportunity that the inclusion of associated latitude figures offers us to assess the accuracy of the owners' geographical awareness alters our initial impression. As we have seen, it emerges on investigation that there were no "standard" latitude figures for locations, and that serious misconceptions are widespread here. To eliminate or reduce such errors, it would clearly have been a useful precaution to check each list of names against a map made according to Alexandrian cartographic principles; but it seems that such checks were not performed, and, of course, unfailing accuracy is not to be expected of such maps anyway.

The owners were evidently not bothered by these flaws and inconsistencies. They already carried a mental map in their heads, one for which Rome's provinces may conceivably have been a framework. The names by which they remembered the provinces—even after the foundation of Constantinople (324 CE), included in all four lists in table 18.1—remained the old forms, without regard in particular to the radical changes in provincial organization and nomenclature introduced by Diocletian.[43] Not so much as a hint remains of how the scientific knowledge that motivated would-be owners to acquire "geographical" portable sundials was spread, remarkable though that circulation seems. One likely possibility, at least, is that owners learned from encounters that we hear of in libraries,[44] as well as from seeing for themselves how others used these enviably intriguing miniature gadgets.[45]

Even so, we may suspect that practical use hardly mattered to most owners who acquired geographical portable sundials. Instead, these objects were valued as showpieces that communicated the owners' (supposed) mastery of scientific principles for computing the time, and—as a glance at the list of geographical names would instantly demonstrate—their pride in a world that

[43] Cf. Racine (2009), 79: "Reading classical poetry closely in school and hearing the grammarian's commentary on place-names mentioned by poets was for the educated Roman the first lens through which he learned to see the wider world, a lens that would be later supplemented but never completely replaced by direct experience, personal contacts and the flow of news."

[44] See Nicholls (2017), 36–39.

[45] The small size of the object surely added to its appeal: owners could feel that they were gaining the chance to hold the Roman empire, indeed the world, in just one hand. Compare the popularity of European pocket globes in the seventeenth to nineteenth centuries: Sumira (2014), index s.v. pocket globes.

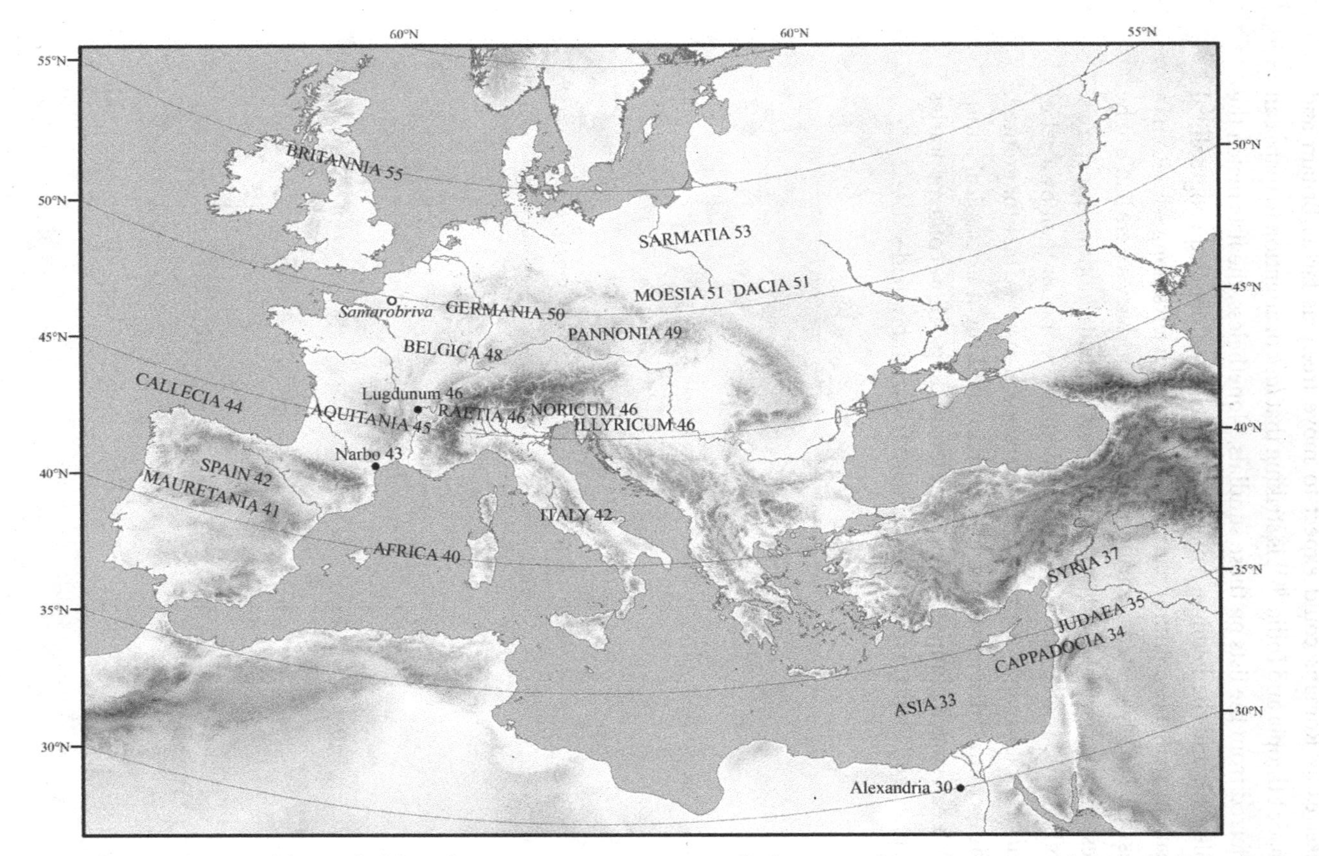

Fig. 18.10 Names on the portable sundial found at Vignacourt, France, marked on a modern map base, each with the latitude stated for it. Accurate longitude is assumed. For clarification, Samarobriva (modern Amiens, France) is added in italics; it is *not* named on the sundial. Map: Ancient World Mapping Center.

Rome dominated in all directions, one through which those who identified themselves as Romans could expect to move freely as far as Britain and Dacia, or Ethiopia and India.[46] It is striking that *lack* of attention to maps can be inferred from the lists on these sundials. Nonetheless, the lists match the Marble Plan and the Peutinger map in their basis of detailed geographical data subordinated to communicating Roman worldviews along with Roman values only loosely related to cartography. These worldviews were liable to be impressionistic, variable, and molded by a mix of mental notions, itineraries, and other lists, as well as some traditional literature, rather than by a set of shared, accurate images comparable to the modern maps we take for granted. Practical communication through maps remained slight in classical antiquity. We can snatch glimpses, but our understanding of how contemporaries conceptualized their surroundings remains frustratingly inadequate.

[46] For a similar conclusion, note now Savoie (2020), 57, 77; likewise Taccola (2021), 97. Jones (2018), 236, on the other hand, envisages greater practical use.

19

Roman Concern to Know the Hour in Broader Historical Context

To contribute to the volume in honor of Alexander Podossinov on his seventieth birthday is a much appreciated opportunity to pay tribute to the longstanding importance of his scholarship. So far as most colleagues in North America and Western Europe (myself included) were aware, the movement to question traditional thinking about ancient mapping and worldview only began during the 1980s, with fresh thinking stimulated especially by the publication of the first volume of *The History of Cartography* edited by Brian Harley and David Woodward in 1987. Unfortunately, however, the barrier of language prevented us[1] from realizing then that, in Russian publications dating to the 1970s, Alexander Podossinov had already initiated such questioning, as his magisterial extended review (coauthored with Leonid Chekin) of *The History of Cartography* volume modestly documented.[2] Ever since, he has continued to publish across an expanding range, and we are all in his debt. I hope that the present contribution will appeal to him as an illustration of how my own work on worldview has led in turn to a focus on time, specifically Romans' consciousness of time in daily life. Here I advocate applying comparative history and social psychology to this aspect of Roman culture, which in my estimation offers untapped potential for such approaches.

Let me acknowledge at once that the topic has already received ample attention from scholars over a long period.[3] Even so, a basic consensus has yet to be reached, although my argument is that by now it should be; at least one lingering obstacle is identifiable, and can be overcome. I proceed to urge that, with this foundation laid, the opportunity (not to say necessity) at once

[1] For Pietro Janni as a striking exception, however, see Janni (1984), 14, and Indice degli Autori Moderni, s.v.

[2] Podossinov and Chekin (1991), 118 n. 31.

[3] Note, for example, Dohrn-van Rossum (1996), 17–28; Laurence (2007), 154–66; Hannah (2009), 134–44; Houston (2015).

arises for setting and re-evaluating Roman consciousness of time in daily life within a broader historical context than has been attempted to date.

My engagement with this topic is serendipitous, arising from the research undertaken for my study *Roman Portable Sundials: The Empire in Your Hand* (2017).[4] The focus adopted there, however, was a counterintuitive one, settled upon in frustrated reaction to scholars' unwavering confinement of their interest in Roman portable sundials to the science behind their design and the accuracy with which they tell the time. Justified though these choices unquestionably are, their narrow scope leaves much else of significance unexplored. So, with the interest in geography, maps, and worldview that I share with Alexander Podossinov, I deliberately opted to focus instead on a distinctive but ignored feature of some of the portable sundials: namely, the inclusion of a list—normally inscribed on the reverse of the bronze disc—of cities or provinces or regions (as many as thirty-six), each accompanied by a latitude figure. This is the latitude to which the sundial should be set if it is to function satisfactorily in the particular location where the user happens to be during a journey. The varied contents of the dozen or so such lists known to date create the impression that most record an owner's personal set of favored locations rather than merely reproducing a workshop's standard selection.[5] Hence may emerge glimpses of individual Romans' spatial awareness and worldview—insights to be linked with others from an intriguing mix of testimony, and all the more precious in light of Romans' evident disinterest in maps.

In reacting to the book, colleagues have reasonably observed that the novel choice of focus marginalizes the opportunity to explore large issues of Roman time-consciousness that the subject by its very nature raises.[6] I concur that further investigation of this aspect is warranted, and welcome the coincidence that an ideal basis for the purpose appeared shortly before my book (though too late to be taken into account): the authoritative catalog and overview edited by Alexander Jones for the memorable exhibition *Time and Cosmos in Greco-Roman Antiquity* at New York University's Institute for the Study of the Ancient World in 2016–2017. In the introduction Jones formulates a key question and then proceeds to offer an unsatisfactory answer in my view (fig. 19.1):[7]

[4] See in this connection essay 18 above.
[5] Talbert (2017), 147.
[6] Note, for example, the reflections of Riggsby (2017) in advance of his monograph on representing information (2019).
[7] Jones (2016), 28–29. For more forthright negative responses to the same question, note Remijsen (2007), 127–30, 140; Schomberg (2018), 322.

Fig. 19.1 Mosaic panel (third century CE?) from the "House of the Sundial" at Daphne, near Syrian Antioch (modern Antakya, Turkey). A man facing a column looks up to check the sundial at the top, his hands anxiously clenched. The pithy Greek ΕΝΑΤΗ ΠΑΡΗΛΑΣΕΝ—"Ninth [hour] has passed"—expresses his dismay at being late for dinner. See Dunbabin (2019), 340–42; cf. frontispiece above and fig. 19.6. Photo: Art Resource, NY (420080).

Confronted as they were with time-keeping devices wherever they turned, we may wonder about the degree to which the ancient Greeks and Romans came to regulate their activities by the numbered hours. The testimony of the surviving Greek and Latin literature and of Greek documentary papyri . . . suggests that, in private life, hours were seldom invoked; the

> conventional "ninth hour" for dinner invitations . . . is perhaps the only important exception.

There is an underlying assumption here that a single answer can serve for the entire long span of Greek *and* Roman civilization. I suggest rather that this is an impossibility even in the most general terms, and that at the very least the two civilizations should be considered separately, because substantial differences are unmistakable. My focus here is confined to Romans. Édouard Ardaillon over a century ago was closer to the mark than Jones, I believe, in the conclusion to his lengthy article "Horologium" for the Daremberg and Saglio *Dictionnaire*:[8]

> La division horaire du jour n'eut jamais chez les Grecs grande influence sur la distribution des occupations. . . . Il n'en était pas de même chez les Romains. La vie publique et privée était reglée heure par heure.

To be sure, the hours referred to by both Jones and Ardaillon are "seasonal" ones of variable length (Greek ὥραι καιρικαί), not uniform periods all the same length—in other words, not the "equinoctial" hours (ἰσημεριναί) that are today's norm. Both Greek and Roman practice was to regard each day from sunrise to sunset as divided into twelve segments or hours each of the same length. Each night from sunset to sunrise was taken to be similarly divided, giving a total of twenty-four hours within each day and following night (or vice versa, νυχθήμερον). Naturally, in consequence each such daylight hour increases or reduces in length day by day, while the nighttime ones correspondingly reduce or increase. From one day or night to the next the difference may be barely noticeable, but from season to season it is marked in the Mediterranean zone and increasingly so at latitudes further north. At Rome itself, for example, the difference in length between a daylight seasonal hour at the winter solstice and one at the summer solstice is about half an equinoctial hour. How severely reference to seasonal hours limited—to our modern way of thinking—the usefulness of timekeeping by this means is an issue to return to. Suffice it to affirm now that, despite the limitations as we may perceive them, Romans' time-consciousness nonetheless remained largely based upon such hours.

[8] Ardaillon (1900), 263.

Jones's conclusion falls short because it is based only on the testimony of literature and papyri, without also taking epigraphic testimony into consideration. In the Roman case with its wealth of relevant inscriptions, the omission is serious and needlessly creates an obstacle to full understanding that can and should be overcome. Moreover, unlike literature in particular, the epigraphic testimony is invaluable for its reflection of both male and female experience at all levels of society.

As I now proceed to present a variety of epigraphic testimony in support of the claim that concern to know the hour made a deep and lasting impact throughout Roman society, it may be instructive to note criteria developed by Paul Glennie and Nigel Thrift for a comparable context—namely, an attempt "to index revolutions in everyday clock-timekeeping practices" in England between 1300 and 1800, as mechanical clocks and watches were invented and developed:[9]

> First . . . we need to ask what statements clocks and other timekeeping devices made through the ways and contexts in which they were used. Second, since we are primarily interested in everyday life, and in how new things became incorporated into "normal" practice, we need to ask now how clocks and other timekeeping devices populated everyday life. Third, since we are interested in embodiment, we need to ask how clock time was taken into the body, for example, through increasing nervousness about timekeeping or punctuality. . . . Last . . . we need to ask what timekeeping practices circulated in which communities and what significance they had. We also need to ask how these practices both related to and differentiated communities of practice from each other.

For certain, a large sundial erected in a prominent location at the expense of an individual keen to secure popular favor was a typical sight in Roman communities and a demonstration of civic pride.[10] According to a damaged inscription,[11] the major port in Campania, Puteoli, had been presented with such a sundial by an emperor (no longer identifiable), and

[9] Glennie and Thrift (2005), 169. A monograph on this topic by them (2009) followed; note Turner's review (2012).

[10] So, for example, the sundial unearthed at Interamna Lirenas (Latium) in 2017; it was erected during the mid-first century BCE at the expense of a local man who was a Roman senator ("The inscribed sundial," www.classics.cam.ac.uk/interamna). Note also another erected at Mevania by its quaestors around 100 BCE, inscribed in the Umbrian alphabet: Crawford (2011), I.122–23.

[11] *CIL* X.1617 with Winter (2013), 509.

Fig. 19.2 Memorable features of Puteoli, Italy, as engraved on a glass flask in the World of Glass museum, St. Helens, UK. The T-shaped feature that may be a large sundial can be seen upper right. Figure: Ancient World Mapping Center, after P. Compton in Painter (1975), fig. 13.

it was later repaired at the city's expense. I identify it as the T-shaped feature depicted on a glass flask—perhaps of third- or fourth-century date—in all likelihood designed and marketed as a tourist souvenir of Puteoli's principal attractions.[12] If STADIVSOLAR placed above this feature on the flask, and AMPITHEAT below, can be taken to provide some indication of where in the city this sundial was erected, the choice of location seems characteristically Roman (fig. 19.2).

How typical it was for public entertainments in Roman communities to be scheduled with reference to hours remains unclear,[13] but there is explicit epigraphic testimony in one striking instance. This is an announcement so matter-of-fact as to leave the impression that neither the magistrates who issued it, nor the general public for whom it was intended, would have regarded reference to hours as out of the ordinary. The occasion is Augustus' *Ludi Saeculares* celebrated in Rome in 17 BCE:[14]

> XVvir. s. f. dic.: Ludos, quos honorarios dierum VII adiecimus ludis sollemnibus, committimus nonis Iun. Latinos in theatro ligneo quod est ad

[12] World of Glass museum, St. Helens, UK, inv. SAHGM 1974.002; photograph in Talbert (2017), 7. The thorough reexamination by Popkin (2018) does not focus specifically on this feature, but see her n. 74.

[13] Cf. Fagan (2011), 209–14, for gladiatorial shows.

[14] *ILS* 5050 lines 155–58 with Bruun and Edmondson (2015), 402, 544.

Tiberim h. II, Graecos thymelicos in theatro Pompei h. III, Graecos asti[cos i]n thea[tro quod est] in circo Flaminio h. I[III].

The *quindecimviri sacris faciundis* declare: We have taken the initiative of adding seven days to the traditional shows, and we are beginning on the Nones of June with: Latin shows in the wooden theater which is by the Tiber at the second hour; Greek dance-shows in Pompey's theater at the third hour; Greek city-shows in the theater which is in the Circus Flaminius at the fourth hour.

It is also unclear how typically Roman public baths announced their opening times in terms of "from hour X to hour Y,"[15] but again a valuable instance—of concern to both sexes—confirms the possibility.[16] This is a regulation of Hadrianic date or later in the second century about one of the services to be provided by a contractor in the settlement near the imperial copper and silver mines in the Vipasca district in Lusitania. The length of time for which the bath here must stay open extends into the evening no doubt in order to accommodate miners who worked in shifts:[17]

Conductor balinei sociusve eius omni{a} sua inpensa balineum, [quod ita conductum habe]bit in pr. K. Iul. primas, omnibus diebus calfacere et praestare debeto a prima luce in horam septim[am diei mulieribus] et ab hora octava in horam secundam noctis viris arbitratu proc. qui metallis praeerit.

The contractor of the bath or his associate is to heat the bath every day and keep it open for use, entirely at his own expense, as stipulated in the lease for the bath that runs until 30 June next, for women from first light to the seventh hour of day, and for men from the eighth hour to the second hour of night, in accordance with the decision of the procurator who runs the mines.

Late first-century BCE Puteoli offers an instance of another service provided by a city—again through a contractor—according to inscribed

[15] Fagan (1999), 326.

[16] Compare the inscribed notice at the late first-century CE library of Pantainos in Athens stating when it would be open (first to sixth hours): *SEG* 21.500 with Nicholls (2017), 37.

[17] *ILS* 6891 lines 19–21 with Bruun and Edmondson (2015), 690. At Puteoli too during the late first century BCE a bath must have remained open beyond the first hour of the night, because a contractor's slave workhands there (see next paragraph) were banned from bathing until then (*AE* 1971.88 II lines 3–4). For Pompeii, note Laurence (2007), 158.

regulations posted publicly that refer to hours: the removal of dead bodies (by slave workhands, evidently during daylight only). While there is no cause to assume that this service was ever typically available in Roman communities, the manner in which one clause is framed takes well-developed time-consciousness (and class-consciousness) among the populace for granted:[18]

> Suspendiosum cum denuntiat(um) erit ead horad is solvend(um) tollend(um) curato, item servom servamve si ante h(oram) X diei denuntiat(um) erit ead die tollend(um) curato, si post X poster(a) d(ie) a(nte) h(oram) II.

> Should instructions be given [to remove] a hanged man, he (the contractor) is to see to their fulfilment and the removal within the hour. In the case of a male or female slave, if instructions be given before the tenth hour of the day, removal is to be effected that day; if after the tenth, on the next day before the second hour.

Within Rome itself, keen attention to the tenth hour of the day can be assumed, because only from then until dawn did a law of Caesarian date (preserved on the *Tabula Heracleensis*) allow most wheeled vehicles onto the city's streets.[19] Similar well-developed time-consciousness is attested at Ostia, albeit in just a single specific context, but an instructive one. It features in the fully preserved inscription on a large marble base for a statue of Publius Horatius Chryseros, deceased *Sevir Augustalis*, erected in 182 CE by his fellow Seviri in gratitude for his benefactions to their *collegium*. These include a capital sum of 40,000 sesterces, with a stipulation:[20]

> . . . <ut> ex usuris semissibus et M′ II s(ummae) s(upra) s(criptae) quodannis Idib(us) Mart<i>is natali suo inter praesentes hora II usque ad asse<m> dividiatur . . .

> . . . that the interest, 6.4 per cent, of the amount stated above be distributed to the last *as*, every year on March 15, his birthday, among those [Augustales] present at the second hour . . .

[18] *AE* 1971.88 II lines 22–23 with Bond (2016), 69–70.
[19] Crawford (1996), no. 24 lines 56–67 (pp. 365, 374, 382), with Laurence (2013), 253–55.
[20] *ILS* 6164 = Gordon (1983), no. 68 with bibliography and photograph (plate 44).

Disqualification of laggards not present at the headquarters of the Seviri Augustales in person by the prescribed hour could be thought Chryseros' twist—the testamentary quirk of a micromanaging freedman upstart (whose area of success in business is not known). That notion appears less likely, however, when the surviving fragment of an almost identically worded inscription dating to the 230s is taken into account.[21] It honors Q. Veturius Felix Socrates, whose birthday fell on the Ides of June. The fact that half a century later the annual distribution of interest from a deceased Sevir's bequest likewise required attendance at the second hour points rather to the *collegium* itself being responsible for the stipulation. How far it predates 182 is impossible to say.

For the conduct of Roman court business with reference to hours, there is epigraphic testimony that can be confidently reckoned to have widespread significance. First, the law (*Lex Coloniae Genetivae*) originally of Caesarian date that gave Urso in Baetica its constitution was assuredly replicated more or less in other colonies. Hence the time-limits, or ones like them, prescribed here for court hearings may be a Roman norm:[22]

> IIvir qui h(ac) l(ege) quaeret iud(icium) exercebit, quod iudicium uti uno die fiat h(ac) l(ege) praestitu<tu>m non est, ne quis eorum ante h(oram) I[23] neve post horam XI diei quaerito neve iudicium exerceto.
>
> A duovir who shall hold a *quaestio* and shall administer a trial according to this statute, insofar as it is not laid down by this statute that the trial is to take place on one day, none of them is to hold the *quaestio* or administer the trial before the first hour nor after the eleventh hour.

Second, and more specifically, *vadimonia* (bail-bonds)—written on wax tablets, and unearthed in the Pompeii area—confirm the location, date, and hour at which the defendant in a private dispute undertakes to appear for a hearing or its continuation, as in this routine example from the mid-70s CE where both plaintiff and defendant happen to be female. The third hour is

[21] *CIL* XIV.431.

[22] Crawford (1996), no. 25 clause 102 (pp. 409, 445). Note that the first day (June 1) of the great conference of Catholic and Donatist bishops at Carthage in 411, with Flavius Marcellinus *iudex in loco principis*, was finally brought to an end after the declaration by Rufinianus *scriba*: "Exemptae sunt horae undecim diei" (*Gesta* I.219 Lancel with Shaw [2011], 580).

[23] L on the stone, but this has to be a slip.

the one most often stipulated in these documents, but instances of the first, second (as here), fourth, fifth, and even ninth are attested:[24]

Vadimonium factum Cala
toriae Themidi in iii Non(as)
Decemb(res) prim(as) [*vacat*]
R[o]mae in foro Augus(to) ante
tribunal praetoris urbani
hora secun[d]a HS M dari
stipulata es[t] ea q[uae] se
Petroniam [*Sp.* f. Iustam]
esse dicat, s[po]po[ndit]
Calatoria [Them]is t(utore) a(uctore) C
Petronio Tel[e]sph[o]ro.
V. f. [C. Petronio Telesphoro].

A *vadimonium* was made against Calatoria Themis for December 3 in Rome in the Forum of Augustus before the tribunal of the *praetor urbanus* at the second hour. She who identifies herself as Petronia Iusta, daughter of [*name lost*], stipulated for, and Calatoria Themis, on the authority of her *tutor* C. Petronius Telesphorus, answered for, payment of 1,000 sesterces.

Only in 2012 was attention adequately drawn by Simeon Ehrlich to the remarkable quantity of Latin inscriptions known—mostly funerary—that either record the length of the deceased's life (or a memorable part of it) down to the number of hours on the last day, or state the hour of death. The first of the two representative instances that follow is from Theveste in Africa Proconsularis, the second from Pisae in Etruria:[25]

D(is) M(anibus) S(acrum): C(aius) Iulius Fortunatianus v(ixit) a(nnos) XVII d(ies) XXV ho(ras) VIII. Post Fabia(m) Fortunata(m) matre(m) pia(m) s(uam) v(ixit) a(nnos) III m(enses) VI d(ies) XXI (h)o(ram). H(ic) S(itus) E(st).

[24] *Tabulae Herculanenses* 14 with Bablitz (2007), 17–18. See further Laurence (2007), 155; Bruun and Edmondson (2015), 310–14.
[25] *CIL* VIII.27884 = Ehrlich (2012), no. 136; *CIL* XI.1458 = Ehrlich (2012), no. 172.

Sacred to the spirits of the departed: Gaius Julius Fortunatianus lived seventeen years, twenty-five days, eight hours. He lived three years, six months, twenty-one days, one hour after his revered mother, Fabia Fortunata, [died]. Here he lies.

D(is) M(anibus) Aufidiae Victoriae coniugi bene mer(enti) fecit P(ublius) Veturius Martialis cum qua vix(it) annis XXV mensib(us) X dieb(us) XV horis VII s(ine) q(uerella) u(lla).

To the spirits of the departed: Publius Veturius Martialis made this for Aufidia Victoria, his well-deserving wife, with whom he lived twenty-five years, ten months, fifteen days, seven hours, without any complaint.

Ehrlich's painstaking searches assembled over 800 such inscriptions, an impressively high total given the low survival rate of inscriptions. When the deceased is a very young child, mention of hours may indeed be intended primarily to underscore the brevity of the life so soon ended. Otherwise it is Ehrlich's persuasive hypothesis that inclusion of hours in these records reflects an obsession with astrology, or more exactly genethlialogy, the science of births and casting horoscopes.[26] From the configuration of the heavens at the time and place of birth, an individual's character and fortunes could be predicted. A record of the hour of death affirmed belief that the deceased was about to be reborn to an afterlife in the Underworld. With knowledge of this hour, a determination of the prospects to be expected there could be made, in particular which divinity the deceased would gain as guardian.

No less important, however, for reference to hours in general are two distinct findings that emerge from Ehrlich's analysis of his material ("sample" in the quotation below). The first is the wide cross-section of society encompassed, as he states:[27]

It will suffice to note that the sample does include tombs both commemorating and set up by soldiers, slaves, freedmen, plebeians, and equestrians; pagans and Christians; men and women; the very young and the very old.

[26] Ehrlich (2012), 84–85.
[27] Ibid., 26.

Also of note, second, is the extensive geographical range across which these inscriptions were erected. The high proportions in Rome, as well as in Italy and North Africa, must be duly recognized, but even so the distribution spans the empire.[28]

In both town and country the hour served as a standard means by which the right to draw water was regulated in Roman communities.[29] As the mid-second-century CE jurist Pomponius[30] pronounced:[31]

> Si diurnarum aut nocturnarum horarum aquae ductum habeam, non possum alia hora ducere, quam qua ius habeam ducendi.
>
> Should I have the right to draw water during day or night hours, I cannot draw it at any other hour than that for which I may have the right to draw it.

Two similar inscribed records stating the quantity that various individuals may draw, and between which hours, are known from Latium.[32] Both are only fragments, with the meaning of much in their texts uncertain, but the way hours are specified is clear enough. The record (fig. 19.3) found in the area of Tibur (and now lost) states the start of one period for which water may be drawn as the first hour of the night (lines 7–8). This makes it conceivable that the tenth hour, when another period ends (line 16), is two hours before dawn rather than in the late evening, but it is impossible to know. The other record—found in a garden on the Aventine hill—presumably preserves timings that relate to some unidentifiable part of Rome (fig. 19.4). At bottom center on the fragment, the right to draw from two channels (*aquae*) from the second hour to the sixth is clear; note also (bottom right) a period from the sixth hour to sunset (*occa[sum solis]*).

Use of hours as a means to regulate the drawing of water is documented far more fully, and for an extensive rural area in a province, in a detailed decree issued by the *municipium* of Lamasba, a thriving frontier settlement in Numidia, not far northwest of Thamugadi.[33] The decree dates to the reign of Elagabalus in the early third century, and affirms decisions made by arbitrators appointed to resolve disputes of a type common among

[28] Ibid., 56.

[29] Cf. Julianus in *Dig.* 43.20.5.

[30] *PIR*² (1998), P 694.

[31] *Dig.* 43.20.2; for drawing water at night regarded as a normal practice, cf. Paulus in *Dig.* 39.3.17 pr.

[32] *CIL* XIV.3676 (Tibur) and VI.1261 (Rome) with Shaw (1982), 74–76.

[33] *BAtlas* 34D2.

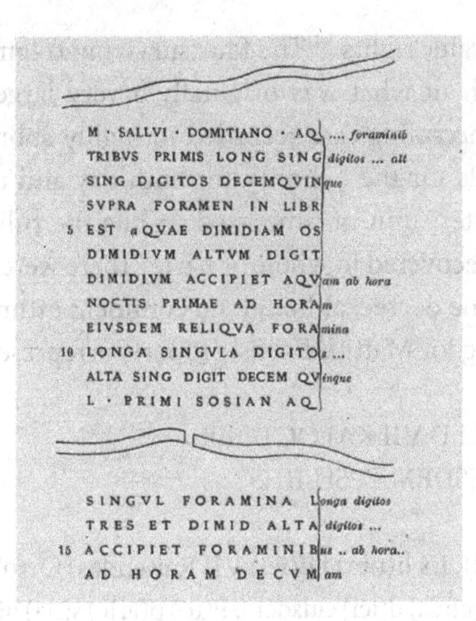

Fig. 19.3 Fragment of an inscribed record stating the quantity of water that various individuals may draw, and between which hours (*foramen* = sluice-gate). Found in the area of Tibur, Latium, and now lost. Figure: *CIL* XIV.3676.

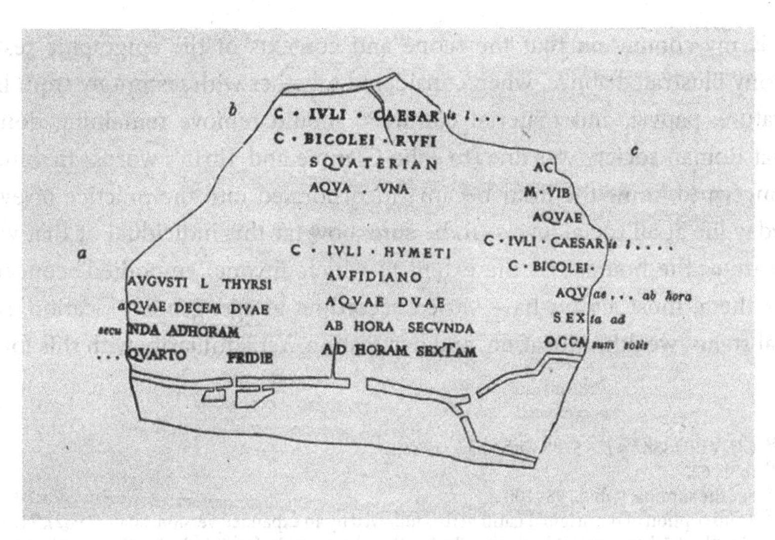

Fig. 19.4 Fragment of an inscribed record stating the quantity of water that various individuals may draw, and between which hours (*aqua* = water-channel). Found in a garden on the Aventine hill, Rome. Figure: *CIL* VI.1261.

landowners over water rights.[34] The four surviving fragments preserve perhaps only one-fifth of what was originally a very large stone inscription (fig. 19.5a, b).[35] The rulings are recorded in highly abbreviated form; they are seemingly made for the fall only, for both day and night hours, with a half hour the shortest unit of time used. While the rulings for eighty-five properties can be recovered in whole or part,[36] there were clearly many more in the lost part of the decree, although no confident estimate of the total can be made. The ruling for Mattius Fortis' property is representative:[37]

KCCCVIII EX.H.I.D.VII KALOCTOBR
INHVS.D.EIVSDEM.PPSH.IIIIS

K(apita) CCCVIII. Ex h(ora) I d(iei) VII Kal(endas) Octobr(es)
in h(oram) V s(emis) d(iei) eiusdem. P(ro) p(arte) s(ua) h(orae) IIII s(emis).

308 Kapita [*land area*]. From the first hour of the day on [*starting-date*] September 25 until the fifth hour and a half of the same day. For his plot, four and a half [*hours*].

*****

It is my contention that the scope and contexts of the epigraphic testimony illustrated above, when considered together with testimony from literature, papyri, and material culture,[38] should remove remaining doubt that Roman society was one (to echo Glennie and Thrift's words) in which concern to know the hour became incorporated into the practice of everyday life at all social levels. To be sure, how far this individual or that was aware of the hours, and the extent to which anyone "embodied" concern for them, must always have varied according to age, gender, location, social status, wealth, education, and occupation. Yet familiarity with this form

[34] *CIL* VIII.18587 = *ILS* 5793 with Shaw (1982).
[35] Ibid., 62.
[36] See the summary, ibid., 98–100.
[37] Transcription of the original Latin here is followed by an expanded version. Shaw (1982), 71, errs in taking the 4.5-hour period here to be during the afternoon. In fact it is during the morning, from the first hour of the day, the point which in Roman reckoning signifies the *end* of the first of the twelve daylight hours (the *beginning* of that hour is *prima lux*, dawn); midday, the sixth hour, is the point reached at the *end* of that hour.
[38] For literary evidence especially, see "The Shape of the Day," in Balsdon (1969), 16–55; note the analysis of *hora* in *TLL* s.v., including the common expression *omnibus horis*, used thrice by Pliny in *NH* 2 for example, sects. 22, 109, 157 (*TLL* VI.3 col. 2956 line 51).

(b)

Fig. 19.5a, b The four surviving fragments— as preserved today (a)—of a Latin inscription on stone regulating the hours at which each landowner in the community of Lamasba (in modern Algeria) may draw water. Detail (b) shows the first five entries in the furthest left column recording each owner's name and allotment, beginning with Mattius Fortis. Photo: Musée Publique National des Antiquités, Algiers (inv. I.S.175-42880 N).

of daily timekeeping—which did not call for literacy—was evidently taken for granted as the norm. As Artemidorus expansively declared during the second century CE:[39]

> ὡρολόγιον πράξεις καὶ ὁρμὰς καὶ κινήσεις καὶ ἐπιβολὰς <τῶν> χρειῶν σημαίνει· πάντα γὰρ πρὸς τὰς ὥρας ἀποβλέποντες οἱ ἄνθρωποι πράσσουσιν.
>
> The sundial symbolizes deeds and actions and movements and the undertaking of business; for people do everything with an eye to the hours.

If the familiarity illustrated here be accepted as an adequate basis on which to proceed, the next step ought to be an inquiry into when and how Roman society adopted this particular form of daily timekeeping. Roman tradition unequivocally acknowledged that it was a foreign practice in origin, one whose introduction was linked to the importation of fixed sundials to Rome. Their arrival was evidently traced in Marcus Terentius Varro's mid-first-century BCE *Antiquitates Rerum Humanarum et Divinarum*, but nothing more than brief references to his account survive in Pliny the Elder's *Naturalis Historia* (70s CE) and Censorinus' *De Die Natali* (238 CE).

Both these authors mention controversy over the identification of Rome's first sundial. Pliny records a claim passed on by an otherwise unknown Fabius Vestalis that a sundial erected in the late 290s BCE was the first; but he then hesitates to believe that, in part because Vestalis supplies so few details, including his source and where the sundial originated.[40] Pliny therefore leans toward acceptance of Varro's claim that a sundial looted from the Sicilian Greek city Catina (modern Catania) early in the First Punic War, and imported during the late 260s, was the first. It was erected in the forum and, according to Censorinus,[41] there was at least no dispute about it being the first placed there. The likelihood of Roman unfamiliarity with sundials at this period seems confirmed by the notorious sequel that, because the one imported during the 260s had been designed for a latitude in Sicily, it failed to function with accuracy at Rome's latitude considerably further north, although not till the 160s—a century later—did the problem attract comment.[42]

[39] *Oneirocritica* 3.66.
[40] *NH* 7.213 with commentary by Beagon (2005), 470.
[41] *DN* 23.7 with commentary by Freyburger (2019), 118–19.
[42] *NH* 7.214 with Beagon (2005), 471.

Censorinus explains, with many examples, that before sundials were brought to Rome, terms for divisions of the day were merely loose for the most part, such as *gallicinium* (cock-crow) and *de meridie* (afternoon).[43] He and Pliny claim that in the collection of early Roman laws, the Twelve Tables, no reference was made to hours, but just to sunrise, sunset, and "before mid-day" (*ante meridiem*).[44] Only in the 240s or later did a Lex Plaetoria refer to an hour.[45]

However, by the early second century the parasite in a fragment of an otherwise lost comedy, attributed to Plautus by Varro, curses whoever invented hours that fragment his life, and he can only expect to amuse the audience with his moans that the town is filled with sundials (*oppletum oppidum est solariis*) which now control his dinner-time.[46] In another of Plautus' comedies, the leading character, Pseudolus, when taunted for the quantity of wine he is capable of consuming in one hour, reacts boastfully "A winter hour, what's more" (*"hiberna" addito*), in other words a shorter one.[47] In his *De Agricultura* (written around 160), Cato twice refers to hours matter-of-factly—when explaining a fattening regime for hens and geese, and specifying the stages in a cure for indigestion.[48] Precise indeed, if from a contemporary (annalistic?) source, is Livy's mention that L. Aemilius Paullus' envoys bringing news of the victory at Pydna in 168 reached Rome "at almost the second hour."[49]

Plausible though it may be that importation of sundials to Rome during the third century led to the adoption of foreign timekeeping with hours, the scope of the account surviving in our scrappy sources is frustratingly limited. Its character foreshadows the modern predilection for assuming uncritically that societies will embrace technology and its advances without hesitation. In fact there could have been no certainty, or even expectation, that the erection of a trophy-sundial or two would lead third-century Roman society to embrace the timekeeping practice that these curiosities made possible. The shift was evidently spontaneous (never enacted by legislation) and established within a century. To explain it just in terms of technology cannot suffice. Hence I perceive the opening for a social-psychological approach to expand

[43] *DN* 23.9–24.6 with Freyburger (2019), 120–22.

[44] Censorinus, *DN* 23.8 with Freyburger (2019), 119–20; Plin. *NH* 7.212 with Beagon (2005), 469. Cf. Crawford (1996), no. 40 I, 7–8 (pp. 579, 592–95).

[45] Crawford (1996), no. 44.

[46] This fragment of *Boeotia* is preserved by Gell. *NA* 3.3.5. Gratwick (1979) argues convincingly that here Plautus is adapting a passage in the original comedy (also now lost) by Menander.

[47] *Pseud.* 1304. The shorter the hour, the more impressive the rate of consumption. Strangely, Hannah (2009), 136 thinks that Pseudolus "lessens the charge" by specifying a winter hour.

[48] 89.1, 156.4, references indicating that he did not resist some Greek practices despite his expressions of hostility to Greek influence in general.

[49] Livy 45.2.2, "hora fere secunda."

the range of possible influences, such as the growth of urban populations perhaps or the pressures of the Punic Wars, not to mention concern for social order and cultural superiority.

Further avenues of inquiry emerge in consequence. It seems that scholarship to date has unquestioningly regarded the fourteenth- and fifteenth-century so-called revolution in daily time-consciousness—stemming from the invention of the mechanical clock which records equinoctial hours—as a unique phenomenon in Europe's social history. For certain, major changes can be attributed to this invention (investigated for England by Glennie and Thrift). Jacques Le Goff's contentions first published in 1960—as summarized by Vanessa Ogle—still hold good:[50]

> Le Goff proposed that the emergence of a commercial market society in fourteenth- and fifteenth-century Europe was accompanied by a growing utility for marking time, hence the proliferation of work bells, tower clocks, and other visible and audible signals denoting various times. Ever since, the emergence of capitalist society has been closely tied to the proliferation of clocks and the honing of time consciousness.

However, I see cause to regard the shift made by Roman society in embracing timekeeping by hours as an earlier, overlooked "revolution" in consciousness comparable to this European experience during the fourteenth and fifteenth centuries. The Roman shift was by no means identical of course, nor did it turn out to act as a springboard for change on the same scale. Even so, beyond inquiry into its causes, there is solid reason to raise the related questions of how the rhythm, behaviors, and cohesion of Roman society came to be affected—and enhanced perhaps—in consequence. The experience of Late Ottoman Beirut around 1900 as explored by Ogle is instructive. There was no external force pressing the ethnically diverse population of this provincial capital to change its ingrained attitudes to timekeeping and time management, yet the proliferation of calendars and watches and the new presence of clock time in public space spurred a transformation:[51]

> Economic gain was not the primary goal in the end but rather, the self-improvement of an entire society and "nation."

[50] Ogle (2015), 11, with reference to Le Goff (1980).
[51] Ogle (2015), 137. For comparable thinking in Japan, see below.

In further evaluation of the Romans' shift, we may fairly seek to identify elements in their society and its attitudes that led them to adopt and maintain a consciousness of hours far more active and pervasive than any found previously in the societies of Greece and Magna Graecia or Egypt or Mesopotamia.[52] Despite the sundial's origin and development in the two latter regions, consciousness of the time of day generated by it evidently remained limited there. In Greek society too, nowhere is there evidence of this consciousness having grown to the level that became the norm in Rome within a century and (to recall Glennie and Thrift's terms) would differentiate it from other communities.[53] Perhaps Romans were energized by the lively concern for time beyond the day that their calendar already reflected,[54] but this cannot serve as a full explanation when the same concern was by no means rare or novel in many Greek societies, Athens especially.

Timekeeping by hours and exceptional preoccupation with it were assimilated among the distinctive practices and values that Romans maintained as they hugely expanded the extent of their control and settlement. Pliny singles out *horarum observatio* as one of the three instances of *gentium consensus tacitus* (along with the Ionian alphabet and shaving!).[55] Tacit agreement among peoples may be too idealistic a view in this instance, however. In reality such *observatio*, especially at the level favored by Romans, was a Roman export, one at risk of not meshing with the established practices of communities brought under Roman control. As is well known, the potential for such tension surfaced often in connection with eras and calendars, and the issue was handled by Roman authority in a variety of ways across the empire according to local circumstances. While Roman practice was promoted, established local eras and calendars were by no means suppressed.[56] Whether comparable tension arose between local daily timekeeping practices and that of the Romans is obscure.[57] Conceivably, struggles did occur of the type explored by Ogle, where the local populace of Bombay stubbornly resisted a shift to the time of Madras (a rival city!) in 1881–1883 or to Indian Standard Time from 1905, exploiting the opportunity to protest not only that it was

[52] For Egypt, note Kadish (2001); Schomberg (2018); for Mesopotamia, Englund (1988).

[53] For the fourth century BCE as the earliest period of Greeks' engagement with sundials, and the earliest at which ὥρα was used to signify "hour" rather than "season," see Hannah (2009), 70, 112–14.

[54] For concise appreciation see, for example, Feeney (2010).

[55] *NH* 7.212; 7.210.

[56] For overview, see Dench (2018), 136–47.

[57] For awareness that the point from which a day is reckoned to begin varies among societies, note Plin. *NH* 2.188, a passage later quoted by Gell. *NA* 3.2.4–11 as Varro's observations.

"'British' time being imposed on colonial subjects" but also that the new daily rhythm would be artificial and unnatural.[58] However, evidence of such a clash involving Rome so far eludes me, and the risk was at least reduced when Rome's means of daily timekeeping remained no more than customary practice, never one legally enacted (unlike the Julian calendar).[59]

Sentiment in Bombay chose to value local interest above one powerful motive for the British authorities to standardize timekeeping in India: the need to facilitate the rapid growth of long-distance communication and travel—especially by rail—in the late nineteenth century.[60] For Romans, by contrast, there was never the need for coordination between widely separated points. Timekeeping remained localized, with its value reduced by sundials' design flaws as well as by the variable length of hours.[61] Mass production of uniform sundials was not attempted;[62] indeed, no two of the 650 or so in the Berlin database are identical.[63] No reading more precise than the hour was typically expected (hence the absence of markers within successive hour-lines on a sundial), as Cetius Faventinus twice insists in his abbreviated version of Vitruvius' instructions for constructing a sundial: "nearly always in their haste everyone merely inquires what hour of the day it is."[64] Even then, the many shoddily produced or poorly sited (seldom in sunlight) sundials would be inaccurate or useless, and it was a standing joke that different sundials did not report the same time.[65]

Yet for all these shortcomings Romans persisted in checking sundials undaunted, and often quite plainly for a variety of practical purposes, not just out of snobbery or love of gadgets or sheer habit.[66] The habit, once "embodied," might indeed prove hard to break.[67] Familiarity meant that the absence of numerals for the hours on most sundials hardly mattered, and

[58] Ogle (2015), 109, 112. The subtitle of Robert Levine's study *A Geography of Time* (1997) underscores his focus on societies' varying daily rhythms: *The Temporal Misadventures of a Social Psychologist, or How Every Culture Keeps Time Just a Little Bit Differently*.

[59] Compare the vicissitudes of sundry twentieth-century efforts (British, Japanese, American) to impose and maintain time-changes in Greece, Turkey, and Korea: Ogle (2015), 94–95.

[60] For this need in broader context, see Ogle (2015), 24–26.

[61] Note discussion by Riggsby (2019), 89, 106, 118, 122–23.

[62] Epigraphic testimony indicates some demand for sundials at military bases: Talbert (2017), 140, 191–201.

[63] http://repository.edition-topoi.org/collection/BDSP. Probably the same can be claimed of the 586 sundials listed by Bonnin (2015), 387–401.

[64] 29.2, cf. 4: "non amplius paene ab omnibus nisi quota [sc. hora] sit solum inquiri festinetur."

[65] Sen. *Apocol.* 2.2.

[66] For possible use by doctors, Galen in particular, see the discussion by Miller (2018). She cautions (121): "it is difficult to say how unique Galen was in privileging hourly schemes within his clinical narratives." Cf. 112–13 and limitations mentioned in note 6.

[67] Melechi (2020), 15, observes that modern trials to understand the effects of extended isolation reveal a strong desire on the part of their subjects to gauge the passing of time.

more specific reference to a half hour is not exceptional.[68] Despite the ubiquity of fixed sundials, there was clearly a demand for portable ones too that could be used on a journey; Vitruvius in the late first century BCE lists these as a distinct type that had attracted plenty of attention from designers.[69] The very wealthy relished having a slave whose function it was to tell them the hour, and Seneca mocks men who succumbed to the extreme of allowing the passing hours to dictate their daily schedule.[70] This said, the importance for guests of punctual arrival at dinner—with the threat of exclusion and its social penalties—emerges as a favorite cautionary tale featured in mosaic panels; new finds from Turkey made within the past decade are especially instructive (fig 19.6).[71] Indeed the tradition transmitted by Ammianus Marcellinus in his fanciful digression about the character of Persian society proudly represents its lack of a fixed dinner-hour as altogether uncivilized:[72]

> Nec apud eos extra regales mensas hora est praestituta prandendi, sed venter uni cuique velut solarium est, eoque monente quod inciderit editur.

> Among them, only at the king's court is there a fixed dinner-hour; otherwise each person's stomach serves as a sundial, and at its prompting they eat whatever is available.

An additional dimension of Roman concern for hours that merits greater recognition is the consciousness shown of nighttime ones—largely, it would seem, by confident instinct, since a sundial would by definition be useless and water-clocks were rare.[73] Even so, nighttime hours

[68] Less rare than Riggsby (2019), 96 n. 36 states. Cf. Remijsen (2007), 138, Ehrlich (2012), 48, and the Lamasba inscription above; possibly also the Frankfurt clepsydra published by Stutzinger (2001), 30 and Abb. 5.

[69] *De Arch.* 9.8.1 with essay 18 above. See Talbert (2017), 9–18, for examples in bone or bronze. Conceivably, more affordable ones were made in wood, too, of which no trace survives. Compare the pocket-size wooden ones (under 4 cm diameter) recovered from the English warship *Mary Rose,* which sank in 1545: Higton (2002), nos. 395–96.

[70] Hannah (2009), 137, with Sen. *De Brevitate Vitae* 12.6.

[71] The wide-ranging analysis in Dunbabin (2019) is invaluable.

[72] Amm. Marc. 23.6.77. Cassiodorus (*Var.* 1.46.3) expresses the same pride in a letter drafted for Theoderic to accompany his gift of a sundial and a water-clock for the Burgundian king Gundobad around 506. For Winston Churchill to demand meals at what he called "tummy time" (when hungry), rather than by the clock, was regarded as an eccentricity and a taxing one for his cook: see Gray (2020), 236, 249, 282.

[73] Wilson (2018), 62. Leslie Dossey (2018), 309–18, discerns that in the eastern Mediterranean during Late Antiquity the rhythm of urban life was modified by a continuation into the evening of what had previously been just daytime activities.

Fig. 19.6 Panel of a damaged floor mosaic in the entrance area to what was perhaps a clubhouse, dating to around 200 CE, discovered at Tarsus, Turkey, in 2012. In the adjacent panel to the left (not shown here), a man is about to dislodge a raven (bird of bad omen) from a sundial by throwing a stone, so that he can then learn the hour from seeing where the shadow falls. Next, in the panel here, with a wreath on his head he approaches a large door which is ajar, saying (above, in Greek): Ο ΘΕΛ[Ε]ΙΣ ΠΟΙ / ΕΙ ΕΙΣΕΛΘΕΙΝ / ΜΕ ΔΕΙ, "Do what you like, I need to come in." From within, a servant in a short tunic thrusts a pole which passes through the door and presses on the man's shoulder. He grips it, while the servant says (below his bare feet): ΠΑΡ ΩΡΑΝ ΗΛΘΕΣ, "You have come past the hour." See Dunbabin (2019), 331; cf. frontispiece and fig. 19.1 above. Photo: Tarsus Archaeological Museum.

are routinely specified in a range of contexts—for stipulating periods within which water may be drawn, for the closing-times of baths, for logging when couriers in Egypt's Eastern Desert arrived and departed,[74] and

[74] Remijsen (2007), 137–39; Cuvigny (2019), 78–79.

for recording times of birth and of death.[75] Also, as noted above, most wheeled vehicles were allowed on Rome's streets only between the tenth hour of the day and sunrise.

In urging that comparative history and social psychology be brought to bear on Roman daily timekeeping practice, I conclude by identifying the experience of Tokugawa Japan as a potentially rewarding case to set against that of Rome. In Japan, too, hours that varied in length according to season (six of light and six of dark every day) were used, a system traditionally thought to have been imported from China by Japan's rulers during the seventh century CE and not abandoned until 1873.[76] Japanese society became one where "the practice of timing even quotidian events [down to the half hour] was . . . an important part of one's daily routine."[77]

The Roman and Japanese experiences differ in two notable ways, however. First, in Japan there was a major effort to make the time of day known aurally and even to synchronize it by this means. Time-bells were rung from temples or towers for the purpose—according to some estimates, as many as 30,000 bells across Japan by 1800—so that one or more could be heard in every urban area as well as far beyond: "Times of work, rest, meals, and leisure—all were structured by the hours announced by the bells."[78] Roman society, by contrast, never developed any such practice. Pliny mentions the announcement of midday at Rome by an attendant (*accensus*) of the consuls before sundials were imported.[79] However, there seems to be no evidence for efforts later on the part of the authorities there or elsewhere in the empire to make any time of day known aurally;[80] hence the importance of sundials.[81] Even the Christian

[75] Varro evidently made some general reference to those born at the third, fourth, or any hour of the night (Gell. *NA* 3.2.1). To record both the birth and the death of an infant as occurring at the sixth hour of the night (i.e. at midnight) could be thought a form of approximation; yet note also the record of deaths at the seventh and tenth hours of the night (*CIL* VI.28044, 13883, 11511 = Ehrlich [2012], nos. 613, 459, 427). Pliny (*NH* 2.229) records the phenomenon of a spring (on Tenedos) overflowing between the third and sixth hours of the night.

[76] See Frumer (2018), 24, 29, 32, for the introduction of the system.

[77] Ibid., 35.

[78] Ibid., 32–35, 36 (quoted). Anna Sherman (2019) reflects on the practice evocatively.

[79] *NH* 7.212 with Beagon (2005), 469–70.

[80] No mention by Bond (2016), 29–40, 45–54, who demonstrates extensive public and private demand for criers (mostly termed *praecones*) in the spheres of "religion, theater, funerals, auctions, civil administration" (31).

[81] Notably not an importance matched later in Arab-Islamic society, where the passage of time was made known aurally by calls to prayer at least five times daily. Sundials could be found in mosques

use of bells to summon worshippers and on other occasions is not reliably attested before the sixth century.[82]

Second, the circumstances in which both societies abandoned their use of seasonal hours differ. So far as I am aware, a methodical attempt to trace the Roman abandonment has yet to be made. In the West at least, it is surely to be associated with the decline of urban life from the fifth century, a transformation that varied considerably in character and pace.[83] Japan, by contrast, did at last adopt the use of equinoctial hours after being aware of this option for over three centuries, but feeling no pressure to pursue it. When first shown mechanical clocks by Europeans in the mid-sixteenth century, the Japanese were fascinated, but saw them as useless.[84] Their sense of cultural superiority for long remained unshaken,[85] and for their purposes the established system of seasonal hours worked satisfactorily. European mechanical clocks continued to impress, however, and so from the seventeenth century Japanese craftsmen who gradually mastered their workings devised means to produce so-called *wadokei* clocks, adaptations which also recorded seasonal hours (fig. 19.7a, b).[86] Yulia Frumer argues that the switch to equinoctial hours was made only once attitudes had shifted in favor of Westernization, and support had grown for instituting a European calendar to replace the outmoded traditional one with its many flaws.[87] The system of seasonal hours, on the other hand, was not considered to be flawed,[88] but the opportunity to replace it too in the spirit of modernization was taken by imperial edict issued in December 1872.[89]

and *madrasas* attached to them, but seldom elsewhere in public places, and there was little demand either for portable ones: see Turner (2019), 110–11, 114–17.

[82] *ODCC* and *ODLA* s.v. bells.

[83] Of the inscriptions studied by Ehrlich (2012), the latest datable ones are from the late fourth and early fifth centuries: see his pp. 30–34.

[84] Frumer (2018), 39.

[85] Ibid., 6.

[86] Ibid., 43–44; Thompson (2004), 142–45. Note the finding of the expert collector Drummond Roberston (1931), 229: "no two clocks are identical, so far as my observation has gone" (for Rome, cf. note 63 above). For invaluable comprehensive guidance, visit Ashley Strachan's site, https://wadokei.org. My thanks to Jakobina Arch for drawing attention to it.

[87] According to contemporary advocates for change, it was "too convoluted and inelegant in its irregularity and was therefore too complex for commoners to understand; and . . . contained instructions for divinational practices, which scholars deemed detrimental to popular morality"; Frumer (2018), 184.

[88] Ibid., 188–95.

[89] Tanaka (2004), 5–17; Sherman (2019), 65–76, 258–63.

Fig. 19.7a, b Japanese copy of a European clock, 43 cm in height to the top of the bell, made around 1700. The maker of this *wadokei* clock intended the two weights suspended either side of the horizontal bar (foliot) to be used for changing the speed of the mechanism, thereby accommodating seasonal hours: it was slowed by shifts outward, and accelerated by shifts inward. The intention of European clock-makers was that only minimal adjustments should be made by these weights; but that was probably unknown to their ingenious Japanese imitators, and of no concern anyway. See further Robertson (1931), 219, figs. 2 and 3; cf. 216, set on a pyramid stand not made for it. Photos: © The Trustees of the British Museum (accession no. 1958.1006.2165).

In contrast to the Japanese experience, no alternative system ever challenged Roman consciousness of time in daily life, but there is still ample scope for instructive comparison of the two societies' practices. In any case, Romans' preoccupation with knowing the hour to an extent unprecedented in the societies of the ancient Mediterranean and Near East demands fresh attention. Here the concern expressed by Anthony Turner in 1985 remains valid:[90]

> We are rich in treatises on how to make and use sun-dials, and well supplied with catalogues of surviving examples in public and private collections. But in historical analysis of this material we are poor in the extreme.

[90] 1985 conference paper published as Turner (1989), 303.

Bibliography

Ancient Maps and Texts

André, Jacques, and Jean Filliozat, eds., trans. 1980. *Pline L'Ancien. Histoire Naturelle*, Livre 6: 46–106. Collection Budé. Paris: Belles Lettres.

Anonymous. 1888. *Peutingeriana Tabula Itineraria in Bibliotheca Palatina Vindobonensi asservata nunc primum arte photographica expressa*. Vienna: Angerer and Göschl.

Arnaud-Lindet, Marie-Pierre, ed., trans. 1994. *Festus. Breviarium*. Collection Budé. Paris: Belles Lettres.

Arnaud-Lindet, Marie-Pierre, ed., trans. 2003. *L. Ampelius. Liber Memorialis*. Collection Budé. Paris: Belles Lettres.

Bayer, Karl, and Kai Brodersen. 2004. *C. Plinius d. Ä., Naturkunde. Lateinisch-deutsch. Gesamtregister*. Sammlung Tusculum. Munich: Artemis and Winkler.

Beagon, Mary. 2005. *The Elder Pliny on the Human Animal*: Natural History Book 7. Oxford: Oxford University Press.

Beaujeu, Jean, ed., trans. 1950. *Pline L'Ancien. Histoire Naturelle*, Livre 2. Collection Budé. Paris: Belles Lettres.

Berggren, J. L., and Alexander Jones. 2000. *Ptolemy's* Geography: *An Annotated Translation of the Theoretical Chapters*. Princeton, NJ: Princeton University Press.

Brodersen, Kai, ed., trans. 1996. *C. Plinius d. Ä., Naturkunde. Lateinisch-deutsch*. Book 6. Sammlung Tusculum. Munich: Artemis and Winkler.

Campbell, Brian. 2000. *The Writings of the Roman Land Surveyors: Introduction, Text, Translation and Commentary*. London: Society for the Promotion of Roman Studies.

Cooley, A. E. 2009. Res Gestae Divi Augusti: *Text, Translation, and Commentary*. Cambridge: Cambridge University Press.

Cooley, M. G. L., ed. 2003. *The Age of Augustus*. London: London Association of Classical Teachers.

Crawford, M. H., ed. 1996. *Roman Statutes*. London: Institute of Classical Studies.

Crawford, M. H., et al., eds. 2011. *Imagines Italicae: A Corpus of Italic Inscriptions*. London: Institute of Classical Studies.

Cumont, Franz. 1925. "Fragment de bouclier portant une liste d'étapes." *Syria* 6: 1–15.

Cuntz, Otto, ed. 1929. *Itineraria Romana*, vol. 1. Leipzig: Teubner. (Reissued with bibliography by Gerhard Wirth, 1990)

De Geoje, M. J. 1889. *Bibliotheca Geographorum Arabicorum*, vol. 6. Leiden: Brill. (Reissued 1967)

Desanges, Jehan, ed., trans. 1980. *Pline L'Ancien. Histoire Naturelle*, Livre 5: 1–46. Collection Budé. Paris: Belles Lettres.

Desanges, Jehan, ed., trans. 2008. *Pline L'Ancien. Histoire Naturelle*, Livre 6: 163–220. Collection Budé. Paris: Belles Lettres.

Desjardins, Ernest. 1869–1874. *La Table de Peutinger, d'après l'original conservé à Vienne*. Paris: Hachette. (Incomplete; only 14 *livraisons* published)

French, D. H. 1981–2016. *Roman Roads and Milestones of Asia Minor*. Ankara: British Institute of Archaeology.

Freyburger, Gérard, and Anne-Marie Chevallier, eds., trans. 2019. *Censorinus. Le Jour Anniversaire de la Naissance*. Collection Budé. Paris: Belles Lettres.

Gallazzi, Claudio, Bärbel Kramer, and Salvatore Settis, eds. 2008. *Il Papiro di Artemidoro (P. Artemid.)*. Milan: LED.

Geyer, Paul, and Otto Cuntz, eds. 1965. *Itineraria et Alia Geographica*. Corpus Christianorum, Series Latina. Turnhout: Brepols.

Jonkers, E. J., ed. 1954. *Acta et Symbola Conciliorum Quae Saeculo Quarto Habita Sunt*. Leiden: Brill.

Labrousse, Mireille, ed. 1995. *Optat de Milève. Traité contre les Donatistes*, vol. 1. Paris: Sources Chrétiennes.

Lancel, Serge, ed., trans. 1972. *Actes de la conférence de Carthage en 411*, vol. 2. Paris: Sources Chrétiennes.

Mannert, Conrad. 1824. *Tabula Itineraria Peutingeriana*. . . . Leipzig: Hahn.

Miller, Konrad. 1888. *Die Peutingersche Tafel*. Ravensburg: Maier.

Miller, Konrad. 1898. *Mappaemundi: die ältesten Weltkarten*, Heft 6. Stuttgart: Roth.

Miller, Konrad. 1916. *Itineraria Romana. Römische Reisewege an der Hand der Tabula Peutingeriana dargestellt*. Stuttgart: Strecker and Schröder.

Müller, Carl. 1901. *Claudii Ptolomaei Geographia: Tabulae XXXVI*. Paris: Firmin Didot.

Mynors, R. A. B., ed. 1964. *XII Panegyrici Latini*. Oxford: Oxford University Press.

Nixon, C. E. V., and B. S. Rodgers, eds., trans. 1994. *In Praise of Later Roman Emperors: The* Panegyrici Latini. Berkeley: University of California Press.

Piccirillo, Michele, and Eugenio Alliata, eds. 1999. *The Madaba Map Centenary, 1897–1997: Travelling through the Byzantine Umayyad Period*. Jerusalem: Studium Biblicum Franciscanum.

Piganiol, André. 1962. *Les documents cadastraux de la colonie romaine d'Orange*. Supplément à *Gallia* XVI. Paris: Editions du CNRS.

Rackham, Harris, trans. 1938–1952. Pliny, *Natural History*. Loeb Classical Library. Cambridge, MA: Harvard University Press. (Vol. 1, books 1–2, 1938; vol. 2, books 3–7, 1942; vol. 3, books 8–11, 1940; vol. 4, books 12–16, 1945; vol. 5, books 17–19, 1950; vol. 9, books 33–35, 1952)

Rees, B. R. 1964. *Papyri from Hermopolis and Other Documents of the Byzantine Period*. London: Egypt Exploration Society.

Roberts, C. H., ed. 1952. *Catalogue of the Greek and Latin Papyri in the John Rylands Library, Manchester*, vol. IV. Manchester: Manchester University Press.

Rodríguez Colmenero, Antonio, et al. 2004. *Miliarios e Outras Inscricións Viarias Romanas do Noroeste Hispánico (Conventos Bracarense, Lucense e Asturicense)*. Santiago de Compostela: Consello da Cultura Galega.

Roller, D. W., trans. 2010. *Eratosthenes'* Geography: *Fragments Collected and Translated, with Commentary and Additional Material*. Princeton, NJ: Princeton University Press.

Roller, D. W., trans. 2014. *The* Geography *of Strabo: An English Translation with Introduction and Notes*. Cambridge: Cambridge University Press.

Roller, D. W. 2018. *A Historical and Topographical Guide to the* Geography *of Strabo*. Cambridge: Cambridge University Press.

Romer, F. E., ed., trans. 1998. *Pomponius Mela's Description of the World*. Ann Arbor: University of Michigan Press.

Rougé, Jean, ed., trans. 1966. *Expositio Totius Mundi et Gentium*. Paris: Sources Chrétiennes.

von Scheyb, F. C. 1753. *Peutingeriana Tabula Itineraria Quae in Augusta Bibliotheca Vindobonensi Nunc Servatur Adcurate Exscripta*. Vienna: Trattner. Reprint 2009 by Fabrizio Ronca et al., *Peutingeriana Tabula Itineraria*. Terni: Arte grafiche Celori.

Schnetz, Joseph, ed. 1940. *Itineraria Romana*, vol. 2. Leipzig: Teubner. (Reissued with index by Marianne Zumschlinge, 1990)

Shipley, Graham, ed., trans. Forthcoming. *Geographers of the Ancient Greek World*. Cambridge: Cambridge University Press.

Stückelberger, Alfred, and Gerd Grasshoff, eds., trans. 2006. *Klaudios Ptolemaios, Handbuch der Geographie*. Basel: Schwabe.

Stückelberger, Alfred, Gerd Grasshoff, et al., eds. 2009. *Klaudios Ptolemaios, Handbuch der Geographie: Ergänzungsband*. Basel: Schwabe.

Wagner, Jörg. 1984. *Weltkarten der Antike / Ancient Maps of the World*, 1.2 *Tabula Peutingeriana. TAVO* B S 1. Wiesbaden: Reichert.

Waterfield, Robin, trans. 1999. *Plutarch, Roman Lives*. Oxford: Oxford University Press.

Weber, Ekkehard. 1976. *Tabula Peutingeriana, Codex Vindobonensis 324*, with *Kommentar* volume. Graz: Akademische Druck- u. Verlagsanstalt.

Wei, Chen, ed. 2014. *Qin jiandu heji. Corpus of Qin Documents Written on Bamboo and Wood*. 7 vols. Wuhan: Wuhan University Press.

Welser, Marcus. 1591. *Fragmenta tabulae antiquae, in quis aliquot per Rom. provincias itinera. Ex Peutingerorum bibliotheca*. Venice: Manutius.

Welser, Marcus. 1598. *Tabula itineraria ex illustri Peutingerorum bibliotheca quae Augustae Vindel. est beneficio Marci Velseri septemviri Augustani in lucem edita*. Antwerp: Moretus.

Winkler, Gerhard, and Roderich König, eds., trans. 1993. *C. Plinius d. Ä., Naturkunde. Lateinisch- deutsch*. Book 5. Sammlung Tusculum. Munich: Artemis and Winkler.

Winterbottom, Michael, ed. 1970. *M. Fabi Quintiliani Institutionis Oratoriae Libri Duodecim*. Oxford: Oxford University Press.

Woodman, A. J., ed., trans. 2004. *Tacitus: The Annals*. Indianapolis: Hackett.

Yanjiao, Yu, et al., eds. 2014. *A Compilation of Silk and Bamboo Books from the Mawangdui Han Tombs in Changsha*. 7 vols. Beijing: Zhonghua Shuju.

Zehnacker, Hubert, ed., trans. 2004. *Pline L'Ancien. Histoire Naturelle*, Livre 3. Collection Budé. Paris: Belles Lettres.

Zehnacker, Hubert, and Alain Silberman, eds., trans. 2015. *Pline L'Ancien. Histoire Naturelle*, Livre 4. Collection Budé. Paris: Belles Lettres.

Modern Scholarship

Adams, Colin. 2001. "'There and Back Again': Getting around in Roman Egypt." In Adams and Laurence, 138–66.

Adams, Colin, and Ray Laurence, eds. 2001. *Travel and Geography in the Roman Empire*. London: Routledge.

Albu, Emily. 2005. "Imperial Geography and the Medieval Peutinger Map." *Imago Mundi* 57: 136–48.

Albu, Emily. 2014. *The Medieval Peutinger Map: Imperial Roman Revival in a German Empire*. Cambridge: Cambridge University Press.

Alcock, S. E., John Bodel, and R. J. A. Talbert, eds. 2012. *Highways, Byways, and Road Systems in the Pre-modern World*. Malden, MA: Wiley-Blackwell.

Aldrete, G. S. 1999. *Gestures and Acclamations in Ancient Rome*. Baltimore: Johns Hopkins University Press.

Andreae, Bernard. 2003. *Antike Bildmosaiken*. Mainz am Rhein: von Zabern.

Appelbaum, N. P. 2016. *Mapping the Country of Regions: The Chorographic Commission of Nineteenth-Century Colombia*. Chapel Hill: University of North Carolina Press.

Ardaillon, Édouard. 1900. "Horologium." In *DarSag* 3.1: 256–64.

Arnaud, Pascal. 1993. "L'*Itinéraire d' Antonin*: Un témoin de la littérature itinéraire du Bas-Empire." *Geographia Antiqua* 2: 33–47.

Arnaud, Pascal. 2007–2008. "Texte et carte de Marcus Agrippa: historiographie et données textuelles." *Geographia Antiqua* 16–17: 73–126.

Austin, N. J. E., and N. B. Rankov. 1995. *Exploratio: Military and Political Intelligence in the Roman World from the Second Punic War to the Battle of Adrianople*. London: Routledge.

Bablitz, Leanne. 2007. *Actors and Audience in the Roman Courtroom*. London: Routledge.

Balsdon, J. P. V. D. 1969. *Life and Leisure in Ancient Rome*. London: Bodley Head.

Barnes, T. D. 2011. Review of Talbert (2010). *Journal of Late Antiquity* 4.2: 375–78.

Beard, Mary. 1998. "Imaginary 'Horti': Or Up the Garden Path." In Cima and La Rocca, 23–31.

Beard, Mary. 2003. "The Triumph of the Absurd: Roman Street Theatre." In Catharine Edwards and Greg Woolf, eds., *Rome the Cosmopolis*, 21–43. Cambridge: Cambridge University Press.

Belke, Klaus. 2008. "Communications: Roads and Bridges." In Elizabeth Jeffreys et al., eds., *The Oxford Handbook of Byzantine Studies*, 295–308. Oxford: Oxford University Press.

Bellón Ruiz, J. P., et al. 2021. "Ianus Augustus, Caput Viae (Mengíbar, Spain): An Interprovincial Monumental Border in Roman Hispania." *JRA* 34.1: 3–29.

Bennett, Alan. 2005. *Untold Stories*. London: Faber.

Bertrand, A. C. 1997. "Stumbling through Gaul: Maps, Intelligence, and Caesar's *Bellum Gallicum*." *Ancient History Bulletin* 11.4: 107–22.

Beutler, Franziska, and Wolfgang Hameter, eds. 2005. *"Eine ganz normale Inschrift" und ähnliches zum Geburtstag von Ekkehard Weber*. Vienna: Österreichische Gesellschaft für Archäologie.

Birdal, Murat. 2010. *The Political Economy of Ottoman Public Debt: Insolvency and European Financial Control in the Late Nineteenth Century*. London: Tauris.

Boatwright, M. T. 2015. "Visualizing Empire in Imperial Rome." In Brice and Slootjes, 235–59.

Bond, S. E. 2016. *Trade and Taboo: Disreputable Professions in the Roman Mediterranean*. Ann Arbor: University of Michigan Press.

Bonnin, Jérôme. 2015. *La mesure du temps dans l'Antiquité*. Paris: Belles Lettres.

Bowden, M. J. 1976. "The Great American Desert in the American Mind: The Historiography of a Geographical Notion." In id. and David Lowenthal, eds., *Geographies of the Mind: Essays in Historical Geosophy in Honor of John Kirtland Wright*, 119–47. New York: Oxford University Press.

Bowersock, G. W. 2005. "The East-West Orientation of Mediterranean Studies and the Meaning of North and South in Antiquity." In W. V. Harris (ed.), *Rethinking the Mediterranean*, 167–78. Oxford: Oxford University Press.

Braund, D. C. 1985. *Augustus to Nero: A Sourcebook on Roman History, 31 BC–AD 68*. London: Routledge.

Brennan, Peter. 1996. "The *Notitia Dignitatum*." In Claude Nicolet, ed., *Les littératures techniques dans l'antiquité romaine: statut, public et destination, tradition*, 147–78. Geneva: Fondation Hardt.

Brice, L. L., and Daniëlle Slootjes, eds. 2015. *Aspects of Ancient Institutions and Geography: Studies in Honor of Richard J. A. Talbert*. Leiden: Brill.

Brilliant, Richard. 1991. "The Bayeux Tapestry: A Stripped Narrative for Their Eyes and Ears." *Word and Image* 7: 93–125. (Reprinted in Richard Gameson, ed., *The Study of the Bayeux Tapestry*, 111–37. Woodbridge, UK: Boydell Press, 1997)

Brodersen, Kai. 1995. *Terra Cognita: Studien zur römischen Raumerfassung*. Hildesheim: Olms.

Brodersen, Kai. 2001. "The Presentation of Geographical Knowledge for Travel and Transport in the Roman World: *Itineraria non tantum adnotata sed etiam picta*." In Adams and Laurence, 7–21.

Brodersen, Kai. 2004. "Mapping (in) the Ancient World." *JRS* 94: 183–90.

Bruun, Christer, and Jonathan Edmondson, eds. 2015. *The Oxford Handbook of Roman Epigraphy*. Oxford: Oxford University Press.

Burnett, D. G. 2000. *Masters of All They Surveyed: Exploration, Geography, and a British El Dorado*. Chicago: University of Chicago Press.

Burton, G. P. 2000. "The Resolution of Territorial Disputes in the Provinces of the Roman Empire." *Chiron* 30: 195–215.

Burton, G. P. 2002. "The Regulation of Inter-community Relations in the Provinces and the Political Integration of the Roman Empire (27 BC–AD 238)." In V. B. Gorman and E. W. Robinson, eds., *Oikistes: Studies in Constitutions, Colonies, and Military Power in the Ancient World, Offered in Honor of A. J. Graham*, 113–28. Leiden: Brill.

Calzolari, Mauro. 1996. *Introduzione allo Studio della Rete Stradale dell' Italia Romana: L'* Itinerarium Antonini. In *Atti della Accademia Nazionale dei Lincei* (Classe di Scienze Morali, Storiche e Filologiche), *Memorie*, ser. IX vol. VII fasc. 4: 369–520.

Calzolari, Mauro. 1997. "Ricerche sugli itinerari romani. L'Itinerarium Burdigalense." In *Studi in onore di Nereo Alfieri. Atti dell'Accademia delle Scienze di Ferrara* 74 supplemento, 127–89.

Calzolari, Mauro. 2000. "Gli itinerari della tarda antichità e il nodo stradale di Aquileia." In Silvia Blason Scarel, ed. *Cammina, Cammina . . . Dalla Via dell'Ambra alla Via della Fede*, 18–41. Aquileia: Gruppo Archeologico Aquileiese.

Calzolari, Mauro. 2003. "L'Edizione di Konrad Miller." In Francesco Prontera, ed. *Tabula Peutingeriana: Le Antiche Vie del Mondo*, 67. Florence: Olschki.

Campbell, Brian. 2005. "'Setting Up True Boundaries': Land Disputes in the Roman Empire." *Mediterraneo Antico* 8.1: 307–43.

Campbell, Brian. 2012. *Rivers and the Power of Ancient Rome*. Chapel Hill: University of North Carolina Press.

Canepa, M. P. 2009. *The Two Eyes of the Earth: Art and Ritual of Kingship between Rome and Sasanian Iran*. Berkeley: University of California Press.

Canto, A. M. 1989. "Colonia Iulia Augusta Emerita: Consideraciones en torno a su fundación y territorio." *Gerión* 7: 149–205.

Carey, Sorcha. 2003. *Pliny's Catalogue of Culture: Art and Empire in the Natural History*. Cambridge: Cambridge University Press.

Cassibry, Kimberly. 2021. *Destinations in Mind: Portraying Places on the Roman Empire's Souvenirs*. Oxford: Oxford University Press.

Chaniotis, Angelos, ed. 2018. *La Nuit: Imaginaire et réalités nocturnes dans le monde gréco-romain*. Vandoeuvres: Fondation Hardt.

Cima, Maddalena, and Eugenio La Rocca, eds. 1998. *Horti Romani: Atti del Convegno Internazionale, Roma, 4–6 maggio 1995*. Rome: Bretschneider.

Cioffi, R. L. 2016. "Travel in the Roman World." In *Oxford Handbooks Online*.

Corbeill, Anthony. 2004. *Nature Embodied: Gesture in Ancient Rome*. Princeton, NJ: Princeton University Press.

Coulton, J. J. 1977. *Ancient Greek Architects at Work: Problems of Structure and Design*. Ithaca, NY: Cornell University Press.

Crawford, M. H., and J. M. Reynolds. 1979. "The Aezani Copy of the Prices Edict." *ZPE* 34: 163–210.

Cressey, G. B. 1934. *China's Geographic Foundations: A Survey of the Land and Its People*. New York: McGraw-Hill.

Cuff, D. B. 2010. "The *Auxilia* in Roman Britain and the Two Germanies from Augustus to Caracalla: Family, Religion and 'Romanization.'" Dissertation, University of Toronto.

Cuvigny, Hélène. 2019. "Le livre de poste de Turbo, curateur du *praesidium* de Xèron Pelagos (*Aegyptus*)." In Anne Kolb, ed., *Roman Roads: New Evidence—New Perspectives*, 67–105. Berlin: de Gruyter.

Daicoviciu, Constantin, and Dumitru Protease. 1961. "Un nouveau diplôme militaire de Dacia Porolissensis." *JRS* 51: 63–70.

Deininger, Jürgen. 1964. *Die Provinziallandtage der römischen Kaiserzeit*. Munich: Beck.

De Lange, N. R. M. 1978. "Jewish Attitudes to the Roman Empire." In P. D. A. Garnsey and C. R. Whittaker, eds., *Imperialism in the Ancient World*, 255–81. Cambridge: Cambridge University Press.

Deluz, Christiane. 2013. "Une image du monde: La géographie dans l'Occident médiéval (Ve–XVe siècle)." In Patrick Gautier Dalché, ed. *La Terre: Connaissance, Représentations, Mesure au Moyen Âge*, 21–40. Turnhout: Brepols.

Dench, Emma. 2018. *Empires and Political Cultures in the Roman World*. Cambridge: Cambridge University Press.

Diederich, Silke. 2018. "The Tabula Peutingeriana (Peutinger Map)." In *OBO Classics*.

Dilke, O. A. W. 1971. *The Roman Land Surveyors: An Introduction to the* Agrimensores. Newton Abbot, UK: David and Charles.

Dilke, O. A. W. 1985. *Greek and Roman Maps*. London: Thames and Hudson.

Dilke, O. A. W. 1987a. "Roman Large-Scale Mapping in the Early Empire." In Harley and Woodward, 212–33.

Dilke, O. A. W. 1987b. "Itineraries and Geographical Maps in the Early and Late Roman Empires." In Harley and Woodward, 234–57.

Dohrn-van Rossum, Gerhard. 1996. *History of the Hour: Clocks and Modern Temporal Orders*. Chicago : University of Chicago Press. (Originally published in German, 1992)

Doody, Aude. 2015. "Pliny the Elder." In *OBO Classics*.

Dossey, Leslie. 2018. "Shedding Light on the Late Antique Night." In Chaniotis, 293–322.

Downs, R. M., and David Stea. 1977. *Maps in Minds: Reflections on Cognitive Mapping*. New York: Harper and Row.

Drew-Bear, Thomas. 1981. "Les voyages d'Aurélius Gaius, soldat de Dioclétien." In *La géographie administrative et politique d'Alexandre à Mahomet*, 93–141. Leiden: Brill.

Drogula, F. K. 2011. "Controlling Travel: Deportation, Islands and the Regulation of Senatorial Mobility in the Augustan Principate." *CQ* 61.1: 230–66.

Dueck, Daniela. 2013. "Geography." In *OBO Classics*.

Dueck, Daniela. 2021. *Illiterate Geography in Classical Athens and Rome*. London: Routledge.

Dunbabin, K. M. D., Işık Adak Adıbelli, Mehmet Çavuş, and Doğukan Alpers. 2019. "The Man Who Came Late to Dinner: A Sundial, a Raven, and a Missed Dinner Party on a Mosaic at Tarsus." *JRA* 32: 329–58.

Eck, Werner. 2002. "Imperial Administration and Epigraphy: In Defence of Prosopography." In A. K. Bowman et al., eds., *Representations of Empire: Rome and the Mediterranean World*, 131–52. Oxford: Oxford University Press.

Eck, Werner. 2010. "Recht und Politik in den Burgerrechtskonstitutionen der römischen Kaiserzeit." *Scripta Classica Israelica* 29: 33–50.

Eck, Werner, Paul Holder, and Andreas Pangerl. 2010. "A Diploma for the Army of Britain in 132 and Hadrian's Return to Rome from the East." *ZPE* 174: 189–200.

Edney, M. H. 2005a. "David Alfred Woodward (1942–2004)." *Imago Mundi* 57: 75–83.

Edney, M. H. 2005b. *The Origins and Development of J. B. Harley's Cartographic Theories*. Toronto: University of Toronto Press.

Edson, Evelyn. 1997. *Mapping Time and Space: How Medieval Mapmakers Viewed Their World*. London: British Library.

Ehrlich, S. D. 2012. "*Horae* in Roman Funerary Inscriptions." Thesis, University of Western Ontario.

Elliott, Thomas. 2004. "Epigraphic Evidence for Boundary Disputes in the Roman Empire." Dissertation, University of North Carolina, Chapel Hill.

Elsner, Jás. 2000. "The *Itinerarium Burdigalense*: Politics and Salvation in the Geography of Constantine's Empire." *JRS* 90: 181–95.

Englund, R. K. 1988. "Administrative Timekeeping in Ancient Mesopotamia." *Journal of the Economic and Social History of the Orient* 31: 121–85.

Fabia, Philippe. 1929. *La Table Claudienne de Lyon*. Lyon: Audin.

Fagan, G. G. 1999. *Bathing in Public in the Roman World*. Ann Arbor: University of Michigan Press.

Fagan, G. G. 2011. *The Lure of the Arena: Social Psychology and the Crowd at the Roman Games*. Cambridge: Cambridge University Press.

Feeney, Denis. 2010. "Time and Calendar." In Alessandro Barchiesi and Walter Scheidel, eds., *The Oxford Handbook of Roman Studies*, 882–94. Oxford: Oxford University Press.

Fireman, J. K. 1977. *The Spanish Royal Corps of Engineers in the Western Borderlands: Instrument of Bourbon Reform, 1764 to 1815*. Glendale, CA: Clark.

Foxhall, Lin, et al., eds. 2006. *The Cambridge Dictionary of Classical Civilization*. Cambridge: Cambridge University Press.

France, Jérôme. 2001. *Quadragesima Galliarum: l'organisation douanière des provinces alpestres, gauloises et germaniques de l'empire romain (Ier siècle avant J.-C.–IIIe siècle après J.-C.)*. Rome: Ecole Française de Rome.

Frei-Stolba, Regula, ed. 2004. *Siedlung und Verkehr im römischen Reich*. Bern: Lang.

Friedman, J. B., and K. M. Figg, eds. 2000. *Trade, Travel, and Exploration in the Middle Ages: An Encyclopedia*. New York: Garland.

Frumer, Yulia. 2018. *Making Time: Astronomical Time Measurement in Tokugawa Japan*. Chicago: University of Chicago Press.

Fugmann, Joachim. 1998. "Itinerarium." In *Reallexikon für Antike und Christentum, Lieferung* 146, 1–32.

Gallazzi, Claudio, and Salvatore Settis, eds. 2006. *Le tre vite del Papiro di Artemidoro. Voci e sguardi dall'Egitto greco-romano*. Milan: Electa.

Gargola, D. J. 2017. *The Shape of the Roman Order: The Republic and Its Spaces*. Chapel Hill: University of North Carolina Press.

Gaube, Heinz, ed. 1986. *Konrad Miller, Mappae Arabicae. TAVO* Beihefte B65. Wiesbaden: Reichert.

Gautier Dalché, Patrick. 2001. "Décrire le monde et situer les lieux au XIIe siècle: l'*Expositio Mappe Mundi* et la généalogie de la mappemonde de Hereford." *Mélanges de l'École Française de Rome, Moyen Âge* 113: 343–409.

Gautier Dalché, Patrick. 2008. "L'héritage antique de la cartographie médiévale: les problèmes et les acquis." In Talbert and Unger, 29–66.
Gautier Dalché, Patrick. 2014. "L'enseignement de la géographie dans l'antiquité tardive." *Klio* 96.1: 144–82.
Gautier Dalché, Patrick. 2015. "Les usages militaires de la carte, des premiers projects de croisade à Machiavel." *Revue historique* 317.1: 45–80.
Gerov, Boris. 1979. "Die Grenzen der römischen Provinz Thracia bis zur Gründung des Aurelianischen Dakien." *Aufstieg und Niederegang des römischen Welt* II.7.1: 212–40.
Geus, Klaus. 2003. "Space and Geography." In Andrew Erskine, ed., *A Companion to the Hellenistic World*, 232–45. Malden, MA: Blackwell.
Geus, Klaus. 2004. "Measuring the Earth and the *Oikoumene*: Zones, Meridians, *Sphragides* and Some Other Geographical Terms Used by Eratosthenes of Cyrene." In Talbert and Brodersen, 11–26.
Gichon, Mordechai. 1972. "The Plan of a Roman Camp Depicted upon a Lamp from Samaria." *Palestine Exploration Quarterly* 104.1: 38–58.
Glennie, Paul, and Nigel Thrift. 2005. "Revolutions in the Times: Clocks and the Temporal Structures of Everyday Life." In D. N. Livingstone and C. W. J. Withers, eds., *Geography and Revolution*, 160–98. Chicago: University of Chicago Press.
Glennie, Paul, and Nigel Thrift. 2009. *Shaping the Day: A History of Timekeeping in England and Wales 1300–1800*. Oxford: Oxford University Press.
Gordon, A. E. 1983. *Illustrated Introduction to Latin Epigraphy*. Berkeley: University of California Press.
Gould, Peter, and Rodney White. 1986. *Mental Maps*. 2nd ed. London: Allen and Unwin.
Grafton, Anthony, G. W. Most, and Salvatore Settis, eds. 2010. *The Classical Tradition*. Cambridge, MA: Harvard University Press.
Grasshoff, Gerd, and Florian Mittenhuber, eds., 2009. *Untersuchungen zum Stadiasmos von Patara: Modellierung und Analyse eines antiken geographischen Streckennetzes*. Bern: Bern Studies in the History and Philosophy of Science.
Gratwick, A. S. 1979. "Sundials, Parasites, and Girls from Boeotia." *CQ* 29.2: 308–23.
Gray, Annie. 2020. *Victory in the Kitchen: The Life of Churchill's Cook*. London: Profile.
Greene, E. M. 2011. "Women and Families in the Auxiliary Military Communities of the Roman West in the First and Second Centuries AD." Dissertation, University of North Carolina, Chapel Hill.
Greene, E. M. 2015. "*Conubium cum Uxoribus*: Wives and Children in the Roman Military Diplomas." *JRA* 28: 125–59.
Greene, E. M. 2017. "The Families of Roman Auxiliary Soldiers in Military Diplomas." In Nick Hodgson et al., eds., *Roman Frontier Studies 2009: 21st International Congress of Roman Frontier Studies*, 23–25. Oxford: Archaeopress.
Gutkind Bulling, Anneliese. 1978. "Ancient Chinese Maps: Two Maps Discovered in a Han Dynasty Tomb from the Second Century B.C." *Expedition* 20: 16–25.
Haley, E. W. 2003. *Baetica Felix: People and Prosperity in Southern Spain from Caesar to Septimius Severus*. Austin: University of Texas Press.
Hamm, Geoffrey. 2014. "British Intelligence in the Middle East, 1898–1906." *Intelligence and National Security* 29.6: 880–900.
Hannah, Robert. 2009. *Time in Antiquity*. London: Routledge.
Harley, J. B. 1987. "The Map and the Development of the History of Cartography." In id. and Woodward, 1–42.

Harley, J. B., and David Woodward, eds. 1987. *The History of Cartography*, 1: *Cartography in Prehistoric, Ancient, and Medieval Europe and the Mediterranean*. Chicago: University of Chicago Press.

Harley, J. B., and David Woodward, eds. 1992. *The History of Cartography*, 2.1: *Cartography in the Traditional Islamic and South Asian Societies*. Chicago: University of Chicago Press.

Harley, J. B., and David Woodward, eds. 1994. *The History of Cartography*, 2.2: *Cartography in the Traditional East and Southeast Asian Societies*. Chicago: University of Chicago Press.

Harris, W. V. 2003. "Roman Governments and Commerce, 300 B.C.–A.D. 300." In Carlo Zaccagnini, ed., *Mercanti e Politica nel Mondo Antico*, 275–305. Rome: Bretschneider.

Harris, W. V. 2005. "The Mediterranean and Ancient History." In id., ed., *Rethinking the Mediterranean*, 1–42. Oxford: Oxford University Press.

Hauken, Tor. 1998. *Petition and Response: An Epigraphic Study of Petitions to Roman Emperors, 181–249*. Bergen: Norwegian Institute at Athens.

Hauken, Tor, and Hasan Malay. 2009. "A New Edict of Hadrian from the Province of Asia Setting Regulations for Requisitioned Transport." In Rudolf Haensch, ed., *Selbstdarstellung und Kommunikation: Die Veröffentlichung staatlicher Urkunden auf Stein und Bronze in der römischen Welt*, 327–48. Munich: Beck.

Heffernan, Michael. 1996. "Geography, Cartography and Military Intelligence: The Royal Geographical Society and the First World War." *Transactions of the Institute of British Geographers* new series 21.3: 504–33.

Henderson, J. B. 1994. "Chinese Cosmographical Thought: The High Intellectual Tradition." In Harley and Woodward, 203–27.

Henderson, J. B. 2010. "Nonary Cosmography in Ancient China." In Raaflaub and Talbert, 64–73.

Herzog, Reinhart, ed. 1989. *Restauration und Erneuerung. Die lateinische Literatur von 284 bis 374 n. Chr.* Munich: Beck = id. 1993. *Restauration et renouveau: la littérature latine de 284 à 374 après J.-C.* Turnhout: Brepols.

Hesberg, Henner von, ed. 1995. *Was ist eigentlich Provinz? Zur Beschreibung eines Bewusstseins*. Cologne: Archäologisches Institut der Universität.

Hiatt, Alfred. 2020. *Dislocations: Maps, Classical Tradition, and Spatial Play in the European Middle Ages*. Toronto: Pontifical Institute of Mediaeval Studies.

Higton, Hester. 2002. *Sundials at Greenwich: A Catalogue of the Sundials, Nocturnals and Horary Quadrants in the National Maritime Museum, Greenwich*. Oxford: Oxford University Press.

Holder, P. A. 2006. *Roman Military Diplomas V*. London: Institute of Classical Studies.

Holliday, P. J. 2002. *The Origins of Roman Historical Commemoration in the Visual Arts*. Cambridge: Cambridge University Press.

Horsfall, Nicholas. 2003. *The Culture of the Roman Plebs*. London: Duckworth.

Houston, G. W. 2015. "Using Sundials." In Brice and Slootjes, 298–313.

Hsu, H.-M. A. 2010. "Structured Perceptions of Real and Imagined Landscapes in Early China." In Raaflaub and Talbert, 43–63.

Hunt, E. D. 1982. *Holy Land Pilgrimage in the Later Roman Empire, AD 312–460*. Oxford: Oxford University Press.

Hunt, E. D. 1998. "The Church as a Public Institution." In *CAH*^{2}, vol. 13, 238–76.

Hunt, E. D. 2004. "Holy Land Itineraries: Mapping the Bible in Late Roman Palestine." In Talbert and Brodersen, 97–110.

Husslein, Gertrud. 1995. "Konrad Miller." *Orbis Terrarum* 1: 213–33 and Tafeln 13–17.

Irby, G. L. 2012. "Mapping the World: Greek Initiatives from Homer to Eratosthenes." In Talbert, 81–107.

Isaac, Benjamin. 1992. *The Limits of Empire: The Roman Army in the East*. Rev. ed. Oxford: Oxford University Press.

Isaac, G. R. 2002. *The Antonine Itinerary Land Routes* (CD-ROM). Aberystwyth, UK: CMCS.

Isaac, G. R. 2004. *Place-Names in Ptolemy's Geography* (CD-ROM). Aberystwyth, UK: CMCS.

Jackson, R. B. 2002. *At Empire's Edge: Exploring Rome's Egyptian Frontier*. New Haven, CT: Yale University Press.

Janko, Richard. 2009. "The Artemidorus Papyrus." *CR* 59.2: 403–10.

Janni, Pietro. 1984. *La Mappa e il Periplo: Cartografia Antica e Spazio Odologico*. Rome: Bretschneider.

Jerphanion, Guillaume de. 1909. Review of Kiepert (1901–1907). *La Géographie* 19: 367–76.

Johnson, S. F. 2015. "Real and Imagined Geography." In Michael Maas, ed., *The Cambridge Companion to the Age of Attila*, 394–413. Cambridge: Cambridge University Press.

Johnson, S. F. 2016. *Literary Territories: Cartographical Thinking in Late Antiquity*. Oxford: Oxford University Press.

Johnston, A. E. M. 1967. "The Earliest Preserved Greek Map: A New Ionian Coin Type." *JHS* 87: 86–94.

Jones, Alexander. 2012. "Ptolemy's Geography: Mapmaking and the Scientific Enterprise." In Talbert, 109–28.

Jones, Alexander. 2018. Review of Talbert (2017). *Classical Philology* 113.2: 232–37.

Jones, Alexander, ed. 2016. *Time and Cosmos in Greco-Roman Antiquity*. Princeton, NJ: Princeton University Press.

Julien, Catherine. 2010. "Inca Worldview." In Raaflaub and Talbert, 128–46.

Julien, Catherine. 2012. "The Chinchaysuyu Road and the Definition of an Inca Imperial Landscape." In Alcock, Bodel, and Talbert, 147–67.

Kadish, G. E. 2001. "Time." In D. B. Redford, ed., *The Oxford Encyclopedia of Ancient Egypt*, vol. 3, 405–409. Oxford: Oxford University Press.

Kiepert, Richard, ed. 1901–1907. *Karte von Kleinasien* (24 sheets at 1:400,000 scale). 1st ed. Berlin: Reimer.

Kim, Nanny. 2012. "Privatizing the Network: Private Contributions and Road Infrastructure in Late Imperial China (1500–1900)." In Alcock, Bodel, and Talbert, 66–89.

Kirsten, Ernst. 1959. "Eine Reise von Hermupolis in Oberägypten nach Antiochia in Syrien zur Zeit Kaiser Konstantins." *Erdkunde* 13: 411–26. Reprinted in id. 1984. *Landschaft und Geschichte in der Antiken Welt: Ausgewählte kleine Schriften*, 263–78. Bonn: Habelt.

Kissel, Theodor. 2002. "Road-Building as a Munus Publicum." In Paul Erdkamp, ed., *The Roman Army and the Economy*, 127–60. Amsterdam: Gieben.

Koeppel, G. M. 1980. "A Military *Itinerarium* on the Column of Trajan: Scene L." *MDAI(R)* 87: 301–306.

Kolb, Anne. 2004. "Römische Meilensteine: Stand der Forschung und Probleme." In Frei-Stolba, 135–55.

Kolb, Anne. 2005. "Reisen unter göttlichem Schutz." In Beutler and Hameter, 293–98.

Kolb, Anne. 2006. "Die Meilensteine von Galicien und Asturien." Review of Rodríguez Colmenero et al. (2004). *JRA* 19: 577–82.

Kolb, Anne. 2007. "Raumwahrnehmung und Raumerschliessung durch römische Strassen." In Michael Rathmann, ed., *Wahrnehmung und Erfassung geographischer Räume in der Antike*, 169–80. Mainz am Rhein: von Zabern.

Kubitschek, Wilhelm. 1917. Review of Miller (1916). *Göttingische gelehrte Anzeigen* 179: 1–117.

Kubitschek, Wilhelm. 1917–1918. Review of Miller (1916). *Zeitschrift für die deutsch-österreichischen Gymnasien* 68: 740–54 and 865–93.

Künzl, Ernst, and Gerhard Koeppel. 2002. *Souvenirs und Devotionalien: Zeugnisse des geschäftlichen, religiösen und kulturellen Tourismus im antiken Römerreich*. Mainz am Rhein: von Zabern.

La Rocca, Eugenio. 2000. "L'affresco con veduta di città dal colle Oppio." In Elizabeth Fentress, ed., *Romanization and the City: Creation, Transformations, and Failures*, 57–71. Portsmouth, RI: Journal of Roman Archaeology.

Laurence, Ray. 2007. *Roman Pompeii: Space and Society*. 2nd ed. London: Routledge.

Laurence, Ray. 2013. "Traffic and Land Transportation in and near Rome." In Paul Erdkamp, ed., *The Cambridge Companion to Ancient Rome*, 246–61. Cambridge: Cambridge University Press.

Laurence, Ray. 2016. "Connectivity, Roads and Transport: Essays on Roman Roads to Speak to Other Disciplines?" Review of Alcock, Bodel, and Talbert (2012). *JRA* 29: 692–95.

Laurence, Ray. 2020. "The Meaning of Roads: A Reinterpretation of the Roman Empire." In Jenni Kuuliala and Jussi Rantala, eds., *Travel, Pilgrimage and Social Interaction from Antiquity to the Middle Ages*, 37–63. London: Routledge.

Le Goff, Jacques. 1980. "Merchant's Time and Church's Time in the Middle Ages." In id., *Time, Work, & Culture in the Middle Ages*, 29–42 (trans. Arthur Goldhammer). Chicago and London: Chicago University Press. (Originally published in French, 1960)

Leclerq, Henri. 1927. "Itinéraires." In id. and Fernand Cabrol, eds., *Dictionnaire d'archéologie chrétienne et de liturgie* VII, cols. 1841–1922. Paris: Letouzey et Ané.

Levick, Barbara. 2000. *The Government of the Roman Empire: A Sourcebook*. 2nd ed. London: Routledge.

Levick, Barbara. 2005. Review of Woodman (2004). *TLS* February 11th, p. 28.

Levine, Robert. 1997. *A Geography of Time: The Temporal Misadventures of a Social Psychologist, or How Every Culture Keeps Time Just a Little Bit Differently*. New York: Basic Books.

Lewis, M. J. T. 2001. *Surveying Instruments of Greece and Rome*. Cambridge: Cambridge University Press.

Leyerle, Blake. 2009. "Mobility and the Traces of Empire." In Philip Rousseau, ed., *A Companion to Late Antiquity*, 110–23. Malden, MA: Wiley-Blackwell.

Lintott, Andrew. 1981. "What Was the 'Imperium Romanum'?" *G&R* 28.1: 53–67.

Littlewood, A. R. 1987. "Ancient Literary Evidence for the Pleasure Gardens of Roman Country Villas." In E. B. MacDougall, ed., *Ancient Roman Villa Gardens*, 7–30. Washington, DC: Dumbarton Oaks.

Löhberg, Bernd. 2006. *Das "Itinerarium provinciarum Antonini Augusti": Ein kaiserzeitliches Straßenverzeichnis des römischen Reiches—Überlieferung, Strecken, Kommentare, Karten*. Berlin: Frank and Timme.

Lozovsky, Natalia. 2000. *"The Earth Is Our Book": Geographical Knowledge in the Latin West ca. 400–1000*. Ann Arbor: University of Michigan Press.

Luzzatto, G. I. 1967. "Provincia. Diritto Romano." *Novissimo Digesto Italiano* 14: 377–82.

MacMullen, Ramsay. 1990. "Some Pictures in Ammianus Marcellinus." In id., *Changes in the Roman Empire: Essays in the Ordinary*, 78–106. Princeton, NJ: Princeton University Press.

Maguire, Henry. 1999. "The Good Life." In G. W. Bowersock et al., eds., *Late Antiquity: A Guide to the Postclassical World*, 238–57. Cambridge, MA: Harvard University Press.

Maslikov, S. J. 2021. "The Greek Portable Sundial from Memphis Rediscovered." *Journal for the History of Astronomy* 52.3: 311–24.

Mattern, S. P. 1999. *Rome and the Enemy: Imperial Strategy in the Principate*. Berkeley: University of California Press.

Matthews, John. 2006. *The Journey of Theophanes: Travel, Business, and Daily Life in the Roman East*. New Haven, CT: Yale University Press.

McCormick, Michael. 2001. *Origins of the European Economy: Communications and Commerce, A.D. 300–900*. Cambridge: Cambridge University Press.

Melechi, Antonio. 2020. "The Sound of Blood Rushing: Exploring the Experimental Science of Isolation." *TLS* July 3rd, 14–15.

Millar, Fergus. 1982. "Emperors, Frontiers and Foreign Relations, 31 B.C. to A.D. 378." *Britannia* 13: 1–23.

Millar, Fergus. 1992. *The Emperor in the Roman World (31 BC–AD 337)*. 2nd ed. Bristol, UK: Bristol Classical Press. (1st ed. 1977)

Miller, K. J. 2018. "From Critical Days to Critical Hours: Galenic Refinements of Hippocratic Models." *Transactions of the American Philological Association* 148.1: 111–38.

Miller, K. J., and Sarah Symons, eds. 2020. *Down to the Hour: Short Time in the Ancient Mediterranean and Near East*. Leiden: Brill.

Mirković, Miroslava. 2007. "Married and Settled: The Origo, Privileges and Settlement of Auxiliary Soldiers." In M. A. Speidel and Hans Lieb, eds., *Militärdiplome: die Forschungsbeiträge der Berner Gespräche von 2004*, 327–43. Stuttgart: Steiner.

Mitchell, Stephen. 1976. "Requisitioned Transport in the Roman Empire: A New Inscription from Pisidia." *JRS* 66: 106–31.

Mitchell, Stephen. 2000. "Ethnicity, Acculturation and Empire in Roman and Late Roman Asia Minor." In id. and Geoffrey Greatrex, eds., *Ethnicity and Culture in Late Antiquity*, 117–50. London: Duckworth and the Classical Press of Wales.

Mitchell, Stephen. 2002. "In Search of the Pontic Community in Antiquity." In A. K. Bowman et al., eds., *Representations of Empire: Rome and the Mediterranean World*, 35–64. London: British Academy.

Moatti, Claudia. 1993. *Archives et partage de la terre dans le monde romain (IIe siècle avant–Ier siècle après J.-C.)*. Rome: École Française de Rome.

Moatti, Claudia. 2000. "Le contrôle de la mobilité des personnes dans l'empire romain." *MEFRA* 112.2: 925–58.

Moreland, Carl, and David Bannister. 1983. *Antique Maps: A Collector's Guide*. Harlow, UK: Longman.

Morris, Jason. 2018. "*Forma Facta Est: Agrimensores* and the Power of Geography." *Phoenix* 72.1/2: 119–42.

Murphy, Trevor. 2004. *Pliny the Elder's Natural History: The Empire in the Encyclopedia*. Oxford: Oxford University Press.

Naval Staff Intelligence Department. 1919. *A Handbook of Asia Minor*. London: HMSO.

Needham, Joseph. 1954. "Note on the Chinese Language." In id., *Science and Civilisation in China*, vol. 1, 27–41. Cambridge: Cambridge University Press.
Needham, Joseph. 1959. "Geography and Cartography." In id., *Science and Civilisation in China*, vol. 3, 497–590. Cambridge: Cambridge University Press.
Nicholls, Matthew. 2017. "Libraries and Communication in the Ancient World." In F. S. Naiden and R. J. A. Talbert, eds., *Mercury's Wings: Exploring Modes of Communication in the Ancient World*, 23–44. Oxford: Oxford University Press.
Nicolet, Claude. 1991. *Space, Geography, and Politics in the Early Roman Empire*. Ann Arbor: University of Michigan Press.
Nylan, Michael. 2012. "The Power of Highway Networks during China's Classical Era (323 BCE–316 CE): Regulations, Metaphors, Rituals, and Deities." In Alcock, Bodel, and Talbert, 33–65.
O'Connor, David. 2012. "From Topography to Cosmos: Ancient Egypt's Multiple Maps." In Talbert, 47–79.
Ogle, Vanessa. 2015. *The Global Transformation of Time 1870–1950*. Cambridge, MA: Harvard University Press.
Östenberg, Ida. 2009. *Staging the World: Spoils, Captives, and Representations in the Roman Triumphal Procession*. Oxford: Oxford University Press.
Ott, Joachim. 1995. *Die Beneficiarier: Untersuchungen zu ihrer Stellung innerhalb der Rangordnung des römischen Heeres und zu ihrer Funktion*. Stuttgart: Steiner.
Painter, K. S. 1975. "Roman Flasks with Scenes of Baiae and Puteoli." *Journal of Glass Studies* 17: 54–67.
Pamir, Hatice, and Nilüfer Sezgin. 2016. "The Sundial and Convivium Scene on the Mosaic from the Rescue Excavation in a Late Antique House of Antioch." *Adalya* 19: 251–80.
Parsons, P. J. 2001. "Rhetorical Handbook." In Traianos Gagos and R. S. Bagnall, eds., *Essays and Texts in Honor of J. David Thomas*, 153–64. Oakville, CT: American Society of Papyrologists.
Pekáry, Thomas. 1968. *Untersuchungen zu den römischen Reichsstrassen*. Bonn: Habelt.
Pferdehirt, Barbara. 2004. *Römische Militärdiplome und Entlassungsurkunden in der Sammlung des Römisch-Germanischen Zentralmuseums*. 2nd ed. Mainz: Römisch-Germanisches Zentralmuseum.
Podossinov, A. V., and L. S. Chekin. 1991. Extended review of Harley and Woodward (1987). *Imago Mundi* 43: 112–23.
Popkin, M. L. 2018. "Urban Images in Glass from the Late Roman Empire: The Souvenir Flasks of Puteoli and Baiae." *American Journal of Archaeology* 122.3: 427–62.
Potter, D. S., ed. 2006. *A Companion to the Roman Empire*. Malden: Blackwell.
Price, S. R. F. 1984. *Rituals and Power: The Roman Imperial Cult in Asia Minor*. Cambridge: Cambridge University Press.
Purcell, Nicholas. 1990. "The Creation of Provincial Landscape: The Roman Impact on Cisalpine Gaul." In Thomas Blagg and Martin Millett, eds., *The Early Roman Empire in the West*, 7–29. Oxford: Oxbow.
Purcell, Nicholas. 2001. "Dialectical Gardening." Review of Cima and La Rocca (1998). *JRA* 14: 546–56.
Quilici, Lorenzo. 2008. "Land Transport, Part 1: Roads and Bridges." In J. P. Oleson, ed. *The Oxford Handbook of Engineering and Technology in the Classical World*, 551–79. Oxford: Oxford University Press.
Raaflaub, K. A., and R. J. A. Talbert, eds. 2010. *Geography and Ethnography: Perceptions of the World in Pre-modern Societies*. Malden, MA: Wiley-Blackwell.

Racine, Félix. 2009. "Literary Geography in Late Antiquity." Dissertation, Yale University.

Rambaud, Michel. 1974. "L'espace dans le récit césarien." In Raymond Chevallier, ed., *Littérature Gréco-Romaine et Géographie Historique* [Mélanges Roger Dion], 111–30. Paris: Picard.

Rapoport, Yossef, and Emilie Savage-Smith. 2008. "The Book of Curiosities and a Unique Map of the World." In Talbert and Unger, 121–38.

Rathmann, Michael. 2003. *Untersuchungen zu den Reichsstrassen in den westlichen Provinzen des Imperium Romanum*. Mainz am Rhein: von Zabern.

Reinhartz, Dennis. 2005. "Spanish Military Mapping of the Northern Borderlands after 1750." In id. and G. D. Saxon, eds., *Mapping and Empire: Soldier-Engineers on the Southeastern Frontier*, 57–79. Austin: University of Texas Press.

Remijsen, Sofie. 2007. "The Postal Service and the Hour as a Unit of Time in Antiquity." *Historia* 56.2: 127–40.

Reynolds, Joyce, et al. 1986. "Roman Inscriptions 1981–5." *JRS* 76: 124–46.

Richardson, J. S. 1986. *Hispaniae: Spain and the Development of Roman Imperialism, 218–82 BC*. Cambridge: Cambridge University Press.

Riess, Werner. 2003. "Die Rede des Claudius über das *ius honorum* der gallischen Notablen: Forschungsstand und Perspektiven." *Revue des Etudes Anciennes* 105: 211–49.

Riggsby, A. M. 2006. *Caesar in Gaul and Rome*. Austin: University of Texas Press.

Riggsby, A. M. 2017. Review of Talbert (2017). *BMCR* 2017.09.56.

Riggsby, A. M. 2019. *Mosaics of Knowledge: Representing Information in the Roman World*. Oxford: Oxford University Press.

Roberts, Michael. 1989. *The Jeweled Style: Poetry and Poetics in Late Antiquity*. Ithaca, NY: Cornell University Press.

Robertson, J. D. 1931. *The Evolution of Clockwork, with a Special Section on the Clocks of Japan*. London: Cassell.

Roby, C. A. 2014. "Land-Surveyors." In *OBO Classics*.

Roldán Hervás, J. M. 1975. *Itineraria Hispana*. Valladolid: Universidad de Valladolid.

Roller, D. W. 2015. *Ancient Geography: The Discovery of the World in Classical Greece and Rome*. London: Tauris.

Romm, James. 1992. *The Edges of the Earth in Ancient Thought: Geography, Exploration, and Fiction*. Princeton, NJ: Princeton University Press.

Roxan, M. M. 1978. *Roman Military Diplomas 1954–1977*. London: Institute of Archaeology.

Roxan, M. M. 1985. *Roman Military Diplomas 1978–1984*. London: Institute of Archaeology.

Roxan, M. M. 1994. *Roman Military Diplomas 1985–1993*. London: Institute of Archaeology.

Roxan, M. M., and Paul Holder. 2003. *Roman Military Diplomas IV*. London: Institute of Classical Studies.

Şahin, Sencer, and Mustafa Adak. 2004. "Stadiasmus Patarensis—Ein zweiter Vorbericht über das claudische Strassenbauprogramm in Lykien." In Frei-Stolba, 227–82.

Şahin, Sencer, and Mustafa Adak. 2007. *Stadiasmus Patarensis. Itinera Romana Provinciae Lyciae*. Istanbul: Ege Yayınları.

Salway, Benet. 2001. "Travel, *Itineraria* and *Tabellaria*." In Adams and Laurence, 22–66.

Salway, Benet. 2004. "Sea and River Travel in the Roman Itinerary Literature." In Talbert and Brodersen, 43–96.

Salway, Benet. 2005. "The Nature and Genesis of the Peutinger Map." *Imago Mundi* 57: 119–35.

Salway, Benet. 2012. "Putting the World in Order: Mapping in Roman Texts." In Talbert, 193–234.

Salzman, M. R. 1990. *On Roman Time: The Codex-Calendar of 354 and the Rhythms of Urban Life in Late Antiquity*. Berkeley: University of California Press.

Sartre, Maurice. 1983. "Les voyages d'Aurelius Gaius, soldat de Dioclétien, et la nomenclature provinciale." *Epigraphica Anatolica* 2: 25–32.

Šašel, Jaroslav. 1974. "Über Umfang und Dauer der Militärzone Praetentura Italiae et Alpium zur Zeit Mark Aurels." *Museum Helveticum* 31: 225–33.

Šašel Kos, Marjeta. 2002. "The Boundary Stone between Aquileia and Emona." *ArhVest* 53: 373–82.

Savoie, Denis. 2020. "Three Examples of Ancient 'Universal' Portable Sundials." In Alexander Jones and Christián Carman, eds., *Instruments–Observations–Theories: Studies in the History of Astronomy in Honor of James Evans*, 45–77. DOI: 10.5281/zenodo.3928498.

Schneider, Helmuth. 1986. "Infrastruktur und politische Legitimation im frühen Principat." *Opus* 5: 23–51.

Schomberg, Anette. 2018. "The Karnak Clepsydra and its Successors: Egypt's Contribution to the Invention of Time Measurement." In Jonas Berking, ed., *Water Management in Ancient Civilizations*, 321–46. Berlin: Edition Topoi.

Scott, J. M. 2002. *Geography in Early Judaism and Christianity,* The Book of Jubilees. Cambridge: Cambridge University Press.

Shaw, B. D. 1982. "Lamasba: An Ancient Irrigation Community." *AntAf* 18: 61–103.

Shaw, B. D. 2011. *Sacred Violence: African Christians and Sectarian Hatred in the Age of Augustine*. Cambridge: Cambridge University Press.

Sherman, Anna. 2019. *The Bells of Old Tokyo: Travels in Japanese Time*. New York: Picador.

Sherwin-White, A. N. 1984. *Roman Foreign Policy in the East: 168 B.C. to A.D. 1*. London: Duckworth.

Sillières, Pierre. 1990. *Les voies de communication de l'Hispanie méridionale*. Paris: De Boccard.

Silverstein, A. J. 2007. *Postal Systems in the Pre-modern Islamic World*. Cambridge: Cambridge University Press.

Sims-Williams, Patrick. 2006. *Ancient Celtic Place-Names in Europe and Asia Minor*. Malden, MA: Blackwell.

Sittl, Carl. 1890. *Die Gebärden der Griechen und Römer*. Leipzig: Teubner.

Smallwood, E. M. 1967. *Documents Illustrating the Principates of Gaius, Claudius and Nero*. Cambridge: Cambridge University Press.

Smallwood, E. M. 1976. *The Jews under Roman Rule from Pompey to Diocletian: A Study in Political Relations*. Leiden: Brill.

Snead, J. E., C. L. Erickson, and J. A. Darling, eds. 2009. *Landscapes of Movement: Trails, Paths, and Roads in Anthropological Perspective*. Philadelphia: University of Pennsylvania Press.

Speidel, M. A. 2017. "Recruitment and Identity: Exploring the Meanings of Roman Soldiers' Homes." *Revue internationale d'histoire militaire ancienne* 6: 35–50.

Speidel, M. P. 1986. "The Soldiers' Homes." In Werner Eck and Hartmut Wolff, eds., *Heer und Integrationspolitik: Die römischen Militärdiplome als historische Quelle*, 467–81. Cologne: Böhlau.

Staccioli, R. A. 2003. *The Roads of the Romans*. Los Angeles: J. Paul Getty Museum.

Stutzinger, Dagmar. 2001. *Eine römische Wasserauslaufuhr*. Berlin: Kulturstiftung der Länder.

Sumira, Sylvia. 2014. *Globes: 400 Years of Exploration, Navigation, and Power*. Chicago: University of Chicago Press.

Sumption, Jonathan. 1975. *Pilgrimage: An Image of Mediaeval Religion*. London: Faber.

Taccola, Emanuele, et al. 2021. "Un orologio solare miniaturistico in avorio da un contesto residenziale tardo-repubblicano di Pisa." *JRA* 34.1: 75–97.

Talbert, R. J. A. 1984. *The Senate of Imperial Rome*. Princeton, NJ: Princeton University Press.

Talbert, R. J. A. 2005. "Rome's Marble Plan and Peutinger's Map: Continuity in Cartographic Design." In Beutler and Hameter, 627–34.

Talbert, R. J. A. 2006. "Meyer Reinhold and Roman Civilization: The Impact of Sourcebooks *sans pareils*." *Classical Bulletin* 82.1: 97–101.

Talbert, R. J. A. 2007. "Konrad Miller, Roman Cartography, and the Lost Western End of the Peutinger Map." In Ulrich Fellmeth et al., eds., *Historische Geographie der Alten Welt: Grundlagen, Erträge, Perspektiven*, 353–66. Hildesheim: Olms.

Talbert, R. J. A. 2008. "Greek and Roman Mapping: Twenty-First Century Perspectives." In Talbert and Unger, 9–27.

Talbert, R. J. A. 2009. "P.Artemid.: The Map." In Kai Brodersen and Jás Elsner, eds. *Images and Texts on the "Artemidorus Papyrus": Working Papers on P.Artemid.*, 57–64, 158–63. Stuttgart: Steiner.

Talbert, R. J. A. 2010. *Rome's World. The Peutinger Map Reconsidered*. Cambridge: Cambridge University Press, with webpage www.cambridge.org/9780521764803.

Talbert, R. J. A. 2016. "Visions of Travel and Their Realization." *Antiquité Tardive* 24: 21–34.

Talbert, R. J. A. 2017. *Roman Portable Sundials: The Empire in Your Hand*. Oxford: Oxford University Press.

Talbert, R. J. A. 2019. *Challenges of Mapping the Classical World*. London: Routledge.

Talbert, R. J. A. ed. 1985. *Atlas of Classical History*. London and Sydney: Croom Helm.

Talbert, R. J. A. ed. 2012. *Ancient Perspectives. Maps and Their Place in Mesopotamia, Egypt, Greece, and Rome*. Chicago: University of Chicago Press.

Talbert, R. J. A., and Kai Brodersen, eds. 2004. *Space in the Roman World: Its Perception and Presentation*. Münster: LIT.

Talbert, R. J. A., and R. W. Unger, eds. 2008. *Cartography in Antiquity and the Middle Ages: Fresh Perspectives, New Methods*. Leiden: Brill.

Tanaka, Stefan. 2004. *New Times in Modern Japan*. Princeton, NJ: Princeton University Press.

Thompson, David. 2004. *The British Museum Clocks*. London: British Museum Press.

Tibbetts, G. R. 1992a. "The Beginnings of a Cartographic Tradition." In Harley and Woodward, 90–107.

Tibbetts, G. R. 1992b. "The 'Balkhī School' of Geographers." In Harley and Woodward, 108–36.

Todd, R. B., ed. 2004. *The Dictionary of British Classicists*. Bristol, UK: Thoemmes.

Trimble, Jennifer. 2007. "Visibility and Viewing on the Severan Marble Plan." In Simon Swain et al., eds., *Severan Culture*, 368–84. Cambridge: Cambridge University Press.

Trimble, Jennifer. 2008. "Process and Transformation on the Severan Marble Plan of Rome." In Talbert and Unger, 67–97.

Trousset, Pol. 1978. "Les bornes du Bled Segui. Nouveaux aperçus sur la centuriation romaine du Sud-Tunisie." *AntAf* 12: 125–78.

Turner, A. J. 1989. "Sun-Dials: History and Classification." *History of Science* 27: 303–18.

Turner, A. J. 2012. Review of Glennie and Thrift (2009). *English Historical Review* 127: 1217–19.

Turner, A. J. 2019. "A Mingling of Traditions: Aspects of Dialling in Islam." In Neil Brown, Silke Ackermann, Feza Günergun, eds., *Scientific Instruments between East and West*, 108–21. Leiden: Brill.

Turner, Brian. 2013. "War Losses and Worldview: Re-viewing the Roman Funerary Altar at Adamclisi." *American Journal of Philology* 134.2: 277–304.

Vaporis, C. N. 2012. "Linking the Realm: The Gokaidô Highway Network in Early Modern Japan (1603–1868)." In Alcock, Bodel, and Talbert, 90–105.

Venture, Olivier. 2016. Review of Wei (2014). *Early China* 39: 255–63.

Veyne, Paul. 1959. "Contributio: Bénévent, Capoue, Cirta." *Latomus* 18: 568–92.

Wang, Zhongxiao. 2015. "World Views and Military Policies in the Early Roman and Western Han Empires." Dissertation, Leiden University.

Weber, Ekkehard. 2006. "Die Spuren des frühen Christentums in der Tabula Peutingeriana." In Reinhardt Harreither et al., eds., *Acta Congressus Internationalis XIV Archeologiae Christianae, Vindobonae, 19–26.9.1999. Frühes Christentum zwischen Rom und Konstantinopel*, 775–82. Vatican City: Pontificio Istituto di Archeologia Cristiana and Vienna: Österreichische Akademie der Wissenschaften.

Wellesley, Kenneth. 1954. "Can You Trust Tacitus?" *G&R* 1: 13–33.

West Reynolds, David. 1996. "*Forma Urbis Romae*: The Severan Marble Plan and the Urban Form of Ancient Rome." Dissertation, University of Michigan, Ann Arbor.

Westrem, S. D. 2001. *The Hereford Map: A Transcription and Translation of the Legends with Commentary*. Turnhout: Brepols.

Whittaker, C. R. 2002. "Mental Maps: Seeing Like a Roman." In Paul McKechnie, ed., *Thinking Like a Lawyer: Essays on Legal History and General History for John Crook on his Eightieth Birthday*, 81–112. Leiden: Brill.

Wilkes, J. J. 1974. "Boundary Stones in Roman Dalmatia." *ArhVest* 25: 258–74.

Wilkinson, K. W. 2012. "Aurelius Gaius (*AE* 1981.777) and Imperial Journeys, 293–299." *ZPE* 183: 53–58.

Wilson, Andrew. 2018. "Roman Nightlife." In Chaniotis, 59–81.

Winchester, Simon. 2013. *The Men Who United the States: America's Explorers, Investors, Eccentrics, and Mavericks, and the Creation of One Nation, Indivisible*. New York: Harper.

Winter, Eva. 2013. *Zeitzeichen: Zur Entwicklung und Verwendung antiker Zeitmesser*. Berlin: de Gruyter.

Witschel, Christian. 2002. "Meilensteine als historische Quelle? Das Beispiel Aquileia." *Chiron* 32: 325–93.

Wittke, Anne-Marie, et al. 2010. *Historical Atlas of the Ancient World. BNP* suppl. 3. Leiden: Brill.

Wolfram Thill, Elizabeth. 2019. "Rome's Marble Plan: Progress and Prospects." In *Mapping the Classical World since 1869*, no. 5. awmc.unc.edu.

Wood, Denis. 1992. *The Power of Maps*. New York: Guilford.

Woodward, David. 1987. "Medieval *Mappaemundi*." In Harley and Woodward, 286–370.

Yee, Cordell. 1994. "Cartography in China." In Harley and Woodward, 35–202, 228–31.

Yee, Cordell. 2001. "A Reaction to the Reaction against Scientism: On the Power and Limits of the Textual Analogy for Maps." In id., David Woodward and Catherine Delano-Smith, eds., *Plantejaments I Objectius d'una Història Universal de la Cartografia/Approaches and Challenges in a Worldwide History of Cartography*, 203–22. Barcelona: Institut Cartogràfic de Catalunya.

Index